Dealing with Challenges in Psychotherapy and Counseling

CHRISTIANE BREMS

Brooks/Cole
Thomson Learning™

Australia • Canada • Denmark • Japan • Mexico • New Zealand • Philippines
Puerto Rico • Singapore • South Africa • Spain • United Kingdom • United States

Counseling Editor: *Eilleen Murphy*
Assistant Editor: *Julie Martinez*
Editorial Assistant: *Annie Berterretche*
Marketing Manager: *Jennie Burger*
Signing Representative: *Tony Holland*
Project Editor: *Marlene Vasilieff*

Print Buyer: *Mary Noel*
Permissions Editor: *Susan Walters*
Copy Editor: *Heidi Marschner*
Compositor: *Pre-Press Co., Inc.*
Cover Designer: *Laurie Anderson*
Printer/Binder: *Webcom*

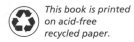

This book is printed
on acid-free
recycled paper.

For more information, contact
Wadsworth/Thomson Learning
10 Davis Drive
Belmont, CA 94002-3098
USA
www.wadsworth.com

International Headquarters
Thomson Learning
290 Harbor Drive, 2nd Floor
Stamford, CT 06902-7477
USA

UK/Europe/Middle East
Thomson Learning
Berkshire House
168-173 High Holborn
London WC1V 7AA
United Kingdom

Asia
Thomson Learning
60 Albert Street #15-01
Albert Complex
Singapore 189969

Canada
Nelson/Thomson Learning
1120 Birchmount Road
Scarborough, Ontario M1K 5G4
Canada

Library of Congress
Cataloging-in-Publication Data
Brems, Christiane.
  Dealing with challenges in psychotherapy
  and counseling /
Christiane Brems.
    p. cm.
  Includes bibliographical references and
  indexes.
  ISBN 0-534-36471-3 (pbk. : alk. paper)
  1. Counseling. 2. Psychotherapy. I. Title.
BF637.C6B723 2000
616.89'14--dc21

99-19539

Contents

Listing of
Figures and Tables

FIGURES

TABLES

To Mark—
who makes it all worthwhile

Preface

Disciples do owe their masters only a temporary belief and a
suspension of their own judgment till they be fully instructed;
and not an absolute resignation nor perpetual captivity.

SIR FRANCIS BACON

This book provides guidance about dealing with challenges in psychotherapy
and counseling not only for mental health practitioners, who are relatively
new to the field, but also for seasoned practitioners, who want a single resource on common challenges they face. The book includes many rules of thumb
and suggestions that have proven useful over the years in managing and dealing
with tough clinical situations as they arise. These guidelines are not to be read as if
they are written in stone. They are suggestions and ideas that may make for better
solutions to some critical incidents in therapeutic relationships. They are not *the*
answers; they are each *an* answer. They are guidelines that are best adhered to most
closely by the less experienced, whereas they may be varied somewhat (or a lot)
by the more-experienced practitioner. Each mental health care provider will, over
the years, develop strategies for dealing with difficult situations. These strategies
and interventions are almost as varied as the practitioners who use them. This book
represents a foundation that is common to all of these variations in intervention
during tough clinical occurrences. In other words, the principles and strategies selected for this book are those that are useful most of the time, with most clients,
from most theoretical orientations, and for most care providers. Following these
suggestions will help the reader cope with these clinical incidents as they occur
and will ensure that the clinician has a model for dealing with challenges that arise.

To respond to crises successfully, it is important to have thought about them beforehand and to have formulated some ideas about how to react and intervene. That is the purpose for this book.

PHILOSOPHICAL UNDERPINNINGS

The types of challenges that I chose to include range from the assessment phase to the intervention phase of therapy and counseling, and represent the types of incidents, crises, or questions that are most likely to occur. They are also situations that require some specific preparation, background knowledge, and intervention. The philosophy that underlies all of these suggestions and guidelines is based upon my personal larger philosophy for psychotherapy and counseling (Brems, 1999a). Specifically, it is a philosophy that reflects humanity, thoroughness, sensitivity, and preparedness. Although it borrows from many clinicians whom I admire, it also adds some unique components that are integrated into a larger whole. The guiding principles that form the foundation of my philosophy for therapy and counseling, and underlie the strategies suggested in this book are as follows:

- Counseling and psychotherapy represent a growth, change, and healing process that is helpful to clients who have experienced events in their lifetimes that have somehow interfered with or challenged the healthy development of a cohesive, orderly, and vigorous self (see Kohut, 1984).

- A contextual approach is important to all work with therapy and counseling clients because the larger environment of a client is generally important to understanding how the person's phenomenological presentation of self has developed over time and through interpersonal influences. Such an interpersonal matrix approach (see Stern, 1985) considers, at a minimum, familial influences, community pressures, cultural values, and societal belief systems that have acted upon the person.

- Therapy and counseling need to take a holistic approach to the client, investigating (assessing) and attending to (intervening with) the person's physical, emotional, cognitive, social, spiritual, behavioral, cultural, and psychological development and well-being (see Brems, 1993).

- Clients are viewed as collaborators in the therapeutic or counseling process as well as during crises. They are perceived as having an enormous amount of knowledge about their needs and as having many answers and solutions within themselves that can be activated through the facilitative and interpersonal presence of the therapist or counselor (see Rogers, 1961).

- The central growth, change, and healing process is the client-clinician relationship and the client's experiences within that relationship. The experiential component of psychotherapy and counseling is of greater importance than the insight component and the emphasis is on processing here-and-now relationships in a holistic, lifetime context (see Kohut, 1984; Teyber, 1997).

■ The provider-client relationship is only one component of a client's journey toward growth, change, or healing and needs to be supplemented with other strategies the client can implement independently of the therapy or counseling process (e.g., family strengthening, social support, environmental modification, nutrition and exercise, physical health) (Brems, 1999a).

To be optimally effective, clinicians understand crises and challenges in such a broad philosophical context and choose interventions that reflect this thorough and sensitive understanding of each individual client. Alexis de Toqueville expressed the belief that "chance is nothing that has not been prepared beforehand." This belief permeates this book and the suggestions within it. Often clinical interventions appear to arise straight from the intuition of the care provider. It is my hypothesis that this intuition is in actuality an expression of preparedness at such a profoundly internalized level that the spontaneous reaction of the clinician is also the empathic and appropriate one. I am not suggesting that being prepared for every single occurrence or event is possible or that one can always have a specific and prescribed strategy of intervention. Preplanning every moment of therapy or counseling is entirely impossible because of the uniqueness and context of each client; general preparedness, however, is not only feasible, it is essential. Mental health care providers are most effective when they have a schema or template of behaviors for situations that involve additional assessment needs for the client (whether they involve substance use issues, medical issues, or complicated psychological issues that require testing) and for situations that involve client behaviors such as suicidality, aggression, or abusive actions toward others. Further, counselors and therapists must consider their clients' perspectives and needs when managing their personal and professional growth. These issues, as independent as they may appear of the clientele the providers will see in a lifetime, are inextricably intertwined with the clinician's clients.

SUGGESTIONS FOR USE OF THIS BOOK

The book is best read from beginning to end at least once, preferably before the clinician begins to see clients for the first time. However, being a pragmatist, I realize that the optimal is not always the practical or realistic. With the exception of Chapter 1, which really is best read first in any and all circumstances, subsequent chapters can be read out of sequence without loss of clarity. I do recommend that clinicians who have read this book once keep it in a place where they can refer to it again and again as a reference manual for critical clinical incidents. I especially urge clinicians to reread relevant chapters whenever they encounter a new client and are required to draw upon specific background knowledge and skills. Clinicians may feel free to copy the many checklists and guidelines provided throughout the book for direct use in client work.

I hope that the guidelines and suggestions provided in the pages that follow will assist the reader in many challenging situations in the future. Unfortunately, they can neither guarantee complete success nor provide simple solutions that will always work. But it is my hope that they will provide a framework that endows the clinician with confidence and a sense of self-efficacy in situations that may otherwise evoke fear and insecurity. Applied with creativity and within a sensitive and individualized context, the principles outlined in this book will guide the reader toward understanding challenges to be opportunities.

In closing, I wish you excitement and enthusiasm as you help your clients recognize problems as incentives for growth and crises as motivators for change. Watching a client grow out of challenge should not be a frightful event, but an honor. Approaching counseling and therapy with such an attitude, when combined with skill, knowledge, and sensitivity, is bound to bring a successful outcome.

ACKNOWLEDGMENTS

Every book is an evolution and this book is no exception. Evolution is not a process that occurs in isolation; instead, it is profoundly influenced by many variables and numerous sources of input. The same is true for this book. I could not have written this book without the invaluable influences and inputs from the many individuals who surround me and guide my professional path and personal growth. I wish to thank them all for their patience, caring, and guidance—in fact, sometimes just for their presence! I owe great debts of gratitude to:

- my most beloved partner in life and love, Mark, whose ever-present love helps me grow and whose profound caring and nurturance during my struggles with a life-threatening illness while I wrote this book kept me on my path; I could not have done this without you and owe my life to you

- my students, whose appreciation for what I have to offer them and share with them encourages me to want to give more and whose support during my personal period of challenge helped me keep alive my hope

- my supervisees, whose responsiveness to their clients' challenges encourages me and whose creativity in problem solving keeps me equally open to solutions of a new and innovative kind

- my clients, whose struggles to survive touch me deeply and help me learn about myself, and whose recovery from their own challenges has modeled graciousness and faith for me

- my family, whose caring and support has helped me become who I am and whose unfaltering closeness keeps me safe in the knowledge that I will never have to face life's battles alone

- Eileen Murphy, Julie Martinez, Heidi Marschner, and Marlene Vasilieff for their wonderful contributions and never-ending support in the development and production of this project

- the following reviewers for their valuable comments and suggestions during the review process: Daisy B. Ellington, Wayne State University; Joan Polansky, Lewis & Clark College; Geoffrey G. Yager, University of Cincinnati; Susan Gray, Barry University; and Dana Schneider, Sonoma State University

Thank you all.

PART I

Challenges Arising During Assessment and Treatment Planning

1

The Challenge of Meeting
a New Client

You can look at every problem you have in your life as an opportunity for some greater benefit. You can stay alert to opportunities by being grounded in the wisdom of uncertainty. When your preparedness meets opportunity, the solution will spontaneously appear.

DEEPAK CHOPRA, *THE SEVEN SPIRITUAL LAWS OF SUCCESS*

Without fail, all therapists and counselors must face the challenge of meeting new clients. The best way to master this task is through preparation and structure that help the clinician remain focused, open to new learning, and empathic toward understanding the client. Preparation and structure are best obtained through attention to a number of critical matters that tend to apply to any client a mental health care provider will ever see. These issues include, at a minimum:

- insight about the clinician's own personal traits and their impact on the meeting

- sensitivity to the cultural factors client and clinician bring to the encounter they are about to have

- awareness of ethical principles that guide the provider's behaviors and actions

- thoughtfulness about the surroundings and structure in which the client is seen

If mental health care providers meet a new client with a firm grasp of all four of these prin-

ciples, they will be able to meet this challenge successfully and keep their clients comfortable and safe. Once these preliminary issues are attended to, clinicians can turn to data gathering.

HELPFUL PERSONAL TRAITS

Therapists and counselors, like all people, possess a number of traits in all life interactions that are either innate, learned, or developed over the years. For counselors to be maximally effective, however, they must develop an armamentarium of traits that reflects a combination of personal and technical competence that is compatible with working with people (Cormier & Cormier, 1999). Personal traits are often difficult to deal with early on in a therapist's career if they are strong but not conducive to therapeutic work; however, they are easy to deal with and a great asset, if they happen to be important therapy ingredients. Technical skills, which are learned and applied strictly in the context of a therapy relationship, include the ability to use reflection, active listening, reframing, empathy, and similar strategies. Although some of these are clearly related to personal traits, they are generally considered therapeutic catalysts and are taught in counseling classes and through specialized textbooks. What is of importance here are the personal traits mental health care providers bring to their education and their relationship with clients, traits that may either help or hinder the therapeutic process. Several of these personal traits will be addressed again in Chapter 9 in the context of preventing burnout and impairment. (Chapter 9 will also give concrete recommendations about how to develop and maintain these traits.)

Self-Esteem

Because clients are not always enamored with a clinician's techniques and may make their displeasure known verbally by attacking the therapist, self-esteem is a particularly important trait for the therapist to have. The insecure therapist may be worn down by such negative feedback, whereas a healthy (and realistic) dose of self-esteem makes the therapist less vulnerable to transferential and otherwise undeserved personal attacks by clients (Brems, 1994). A therapist who cannot deal with discomfort during a session is bound for failure or burnout (Kottler & Brown, 1992).

Self-Respect

Relatedly, a clinician's self-respect helps set appropriate therapeutic boundaries with the client. Only self-respecting clinicians are able to adhere to the agreed-upon time frame for a session without feeling guilty. They are also able to request payment and do not accept unnecessary calls or contacts from the client between sessions. Such appropriate setting of boundaries is important not only for clients, in that it communicates control and safety of the therapeutic setting, but it is also necessary for mental health care providers because it takes care of their need for a predictable schedule and private life outside of the clinic (see Herlihy & Corey, 1997, for a complete discussion of boundary issues).

Cognitive Capacity

Therapists and counselors must have the cognitive complexity required to conceptualize client cases and to think on their feet. They must be knowledgeable, have the desire and ability to assimilate new information quickly, and be able to reason abstractly. Only a bright clinician can respond quickly enough to the outpouring of data that can occur in therapy and keep up with the client in terms of conceptualizing and revising treatment plans effectively (Cormier & Hackney, 1987). Striving for excellence is how Kottler and Brown (1992) refer to the process of constantly learning and increasing the therapist's awareness and therapeutic skill. Each new client is viewed as a learning opportunity that will stretch therapists' limits and broaden their horizons. Additionally, continuing to learn about therapy and related strategies is essential because the field continually changes (Egan, 1994). The clinician must also be fully competent in the ethical sense. Most, if not all, mental health professions have a code of ethics that requires care providers to be fully trained, up-to-date, and generally competent at what they do (Swenson, 1997). Counselors must know their professional limits and know when to refer to another provider if these limits have been reached by the demands of a case or client.

Self-Awareness

A therapist needs to possess a high level of self-awareness and maturity, lest confounding feelings and attitudes enter into the treatment process (see Knobel, 1990). Self-awareness helps prevent inappropriate countertransference reactions and helps the therapist respond out of concern for the client, not for the self. Self-awareness is also necessary to help therapists recognize when their personal needs are mobilized by a therapeutic relationship and to guide them in keeping those needs out of the therapy room. The counselor may have to address these needs, but this is never done in the presence of the client (Kottler & Brown, 1992). The competent therapist is willing to seek supervision and consultation to improve and enhance self-awareness, not only when personal needs and limitations rise to the surface (Strupp, 1996), but on a regular basis to prevent inappropriate expression of personal needs in therapy sessions (Basch, 1980). Gaining self-awareness and working to maintain it also means that clinicians practice what they preach, looking at themselves regularly with caring and realistic criticism (Egan, 1994, chap. 9).

Open-Mindedness

Open-mindedness is another essential trait that helps clinicians welcome even those clients whose values may differ from their own (Cormier & Cormier, 1999). Open-minded therapists will not inadvertently or deliberately force personal values onto clients, especially clients who grew up in an environment significantly different from the therapist's (Castillo, 1997; Pinderhughes, 1983). A difference in backgrounds will not threaten treatment as long as the mental health care provider remains open minded and is able to see the client's life from the client's unique perspective. The clinician may need to understand that some of the client's behaviors that would be considered maladaptive or questionable in the therapist's personal background may have had great adaptive value in the client's environment. Clearly, counseling cannot be entirely free of values and value judgments (Lewis & Walsh, 1980; Pinderhughes, 1997). However, the counselor is encouraged to be as flexible and open-minded as possible and to recognize personal values and how they may collide with the client's values. In other words, although therapists may be unable to refrain from some value judgments, they must not allow them to cloud the work that needs to be done. Remaining as nonjudgmental and compassionate about clients' realities as possible both facilitates the therapeutic relationship and enhances the likelihood of helping clients help themselves (Reid,

1998). It also suggests and requires respectfulness vis-à-vis the client as well as a stance of egalitarianism that eliminates the power differential between the client and provider (Land, 1998).

Empathy

Each successful therapist strives to understand how a client feels in a given situation based on that client's specific and unique experiences, history, and background (cf., Kohut, 1984; Shulman, 1988). Empathy, thus defined, requires the therapist to listen carefully and to hear or see, not only the overt content of what is being expressed either verbally or behaviorally, but also to listen to the latent message contained within the client's expression. Such empathy, also termed vicarious introspection (Kohut & Wolf, 1978), is more than the warm, fuzzy feeling of caring: it is an artful and scientific approach to better understanding. Empathy is incomplete if it ends with the internal or private understanding of the client by the therapist. Empathy serves a positive therapeutic purpose only if the therapist is able to communicate understanding back to the person (Brems, 1999a). Once counselors have listened carefully and believe they have empathically understood the client, they can then communicate this understanding back to the client. The interpersonal cycle of empathy is complete when the client receives the message of understanding and feels the therapist's empathic concern (cf., Barrett-Lennard, 1981; Brems, 1989).

Flexibility

Tolerance and flexibility are two further crucial therapist traits (Choca, 1988; Cormier & Cormier, 1999; Knobel, 1990). In all therapies, new information constantly emerges as work progresses. This results in positive revisions of treatment plans and conceptualizations and requires meaningful adaptability of treatment strategies (Morrison, 1995a). Not all human beings are capable of functioning in such an environment of ambiguity and tentativeness. Some will attempt to make counseling or therapy fit a rigid model, forcing the understanding of the client into a mold. No client, no family, no therapy fits a specific mold (Land, 1998). In fact, the whole therapeutic process relies upon change, upheaval, tentativeness, and ambiguity. Related to the concept of flexibility is the idea that a good therapist must have patience (Strupp, 1996). Counseling cannot be rushed and counselors cannot expect clients to follow a particular time line.

Intuition and Risk Taking

In the same way that counselors and therapists need to be flexible, they also have to deal with unknowns and be willing to take risks and explore new grounds. The ability to deal with ambiguity is a critical skill of every counselor (Kottler & Brown, 1992). All therapists have to be capable of "epistemological feeling" (Knobel, 1990, p. 61), that is, they must have the ability to listen empathically and to alter their assessments of a client's situation flexibly and appropriately in changing contexts (Cormier & Cormier, 1999). Unwillingness to follow intuitions can result in leaving facets of the client undiscovered that may otherwise prove crucial to growth and change. Although this risk taking has to be weighed against the possible consequences of making a mistake, it is rare that one failed or inappropriate treatment intervention derails the entire therapeutic process. Indeed, some clinicians believe that the occasional empathic failure of the therapist is crucial to successful treatment (Kohut, 1984; Wolmark & Sweezy, 1998). Only repeated failures, not one unfortunate choice of wording or behavior, are likely to have an impact. It is often preferable for the counselor to follow intuition and risk a new intervention rather than to adhere rigidly to one that has already proven less than successful.

Authenticity and Genuineness

In facilitating therapeutic process, it is important that counselors allow their own individualities to come through, to be authentic (Knobel, 1990) and genuine (Egan, 1994). Since personality can neither be hidden nor camouflaged, it is quite impossible for clinicians to deny who they are outside of the therapy room (Chrzanowski, 1989). It is impossible for counselors to deny who they are in the way they greet their clients, sit in the room, and interact throughout the session. Clinicians express their humanity when they can be themselves, and this allows them to extend common human courtesies to their clients. Authenticity, however, does not equal self-disclosure within the therapy room. The therapy is for the client, not for the therapist to self-disclose or deal with personal psychological or emotional issues. However, all therapists and counselors have a certain interpersonal style: some are extroverted and active; others are introverted and observing. This general pattern shows in the type of interventions a clinician chooses (cf., Keinan, Almagor, & Ben-Porath, 1989; Kolevzon, Sowers-Hoag, & Hoffman, 1989), and finding a way of doing therapy that fits with the clinician's general style of being and life values is critical (Dorfman, 1998). Only if the style fits the clinician will she or he be able to muster the enthusiasm that is so crucial to the treatment of any client.

Adapting therapeutic style and technique to personal traits is not the same as abandoning the therapeutic neutrality. It is still important to be nonjudgmental and to avoid imposing certain opinions or outcomes on a client (Kottler & Brown, 1992). However, a clinician takes the neutrality issue too far if it translates into sterile and impersonal ways of relating to clients. Even psychoanalytic therapists have come to recognize that remaining anonymous does not mean being nonresponsive (cf., Basch, 1980; Wolf, 1988) or having no personality. It merely means maintaining clear boundaries and retaining the focus of the session on the client, not the therapist (Brems, 1994; Morrison, 1995a). The therapist's immediate needs are kept out of the therapy session. For example, if clinicians are hungry or upset, they delay gratification of these needs and feelings while with the client (Kottler & Brown, 1992). This may be one of the greatest challenges for the new practitioner, but it becomes second nature with experience. For this reason, self-care outside of the therapy room is an important component of therapists' lives so as to foster individuals who are self-aware and who recognize their personal needs and feelings.

Mental Health

Self-care is also crucial to maintaining personal mental health. Mentally healthy people are capable of empathic attunement: they have the wish and the ability to understand both the needs of others and their own needs; they are capable of delaying their own needs to meet the needs of others; they are realistically self-confident with a clear acceptance of personal flaws and shortcomings; they have little fear of rejection or humiliation; and they possess a certain amount of creativity, a sense of humor, and wisdom (Rowe & MacIsaac, 1986).

The latter three concepts beg some definition. Creativity refers to a person's ability to derive pleasure from problem solving. This pleasure in finding solutions to tough circumstances rewards clinicians, propelling them toward finding options and alternatives, both in their own lives and in the lives of clients. Creativity is one of the most important traits in successfully dealing with the challenges of clients and daily living. A sense of humor is the ability to laugh at oneself. It excludes biting sarcasm or vicious irony, instead it is the capacity to make light of past failures and minor imperfections and to avoid taking life or oneself too seriously. Finally, wisdom is characterized by the acceptance of personal limitations and frailties. Wise persons can forgive mistakes and lack of ability, accepting self and others fully. Wisdom dictates that parents, teachers, and clinicians can be forgiven and accepted despite past mistakes, and that the person can maintain a respectful and caring relationship nevertheless. True wisdom is achieved if this capacity is present when relating to others and the self.

CULTURAL ISSUES

Closely related to personal traits, the cultural sensitivity of the successful counselor is embedded in a larger context of understanding cultural issues and developing cultural awareness. Without attention to cultural issues, therapeutic rapport and process is not possible. The best counselors and therapists evaluate their level of cultural sensitivity and challenge themselves to think about their perceptions of other cultural groups. Gaining the skills and knowledge necessary for dealing with a racially, ethnically, and culturally diverse clientele is as important to a therapist's education as gaining the basic skills and knowledge of assessment and treatment itself (Iijima Hall, 1997; Schlesinger & Devore, 1995).

Before discussing cultural sensitivity, the labels *race*, *ethnicity*, *culture*, and *minority* need to be defined. The term *race* refers to a biological classification that is based on physical and genetic characteristics, with only three primary races identified (Caucasoid, Mongoloid, and Negroid). *Ethnicity* refers to a shared social and cultural heritage (for example, Asian Americans or Alaska Natives). Finally, *culture* refers to learned behavior that is shared and transmitted within a group. Such transmission can occur across generations through the teaching of shared values and rules, or it can be initiated with new members, as frequently occurs in gay and lesbian cultures. To give some examples: members of the Jewish faith constitute an ethnic group with a shared social, cultural, and religious heritage; however, they do not constitute a race. White members of society also constitute an ethnic group in the United States. Like most ethnic groups, this group contains a number of cultures within it (Irish Americans, Italian Americans, German Americans, and so forth), which each share a learned set of behaviors. Although ethnic or cultural status often overlaps with minority status of a group of people, this is not always so. A comprehensive approach to multigroup or multicultural sensitivity encompasses not only ethnicity and culture, but also minority status. Minority status as it relates to therapy has nothing to do with the actual number of people within a specific group. Instead, a minority group is defined as

> A group of people who, because of physical or cultural characteristics, are singled out from others in the society in which they live for differential and unequal treatment, and who therefore regard themselves as objects of collective discrimination. . . . Minority status carries with it the exclusion from full participation in the life of the society (Wirth, 1945, p. 347).

This characterization applies to a number of groups in American society who experience oppression and, as a result, are not able to participate fully in society as a whole. It separates the identification of a minority from the numerical concept. For example, in many cultures, women suffer oppression at the hands of men, rendering them a conceptual minority despite the fact that they are a numerical majority. Using this definition, other minorities include individuals with physical disabilities, the elderly, gays and lesbians, and individuals who are economically disadvantaged. Thus, the therapist in a culturally diverse society works with individuals who vary, not only in terms of ethnic or cultural background, but also in terms of other avenues of oppression (Lum, 1995; Schlesinger & Devore, 1995).

The term *race*, properly defined, is purely biological in nature and hence does not have direct psychological consequences. It is often perceived as derogatory, carrying significant stereotypes with it and generally has fallen into disuse. Ethnic and cultural differentiations, as well as minority status, on the other hand, have social and psychological relevance in that they contribute greatly to the self-development of the group's members. Often members of a specific ethnic, cultural, or minority group cannot be understood properly if taken out of their contextual or multicultural matrix of relationships. Thus, ethnic, cultural, and other group variables have a great impact on social and psychological issues, and their implications for diagnostic and therapeutic work are significant.

The American Psychological Association (APA), the National Association of Social Workers (NASW), the American Counseling Association (ACA), and other professional mental health organizations recognize the diversity of the world's population and acknowledge the need to provide mental health services adequately and appropriately to all ethnic, cultural, and minority members of a given society. They support the need for therapists to be culturally sensitive and the need for training programs that will help meet this need. For example, the APA's ethical guidelines clearly state that "psychologists are aware of cultural, individual, and role differences, including those due to age, gender, race, ethnicity, national origin, religion, sexual orientation, disability, language, and socioeconomic status" (American Psychological Association [APA], 1992, p. 1599). Similarly, the need for including cultural issues in all therapists' training was advanced by the National Conference on Graduate Education in Psychology in their statement that "psychologists must be educated to realize that all training, practice, and research in psychology are profoundly affected by the cultural, subcultural, and national contexts within which they occur" (APA, 1987, p. 1079). Ever-increasing pressure from professional organizations, as well as from individual practitioners, urges therapists to become culturally sensitive to meet the needs of a culturally diverse population (see Iijima Hall, 1997; Ponterotto, Casas, Suzuki, & Alexander, 1995). The code of ethics for the NASW indicates that not only do mental health care providers need to avoid cultural insensitivity in themselves, but they also must not be inactive witnesses to such practices imposed upon clients by others (National Association of Social Workers [NASW], 1993).

Empathy and open-mindedness about values and standards are necessary therapist traits (discussed above) that enhance a counselor's sensitivity to ethnic differences (Ponterotto, Casas, Suzuki, & Alexander, 1995). It is important to keep in mind that although there is significant variation across ethnic groups, there is also significant variation within each ethnic group (Johnson, M., 1993). An open-minded and empathic clinician never assumes that because clients are from a certain ethnic group, they will display certain attitudes or behaviors. This attitude, often misrepresented as cultural sensitivity, is actually one of the worst manifestations of prejudice. Sensitivity to and respect for cross-cultural differences are not to be used to categorize people along certain dimensions (Egan, 1994; Steiner & Devore, 1983). Uniqueness exists in every person, whether that person is Asian American, African American, or European American. No therapist would ever assume that all White clients have the same traits and concerns. However, unfortunately it is still common to hear therapists talk in such global terms about clients from other ethnic backgrounds (Lum, 1995; Namyniuk, Brems, & Clarson, 1996).

Another aspect of cross-cultural sensitivity is that language must never be pejorative (Strupp, 1996). All sexist and racist terminology must be eliminated from a clinician's verbal repertoire (Johnson, M., 1993; Land, 1998). Terms that are openly racist (such as *spic* or *nigger*) are easily avoided because of widespread awareness about the inappropriateness of such language. However, more subtle terms of racism, sexism, and homophobia continue to prevail. For example, terms such as *Indian giver* remain in common use despite their clear racial prejudice. Further, though the speaker may not intend for them to be, word choices are often sexist. For instance, using *he, his,* and *him* as generic pronouns is unacceptable in treatment because this language excludes women. Labels such as *mankind, chairman, policeman,* and so forth are equally biased, especially given that less sexist options have been coined (e.g., *humankind* or *humanity, chair, police officer*). Sexist assumptions about professions must be monitored as well, with the clinician making conscious word and pronoun choices; for example, the clinician should not refer to physicians, politicians, or construction workers as men or to nurses, secretaries, or teachers as women. These are only a few common examples of sexist assumptions.

Cultural sensitivity and its components of awareness, knowledge, and skills can be learned.

Awareness is gained through self-reflection and respect for others as well as through acceptance of the notion that difference does not equal deviance (Namyniuk, 1996). Knowledge can be accumulated via familiarization with cultural, anthropological, historical, and related events involving or affecting all cultural and ethnic groups with whom a clinician anticipates working (see Ponterotto, Casas, Suzuki, & Alexander, 1995, chap. 5–9). Skill is developed through learning about alternative approaches to intervention, reduction in prejudicial or stereotyped use of language, and political activism (Ivey, 1995). Therapists who strive to be culturally sensitive claim that all three of these traits are necessary parts of their repertoires. Cultural sensitivity is defined in more detail in Table 1-1 (adapted from Brems, 1999a, and Johnson, M., 1993).

Cultural issues are also inherent in the diagnostic process (Solomon, 1992). Cultures vary greatly in what they consider a problem or an appropriate strategy for coping within a given situation (cf., Castillo, 1997; Dana, 1993; Iijima Hall, 1997). What may constitute abnormality in one culture, may be acceptable, if not mainstream behavior, in another. Different cultures may express the same type of problem in different ways, choosing different idioms to describe an essentially identical emotional intensity and type of pain (Matsumoto, 1994). For example, depression among mainstream White clients tends to conform to the criteria outlined in the *Diagnostic and Statistical Manual of Mental Disorders*, Fourth Edition (DSM-IV), whereas depression among the Chinese manifests itself through a different set of highly somatized symptoms (such as constipation, loss of appetite, and fatigue), with little expressed dysphoric affect (Castillo, 1997; Dana, 1993). Some disorders appear to be culture-bound, appearing predominantly in some, but not all, cultures (American Psychiatric Association, 1994; Suzuki, Meller, & Ponterotto, 1996). This phenomenon can be explained by the observation that different cultures literally "cultivate" different traits and behaviors. Because any traits or behavior taken to an extreme may result in pathology, different cultural groups will have different manifestations of pathology based on the types

TABLE 1-1   Traits of a Culturally Sensitive Mental Health Care Provider

SENSITIVITY COMPONENT	SPECIFIC TRAITS
Cultural Awareness	■ is aware of and sensitive to personal own cultural heritage
	■ is conscious and embracing of all minority groups
	■ values and respects cultural differences
	■ is aware of personal values and biases and their effect on therapy
	■ is sensitive to neither over-emphasizing or under-emphasizing clinician-client cultural differences
	■ feels comfortable with cultural differences between self and client
	■ demonstrates sensitivity to situations which may require referral of a minority client to a member of the same cultural heritage
Cultural Knowledge	■ understands how the sociopolitical system in the United States treats minorities
	■ knows about the presence and various manifestations of racism, sexism, and heterosexism and their effects on minorities
	■ is familiar with the history of mental health treatment for minorities and potential biases of traditional psychotherapy theories
	■ is aware of cultural definitions of mental illness and perspectives on mental health services
	■ is knowledgeable about cultural and minority groups in the United States
	■ possesses specific knowledge about particular groups with whom the clinician is working
	■ has clear and explicit knowledge and understanding of the generic characteristics of therapy
	■ is familiar with cross-cultural applications of psychotherapy skills
	■ is aware of the effects therapy setting and office can have on minority clients
	■ is knowledgeable about institutional barriers that prevent minorities from using mental health services
Cultural Skillfulness	■ is adept at adjusting communication and therapeutic style to match individual client's needs
	■ knows how to place the appropriate amount of attention to the role of culture
	■ does not categorize individuals according to stereotypes and prejudices
	■ is respectful and flexible in providing services to meet the individual needs of clients
	■ exercises intervention skills as needs of the client dictate and as appropriate to the client's personal contextual background
	■ acts as a social change agent to help reduce or eliminate racism, sexism, and heterosexism
	■ uses language devoid of prejudice and bias

of traits they emphasize in their healthy population (Alarcon & Foulks, 1995; Iijima Hall, 1997; Land, 1998).

For example, eating disorders appear to be largely a phenomenon of industrialized, Western cultures with a strong focus on being thin (especially for women) and with an abundance of food (American Psychiatric Association, 1994;

Castillo, 1997). *Frigophobia*, a disorder characterized by an irrational fear of being exposed to the cold and freezing to death, is identified only among the Chinese where somatic concerns are prevalent in the culture (Dana, 1993). *Amok*, a disorder characterized by a period of brooding, followed by a violent outburst, and usually ending with exhaustion and amnesia for the event, appears limited to Southeast Asian men (Castillo, 1997). *Ataques de nervios* is a disorder that includes symptoms such as hyperemotionality, accompanied by trembling, palpitations, and general hyperkinesis, it occurs primarily among Puerto Rican women who are culturally encouraged to express distress through somatization and who are encouraged toward high emotionality (Rivera-Arzola & Ramos-Grenier, 1997).

Personality disorders are particularly vulnerable to cultural influences because they are even more bound to environmental factors and by definition are interpersonal in nature (see Castillo, 1997; Fabrega, 1992). What complicates a universal standard for diagnosis, especially of personality disorders, is "the observation that different cultures have tended to emphasize different traits of personality as ideal" (Alarcon & Foulks, 1995, p. 6). This tendency may lead to behaviors that are acceptable, even admired, in some cultures and considered inappropriate, or even pathological, in others. For example, persons of Mediterranean and Latin descent are most commonly mislabeled as histrionic due to their cultural emphasis on emotionality, dramatic interpersonal style, novelty-seeking, and their tendency toward somatization; Asian and Asian-American clients are frequently mislabeled as dependent because their culture stresses politeness, deference, acceptance of others' opinions, and passivity (Castillo, 1997; Johnson, F., 1993); Talmudic scholars may be misdiagnosed as obsessive-compulsive because of their cultural encouragement to be conscientious and scrupulous with regard to morality, restricted affective expression, and striving for perfectionism; Hindu yogis may be mislabeled as schizoid because of their choice to withdraw from interpersonal contacts for extended periods of time (Castillo, 1997; Witztum, Greenberg, & Dasberg, 1990).

LEGAL AND ETHICAL ISSUES

Every clinician must learn about ethical and professional issues that are relevant to psychotherapy and counseling. Ethical treatment of clients begins when the clinician obtains informed consent that outlines the responsibilities of the clinician and the rights of the client. The clinician must also clarify the expectations a client can have about treatment. A sample of an informed consent form is provided at the end of this chapter. The "elements of legally adequate informed consent are competence, voluntariness, and knowledge" (Crawford, 1994, p. 56). In this definition, *competence* means that the person rendering consent has to be legally competent to do so (he or she must be over 18 years old and must not be developmentally disabled or otherwise legally "incompetent"). *Voluntariness* means that therapy is entered voluntarily and that informed consent is not rendered under duress. *Knowledge* refers to the fact that the counselor must ascertain that the client is knowledgeable of and understands the types of rights covered under an informed consent to psychological treatment (Canter, Bennett, Jones, & Nagy, 1994). Several pieces of information are covered with each client in person during the first meeting:

- client confidentiality (i.e., the guarantee that the therapist will not disclose information obtained in the therapeutic relationship without proper authorization)
- exceptions to confidentiality (such as the duties to warn, protect, and report)
- limitations of privileged communication (e.g., court orders, diagnostic disclosure to insurance companies to ensure payment)
- information about release of information authorized by the client
- voluntary nature of therapy
- audio- or videotaping policies
- cancellation policies
- fee and payment schedules
- dual relationship issues
- possible alternatives to therapy for the client's presenting concern
- credentials or educational background of the service provider and her or his supervisors
- supervision or consultation arrangements

- clinic procedures for case staffings and treatment plan meetings

This information is covered explicitly, directly, verbally, and in writing with all clients during the initial session, along with the following additional points (see Swenson, 1997). Clients are best made aware:

- that therapy or counseling cannot guarantee positive outcomes and cannot guarantee definite solutions to presenting problems

- that therapy or counseling can have some inherent risks, depending on the techniques employed by a counselor or therapist (e.g., flooding techniques may lead to panic reactions)

- of where they can turn should they have complaints or grievances about the treatment they receive

- that therapy and counseling can have negative side effects as the client changes and possibly outgrows current relationships or behavior patterns

Confidentiality is the cornerstone of counseling and therapy. "For therapy to work the client must trust the therapist. For the client to trust, the therapist must keep promises about secrecy" (Swenson, 1997, p. 70). Even Hippocrates, who practiced before 400 B.C., was convinced of the importance of confidentiality, stating, "Whatever I shall see or hear in the course of my profession . . . I will never divulge, holding such things to be holy secrets." Important exceptions have been made to confidentiality through laws, legal precedent, and ethical codes. These exceptions, which must be covered in the informed consent, are as follows (Arthur & Swanson, 1993; Swenson, 1997):

- if the client requests or allows disclosure by signing a release of information

- if disclosure is necessary to prevent certain crimes

- if the client presents a danger to self or others

- if the client discloses or arouses reasonable suspicions about child abuse or neglect

- if the therapist receives a court order or *subpoena duces tecum* for disclosure to serve the cause of justice

- if there is a criminal or civil legal action related to sanity or competence

- if the client has initiated a legal action or ethical charge against the therapist (e.g., a malpractice lawsuit)

- for the counselor's supervision purposes

- for clerical purposes (i.e., for secretary typing reports or filing insurance paperwork)

- for intraagency sharing that is part of treatment (e.g., case staffings or treatment plan meetings)

Swenson (1997) calls the inclusion of these exceptions in an informed consent the equivalent to a psychological Miranda warning, an apt comparison: "A psychological Miranda warning lets clients know under what circumstance the therapist will share potentially damaging information so that clients will not be misled into incriminating themselves" (Swenson, 1997, p. 72). He also notes that many clinicians (in fact, more than half!) fail to give such a warning to their clients, thus shortchanging their clients and putting themselves at risk for lawsuits.

Finally, in this day and age of managed care and health maintenance organizations, the issue of confidentiality and privileged communication arises because it has become difficult to determine what type of client information is required of the clinician for third party reimbursement purposes (Swenson, 1997). Clients need to be informed that, at a minimum, insurance companies require diagnostic information and, at a maximum, detailed information about symptoms and session progress to continue making payments. Clients need to give informed consent to these procedures if they are required, otherwise they may need to decide not to use third-party payment options if they do not agree with the disclosures required by their third-party payer. Most importantly, therapists cannot assume that clients are willing to have this information disclosed merely for the purpose of being financially reimbursed for the cost of treatment (Arthur & Swanson, 1993). Clients waive their right to privacy vis-à-vis their insurance company when they sign their claims form. Many clients are not aware of this fact, and it should therefore be mentioned in the informed consent. Disclosing excessive information or disclosing any information to the third-party payer without receiving the client's consent to do so may well be considered a breach of confidentiality (Okun, 1997) even if the client signed the insurance claims form. Clients are best encouraged to check with their personal insurance carrier about what type of disclosures will be required of the clinician before making a final commitment to reimbursement and treatment.

Although all of this ethical and legal information is dealt with nonverbally through an informed consent form that was signed by the client before the session, it is the mental health care provider's responsibility to ask if the client read and understood the informed consent. A good way to do this is to begin the initial session with, "Before we get started, I want to talk to you about the informed consent form you read and signed before we came in here. I just wanted to reiterate that . . ." Clients are then familiarized with their consumer protection rights. These consumer rights have been amply covered earlier in this chapter and in other literature (e.g., through ethical guidelines provided by relevant associations such as the APA [1992], the NASW [1993], and the ACA [1995] and through books written on the topic [Anderson, 1996; Corey, Corey, & Callanan, 1988; Herlihy & Corey, 1992; 1996; Swenson, 1997]). The client should be invited to ask questions about the informed consent, and often it is a good idea to probe with a couple of questions to make sure the client truly read and understood what she or he signed (e.g., "Did the information about the fact that we cannot guarantee successful outcomes for therapy make sense to you?" "Do you have any questions about my supervisor/consultant?" "Do you know what information your insurance carrier requires from me to make reimbursements to you?").

CLINIC FEATURES

Developing an atmosphere conducive to psychotherapy and counseling is a deciding aspect in designing a therapy office. This is true regardless of where a counselor's office is housed. An atmosphere that is conducive to treatment will communicate several impressions as soon as a client enters the office, most importantly impressions of confidentiality, privacy, comfort, and caring. With regard to confidentiality, for example, no client records should be visible in the lobby and waiting area. Unlike in many physicians' offices, where patient charts are kept in a shelf or file cabinet behind the receptionist's desk, such a practice is inappropriate for a counseling office. Clients are best greeted discretely by the receptionist without loud announcement of their names and no use of last names. Clinic staff need to be well versed in the basics of client confidentiality and interpersonal skills that help make clients comfortable. Chitchat and social conversation between staff and clients is best kept to a minimum because it is often impossible to predict the mental state of the client (Giordano, 1997). An innocuous social question by a staff person could trigger a flood of emotion in a client and may prove embarrassing. The environment also must be quiet and void of any loud music, noisy staff, or loud equipment (e.g., typewriters, shredders).

The waiting room needs to be large enough to allow clients to fill out paperwork without fearing that anyone may be able to read over their shoulders. Preferably, the waiting area is located away from the treatment area to maximize privacy (Wolf, 1988) and to ensure that clients in the waiting area cannot overhear clients in the treatment room who may either be speaking loudly, weeping, or otherwise making noise (e.g., children playing in play therapy). If this is not possible, playing soft, neutral music (e.g., classical music) in the waiting area may be used to cover any sounds coming from the treatment area. The waiting area is best equipped with comfortable seating that includes couches and individual chairs. Reading materials in the waiting room can be diverse but tasteful and appropriate, avoiding controversial or potentially offensive topics that may alienate some (see Brems, 1999a). Relevant pamphlets (e.g., advertisements/announcements of mental health resources in the community, informational brochures about mental health issues) can be displayed in the waiting area with extra copies provided for clients to take home. The waiting room also reflects the level of thoughtfulness of the clinic staff. Materials (such as magazines, wall coverings, art) that are offensive to any gender, cultural, or ethnic groups are inappropriate and should not be used.

The therapy rooms themselves must be appropriately equipped and soundproof (Wolf, 1988). Proper equipment includes comfortable seating for therapist and client at a reasonable distance that helps both therapist and client feel comfortable (Basch, 1980). A small table that holds appointment cards, clinic brochures, extra forms (e.g., releases of information), tissues (very important), and similar items is useful, as is a shelf that holds extras of all of these items. A small wastebasket for used tissues needs to be provided. Wall and other decorations are best neutral and tasteful, not reflecting any specific message. Lighting for the room should be adjustable, and table or floor lamps are preferable to overhead lighting. Fluorescent lighting is not conducive to helping create a warm atmosphere and is best avoided. Carpeted floors assist with noise absorption and create a warmer atmosphere than tile or linoleum.

STRUCTURAL ISSUES

Many people come to counseling because of chaos and disorder in their lives; all come with various sets of expectations and misperceptions that have to be met and addressed by the therapist. One way to help introduce and model structure (to decrease some of the experienced chaos) is to provide explicit guidelines about the therapeutic process (Cormier & Hackney, 1987). It is helpful for clients to know what will be ex-

pected of them and how various situations (e.g., payment and scheduling) will be handled. This prior knowledge protects clients from misunderstandings and assists therapists in maintaining a healthy boundary between their professional and work lives (Kottler & Brown, 1992). Following are some suggested structures and guidelines that would meet these purposes.

Structuring the Interaction

It is important for counselors and therapists to set clear boundaries about their interactions with their clients (Kottler & Brown, 1992). Therapy generally has a fixed structure that most superficially and traditionally consists of a 45- or 50-minute hour that begins and ends on time and is scheduled on a regular weekly basis, usually at the same time each week. This schedule provides predictability and allows client and clinician to plan schedules around the session for weeks in advance. It is important for the client to understand the limits of therapy and that lateness for sessions is not accommodated. In other words, the counselor will end at the same time as always. This is necessary not only to model structure for the client but also to protect the therapist from undue stress. Imagine running over with a late client and then having to run late for all subsequently scheduled sessions; or imagine running late with a client only to have the therapist who is scheduled next for a given therapy room intrude on the session. These situations not only interfere with the work with the client, but they also increase the stress perceived by both client and counselor. Clinicians do vary in opinions about how long to wait for a late client: some wait 15 minutes and some up to 30 minutes; some wait the entire scheduled time. The important question to ask is how much time is needed to get some therapeutic work done. If that much time is left in the session when the client arrives, the remainder of the session can be used. If clients arrive with less than that amount of time left in the session, they are best asked to return for the regularly scheduled appointment during the next week.

Another relatively agreed-upon structure is requiring payment at the time services are provided (unless client and provider have explicitly and collaboratively made other prior arrangements). Payment is best accepted before the session starts and needs to be acknowledged with a written receipt. One reason for taking payment at the beginning, not the end, of the session is that the client cannot prolong a session by delaying the process of writing a check or finding money in a wallet. Further, it protects the client from potential embarrassment should a particularly difficult session leave the client emotionally upset at the end of the session. Having to deal with a receptionist or with payment in front of others at such a time may be difficult.

The best therapists and counselors avoid social chitchat with clients even during sessions, but they especially refrain from such conversation in waiting or public areas where others may be present (Basch, 1980). The socially acceptable "How are you?" is not an appropriate greeting with a client in a public area (Brems, 1999a; Giordano, 1997). What if the client feels terrible and bursts into tears at the mere inquiry into her or his emotional state? What if the client begins to self-disclose before the therapy room door has been shut? These violations of the client's confidentiality are easily avoided by greeting the client with a friendly "Hello. Let's go . . . (to the room, on back, etc.)." Once a session is over, it is best to let the client walk out of the room and clinic alone, thereby avoiding prolonged confidential conversation in the hall or in a public area. Ending the session before the office door is opened with a friendly emphasis on the client's return the following week will serve this purpose well. The therapist can simply say, "I'll see you again next week, same place, same time. . . . Take care." Then the door is opened, the client leaves, and the therapist hangs back, perhaps straightening the room for the next person. It is good practice to empty wastebaskets between sessions to avoid the accumulation of visibly used tissues. Also, this may be a good time to make sure an ample supply of tissues is on hand. This is particularly true when several therapists share a single office.

Finally, if the client is seen in a clinic where either video- or audiotaping is standard procedure, the equipment is best started before the client enters the room. In this manner, the clinician's attention is not distracted once in the presence of the client. When therapists video- and audiotape sessions, they do not need to rely on note taking, even during intake sessions involving the gathering of a large amount of information. Note taking can be a distancing behavior that is not recommended with most clients even in settings where taping is not commonly practiced (Brems, 1999a; Seligman, 1996). In such situations, counselors are urged to do their case-note writing as soon after a session as possible, while the content and process of that session is still fresh in their mind. This issue will be addressed in detail later in this chapter.

Structuring Potentially Challenging Situations

The initial interaction with the client may pose some difficulties for the novice clinician. In order to be fully prepared, the new counselor needs to make several decisions about these potentially challenging occurrences that are common to initial sessions. Although none of these situations is truly difficult, any one can become problematic if the mental health care provider is not prepared for it. The most common of these situations will be presented in the sections that follow. For a more thorough discussion the reader is referred to Brems (1999a).

Opening of the First Session. Once issues of informed consent have been dealt with, the client needs to be prepared for what will follow. A good introduction about what to expect from the first and subsequent sessions helps clients relax tremendously and often increases their level of cooperation during the initial interview. The therapist explains the nature of the intake so that the client can understand why all the questions will be asked (Morrison, 1995a; Seligman, 1996). This is particularly important because the counselor will request a lot of highly personal and sensitive information from the client. If the client does not understand why these questions are being asked, and if the questions are not formulated carefully and respectfully, the client may become resistant or even lose interest in the therapeutic process (Choca, 1988). A sample of an introduction to the intake and counseling process is presented in Brems (1999a).

Asking clients why they decided to come to the clinic follows the introduction to the first session. The question "Why did you decide to come for therapy at this time?" is recommended over all others because it emphasizes the client's choice in the decision-making process, and it does not imply anything that cannot be guaranteed by the clinician (as does "How can I help you?" the opening question used by some). Further, it generally prevents wisecracks sometimes elicited by the question "What brought you here?" (The client might answer, "My car."). Other opening questions may be perceived as rude (e.g., "What is your problem?"), as challenging or uninviting (e.g., "Why are you here?"), or simply as too vague (e.g., "Tell me about yourself."). Once the opening question has been asked, the practitioner must be flexible enough to follow the client's lead, yet structured enough to gather the data necessary to arrive at a (at least preliminary) conceptualization and treatment plan (more about this later).

Closing the First Session. Although it is the counselor's role to pay attention to the client's emotional state and to ensure that the client feels comfortable throughout the initial interview, this is especially true during the closing of the first session (Kottler & Brown, 1992). Before beginning the closure of the session, the therapist needs to help the client regain composure should the client be upset in some way or another (Kottler & Brown, 1992). Once the client is in a receptive frame of mind, the closing itself starts, with the counselor's brief summary of what has been talked about in the session. The clinician states his or her understanding of the client's

problem and reiterates the primary issue to be worked on, as perceived by the client. The therapist asks whether the client feels comfortable with the clinician or whether the client might prefer to work with someone else (Choca, 1988). The practitioner also gives the client an opportunity to ask questions about any content that emerged during the session or about counseling in general. If the client has sought prior therapy or is currently under the care of a physician for what might be related symptoms, proper releases of information (ROIs) are collected at the close of the initial session.

Once counselors have concluded the summary, checked in with the client, obtained ROIs, and made preliminary treatment recommendations, they schedule the standing appointment that they and the client will have for the next few months. The clinician stresses that this 45- or 50-minute session will take place weekly, at the same time and in the same place. The client must commit to an appointment time that is convenient and appropriate for the client. Careful discussion of this issue now may prevent client no-shows and cancellations later. Finally, the therapist may want to verify again that the client is knowledgeable about fees and cancellation policies, the lack of a guarantee for success, the possibility of negative emotions and side-effects caused by therapy, and the requested commitment to therapy (Cormier & Cormier, 1999; Hutchins & Vaught, 1997; Morrison, 1995a).

Note Taking. Although this is a common practice, it is not recommended (Brems, 1999a; Seligman, 1996). Even Freud suggested that note taking creates an emotional distance between client and therapist, often drawing the therapist's attention away from the client. The therapist becomes so concerned with getting things down on paper, that essential emotional nuances about the client (especially those that are communicated nonverbally) can be missed. Note taking is appropriate only in settings where therapeutic rapport is not an issue; when a relationship is to be established, note taking interferes. If a counselor is worried about not remembering the content of an intake session with a client, taping the session and reviewing the tape while writing case notes or reports can be helpful. Writing detailed notes after sessions, and especially writing a report based on the history gathered during the initial session, is critical for several reasons. First, it forces the therapist to organize the information in a way that facilitates case conceptualization. Second, it clarifies where the counselor has missed essential bits of information, thus providing a stimulus for further exploration in later sessions. Third, clinicians can refer back to the notes and reports to refresh their memory about essential aspects of the client. Such refueling of a therapist's memory is extremely helpful since clients appreciate counselors who can remember what was previously discussed. When the client feels heard and understood, rapport and trust are strengthened.

Food and Drink in Sessions. Food and drink greatly distract from therapeutic work and are strongly associated with social interaction: family members eat together, and friends and acquaintances join one another for meals. A therapeutic relationship needs to be set apart from social and familial interactions and, hence, food and drink are rarely acceptable, especially among novice clinicians who are still struggling to set and maintain appropriate therapeutic boundaries. Certainly, if at all possible, a therapist would never eat or drink in the client's presence. Food or drink are also excellent means for clients to distract themselves from therapeutic work or difficult affects. Given the social connotations and affect-reducing properties of food and drink, they generally have no appropriate role in counseling or therapy.

Are there exceptions to this rule? Perhaps. With some child clients, the sharing of food can be therapeutic (e.g., see Brems, 1993); however, if therapists share food, they must be clear that they are doing so for therapeutic, not social, reasons. Sometimes, offering a drink of water may be appropriate if there is a medical reason (e.g., the

clinician or client has a cough that can only be alleviated by drinking). To reiterate, if a clearly defined therapeutic (or a compelling medical) reason is available, food or drink may be used on a specific occasion; however, overall, food or drink seldom have a place in psychotherapy or counseling.

Related to the discussion of permitting food and drink is the issue of smoking. More than any other behavior, smoking can serve as an anxiety-reducer that distracts the client; it should therefore never be permitted. Smoking by the therapist is equally distracting and interferes with the type of empathic relationship crucial to successful psychotherapy. Because of the unique relationship of therapy, it can never be assumed that a client's permission allowing the therapist to smoke is truly informed consent. Further, the profound health concerns of smoke in a closed room must be considered and must never be imposed on clients, even if they agree to the therapist's smoking or if they themselves smoke. Given these considerations, there can be no exceptions to the no smoking rule in therapy.

Personal Questions. Personal questions can range from simple requests for information to intrusive inquiries about the clinician's personal life. A few types of personal questions are appropriate and may be answered. Such appropriate questions most commonly have to do with a care provider's credentials, experience, and theoretical orientation; they can be answered matter-of-factly, directly, and nondefensively. Clinicians most vulnerable to mishandling these requests for information are novices who still have doubts about their abilities or who believe that they are not satisfactorily credentialed. It is important for these practitioners to be honest in their responses to the client's questions and to remain nondefensive. Clinicians do not have to justify their background in response to these questions; they merely have to state their credentials and experiences, allowing the client to decide whether the answers meet with the client's expectations, needs, and approval.

Most personal questions, however, have a different purpose and are best not answered. Instead, they should be responded to in the context of their meaning to the current therapy. In these situations, the clinician acknowledges the client's curiosity, validates that this curiosity is natural and common, and then returns the focus of the session to the client. If this simple approach suffices to return therapy to its intended purpose (i.e., shifts the attention back to the client's life), no further intervention is necessary. However, at times, clients are persistent in their inquiries and are not easily redirected in their focus. If this is the case, client and therapist should consider the purpose or underlying need fueling the client's questions. The counselor then responds to the underlying concern without answering the question. For example, a client may ask a clinician whether she or he has children. If the therapist is unable to redirect the focus through simple acknowledgment of the client's curiosity, she or he will need to try to find out why the client is asking this question at this time. Because most personal questions arise in a broad therapeutic context that guides the counselor toward understanding why the question emerged, this process sounds more difficult than it really is. Returning to the example, the therapist may note that the client had been talking about her own children and her self-doubts with regard to parenting. The clinician may deduce that the client is really asking about the therapist's expertness in the area of parenting. A response would therefore be directed toward this presumption and may be something like "You are wondering whether I will understand the complexity of issues involved in parenting three children." If the client acknowledges this as the truth, counseling can continue with a consideration of the client's concerns about the therapist's understanding and expertness. The issue may reemerge and will possibly need to be dealt with further, but clearly the question was not about whether the therapist has children but about the therapist's ability to understand the client; it is this latter topic that needs to be dealt with in the therapy.

Sexual Advances and Seductiveness. Sexual relationships may not be the only dual relationships clients attempt to establish with their therapists; however, they are certainly the most potentially damaging and destructive. All dual relationships must be avoided with clients. Mental health care providers must learn to resist offers of social interaction with clients to make sure that treatment remains nonexploitative and objective. Sexual advances and seductive behaviors are only the extreme end of the dual relationship continuum, but they serve well as examples of how to deal with clients' inappropriate dual relationship suggestions toward clinicians.

For some clients, seductive behavior is a normal pattern of relating, which they have developed over a lifetime of experiences in interpersonal contexts. Most seductive behavior can be viewed this way and hence can be understood as transferential in nature. Keeping this in mind may help tempted therapists recognize that they are not truly the target of admiration, but rather they are only convenient objects in the client's environment with whom to act out ingrained behavior patterns. At all times and under all circumstances, the therapist must neither give in to a client's sexual advances directly (by establishing a sexual relationship) nor indirectly (by allowing a mutually sexualized or seductive relational pattern with the client). The clinician draws and maintains firm, professional boundaries and does not respond to the sexual advances of the client. Instead, the clinician understands the client's sexualized behavior from its etiological or historical perspective and deals with it like any other relational pattern or resistance that emerges during treatment.

To do so, the counselor empathizes with the client's need for the behavioral pattern, communicates this understanding, accepts the client's needs and their developmental relevance, but does not gratify them. Instead, the clinician helps the client understand how and why the behavior developed, what function it has served and continues to serve, and what consequences it may have for the client's relationships. This process, while presented here in very few sentences, may well require many months of therapeutic work. Throughout this period, the clinician consistently sets firm bound-aries with the client around the behavior, remains committed to preventing a dual relationship with the client, avoids becoming seductive or sexual in personal reactions to the client, and consistently maintains firm and clear therapeutic boundaries.

Mandated Clients. Some clients seek therapy not by personal choice but because of external pressures. Child protection agencies, the criminal justice system, or employers may refer such clients. Frequently they will have been given a choice of either appearing for treatment or suf-fering a variety of consequences (e.g., incarcera-tion, removal of a child, termination of employ-ment). Additionally, clients may be pressured into treatment by spouses, family members, or signifi-cant others who have given the client some type of ultimatum. What all of these clients have in common is an external, rather than internal, mo-tivator for treatment that serves to intensify the natural ambivalence most clients feel about en-tering counseling or psychotherapy. Often these clients would rather not talk about themselves, do not trust the clinician's motives, and may take a hostile stance toward treatment (Patterson & Welfel, 1993). To these clients, therapy is not an act of freedom but rather an act of coercion, and "accepting the reluctant client involves accepting the client's reluctance as part of the agenda" for therapy (Patterson & Welfel, 1993, p. 207).

Dealing with mandated clients usually pre-sents a significant challenge, especially in the early stages of treatment when the client is de-ciding about level of commitment to therapeutic work. With some mandated clients, counseling may never be possible. However, there are a few interventions the mental health care provider can attempt to pull the client into the therapeutic process. First of all, the counselor acknowledges to the client that it is the client's right not to want to be in therapy and not to want to talk. The client's reluctance is addressed in a caring and concerned manner, communicating to the client that the therapist is not colluding with the external pressures that have forced the client into treatment. The client's reluctance and emotions

can be put into the larger context of the client's general experience of life and thus can represent a point of empathic entry into a positive rela-tionship with the client. Much of this early ther-apeutic work can center around the client's feel-ings about being forced into therapy, and the practitioner can join the client in recognizing and acknowledging the client's tough situation of being asked to perform a task that is not intrinsi-cally motivated.

Additionally, or alternatively, the clinician can attempt to shift the therapeutic focus to an area that may be more motivating for the client. The clinician may tell the client that, although the client was referred for a particular reason, there may be some other agenda that is important to the client and that can be integrated into the current therapeutic work. In other words, the therapist may identify a presenting problem that is genuinely bothersome to the client, not the re-ferring agent, thereby making the content rele-vant and interesting to the client. If this route is taken, both client and therapist must acknowl-edge that they may also have to work on the spe-cific referral requests made by the mandating agency. However, at least the client will have some input into the agenda and may feel more in control of a situation that initially seemed en-tirely externally controlled.

DATA GATHERING

To summarize, every counselor and therapist best explores certain personal and environmental factors before meeting the first client. These explorations, at a minimum, include a review of the environment in which the client is to be seen, an open and honest self-evaluation with re-gard to helpful personal traits, the knowledge of ethical and legal issues, and a clear appreciation about cultural issues. The clinician who has be-come familiar with these aspects of counseling and therapy has obtained a certain readiness for meeting the first client. Assuming that the clini-cian has ensured that personal traits are positive, the environment is conducive, ethical issues are observed, and cultural issues are attended to, a successful interaction with the client is possible. To make the first interaction with the client not only pleasant but also productive, the clinician also has a game plan for the process of the interview. This plan evolves as the clinician gains clarity about what questions a new client needs to be asked in order to be best understood and in order for the clinician to develop an optimal treatment plan. Fortunately, this process of data gathering is relatively straightforward and general. If applied with sufficient caring, flexibility, sensitivity, and empathy, even a novice mental health care provider can easily master the challenge of the first meeting with a client. The framework for gathering information is presented next.

There are several aspects about the client's life that are explored during the initial interview. Most importantly, the mental health care provider needs to gain a thorough understanding of the presenting problem that brought the client to treatment. To understand the essence and context of the presenting concern, additional data are collected. These are derived from the client's family history, sexual history, social history, academic and professional history, developmental and health history, substance use history, and nutritional and exercise history. Further, the counselor notes behavioral observations about the client and pays attention to his or her strengths. In gathering these data, the clinician takes care to use all of the therapeutic skills that have been learned through courses and textbooks in counseling or psychotherapy skills. It is beyond the scope of this book to discuss the acquisition of the microskills essential to the counseling process; it is assumed that the reader of this book already has basic knowledge in this regard. For a review of issues such as rapport building, questioning, empathy, reflection, and similar skills the reader is referred to Brems (1999a), Cormier and Cormier (1999), Egan (1994), Ivey (1994), and Teyber (1997).

Suffice it to say here that throughout the data gathering process, the clinician remembers that rapport building is as important as data gather-

ing. The essential components of therapeutic rapport that are established include:

- a communication of commitment by the therapist to the client and vice versa
- trust by the client in the therapist and a recognition that the therapist is a caring, knowledgeable expert in mental health issues
- stimulation of hope for change and improvement in the client that is shared by therapist and client, but that is felt foremost by the client
- belief by the client in the therapist's goodwill and basic benevolence
- the client's sense of safety that the therapist will not violate the client, either through breaches of confidentiality or through value judgments
- empathy on the part of the therapist for the client's situation and self-expression; understanding on the part of the therapist of the client's unique situation
- agreement on a structure for the relationship that keeps healthy boundaries and limits and that facilitates growth and change

Given the importance of rapport, clinicians maintain good relationships with the clients by using their clinical judgment about when to abandon data gathering during intakes. For example, if painful contents are discussed, it may be important to slow down and make sure that the client is not overwhelmed with affect or racing thoughts. An empathic listening and questioning style is essential and cannot be overrated. Developing all components of therapeutic rapport in a single session is impossible. A therapeutic relationship develops over a long period of time; however, often the foundation is laid in the very first hour of client contact, while the client's feelings of vulnerability are heightened by the newness of the situation. Nevertheless, the counselor structures the initial session more thoroughly than future therapy sessions because the counselor has a clear agenda, namely the gathering of information that will make it possible to concep-

TABLE 1-2   Questions to Be Answered about the Presenting Concern

What is the overriding presenting concern?

What are the circumstances?
- When does the problem arise?
- Where does the problem arise?
- How does the problem arise?
- With whom does the problem arise?
- How often does the problem arise?
- How intense is the problem?

What is the history of the presenting concern?
- How long has the problem been occurring?
- How has the problem changed over time?
- When is the first time the problem was noticed?
- When is the last time the problem was not at all present?

Why is the client seeking treatment now (what was the precipitating event)?

---

tualize the presenting concern of the client. A certain amount of directiveness and obvious clarity about the focus of the sessions is absolutely essential and conducive to the purpose of this process.

First and foremost, the counselor must determine why the client is seeking treatment. A clear definition of the presenting problem is an essential outcome of the initial meeting. A number of questions need to be answered once the initial interview is over. These are presented in Table 1-2 for quick and easy review. Table 1-2, as with all other tables included in this book, can be used by the clinician as a checklist to ensure that all the necessary information was gleaned during the initial interview. (It is best to memorize this list because it is not recommended that the clinician bring writing or written materials into the therapy room.) Once the presenting concern is clear to the clinician, a context is developed for understanding why and how the concern developed in the client's life. To create this context, a variety of additional data points are gathered. Often it is most logical, based on the types of questions that have already been asked, to move into a series of questions about the client's family. Most clients expect that they will be asked about their family and often volunteer informa-

tion, recognizing that their presenting concern is connected to the people to whom they are closest. Both family-of-origin and nuclear family issues are explored. The kind of information that is elicited through this line of questioning is presented in Table 1-3.

Once information has been gathered about the client's family, it is usually relatively easy to segue into the client's sexual and/or social history. With regard to questioning a client's sexual history, novice therapists can rest assured that there is no need to be shy or embarrassed when talking about these topics. There is often a certain hesitation among new counselors to broach the topic of sexuality, but most clients actually expect that they will be questioned about their sexuality. Popular media have assisted in this process through emphasis on sex as an important therapy topic and through the open discussion of sexual abuse in families. It is helpful to begin by asking a client about his or her first sexual experience. This line of questioning can lead to the revelation of childhood sexual abuse if it is present. Social history is a logical follow-up to sexual history since the exploration of the client's sexual life most certainly will involve significant others. Social history then moves beyond significant relationships to include social support

TABLE 1-3 Family History

TYPE OF FAMILY	MAIN ISSUE	SUBISSUE
Family-of-Origin	identification of all family members with whom client interacted in childhood	
	family interactions during childhood and adolescent years	
		interactions with siblings, stepsiblings, half-siblings, etc.
		interactions with parents, stepparents, foster parents, etc.
	structure of family (persons, relationships, communication, etc.)	
		generational boundaries
		coalitions
	family experiences during childhood and adolescent years	
		parenting styles experienced
		communication styles and patterns
		memories in the family setting
		family trauma (history of abuse, witnessing domestic violence)
	parental family background	
		parental family trees
		parental family experiences
		parental experience of childhood and adolescent trauma
		parental medical and psychiatric history
	current interactions with family of origin	
		genogram or family genealogy (optional)
Nuclear Family	identification of nuclear family members in client's adult life	
	nuclear family interactions	
		interactions with significant others (SOs) and former SOs
		interactions with children, stepchildren, foster-children, etc.
		interactions with family of SOs
	structure of family (persons, relationships, communication, etc.)	
		generational boundaries
		coalitions
	nuclear family experiences	
		parenting styles exercised with own children
		memories in the family setting
		family trauma (domestic violence, perpetration of abuse)
		communication styles and patterns
	functionality of the family	

networks, interests, and the many other issues listed in Table 1-4.

After the counselor has gained a thorough understanding of the client's social and economic circumstances, an exploration of the person's academic and professional history is rather logical and the counselor can smoothly segue into this set of questions. The content that needs to be ex-plored in this regard is shown in Table 1-5. The next shift can be somewhat more awkward, for the therapist now focuses attention on health-related and developmental issues. It is best to introduce the client to the shift by saying something to the effect of, "Now let's shift gears a little bit. Often there can be a strong connection between how a person feels emotionally and

TABLE 1-4   Sexual History, Social/Economic History, and Professional/Academic History

CATEGORY	ISSUES TO BE ADDRESSED	SUBISSUES TO BE ADDRESSED
Sexuality	sexuality in current intimate relationship	
		quality
		frequency
		enjoyment, compatibility
	sexuality in other relationships	
		quality
		frequency
		enjoyment, compatibility
	first sexual experience (as a possible lead in to sexual abuse)	
	later sexual experiences	
		quality
		frequency
		masturbation
		enjoyment, compatibility
	sexual abuse (incest, molestation, rape)	
		perpetrator(s)
		specifics about the abuse (type, form)
		age, frequency, and duration
		events surrounding the abuse (e.g., where, when, threats made)
		presence of a protector/confidant
Social History	number and description of close friends	
		current
		past
	acquaintances and colleague relationships	
		current
		past
	interests, hobbies, recreational activities and interests	
		current
		past
	number and level of involvement in community groups (e.g., environmental protection groups, sports leagues, special interest groups, hiking clubs, collectors clubs, professional associations, reading circles)	
		current
		past
	religion and spirituality, including church memberships	
	sociocultural issues	
		socioeconomic variables
		cultural background
		ethnic background
		minority status
		level of acculturation
		context of acculturation (forced versus voluntary)
		world views

Continued

TABLE 1-4   Sexual History, Social/Economic History, and Professional/Academic History (continued)

CATEGORY	ISSUES TO BE ADDRESSED	SUBISSUES TO BE ADDRESSED
Professional History	description of current employment	
	career plans and aspirations	
	jobs and/or occupations in the past	
Academic History	adult academic or vocational preparation background	
		degrees or certificates
		performance (i.e., grades, level of success)
		problems (e.g., learning disabilities or physical impediments)
	school (K to 12) background	
		graduation
		performance (i.e., grades, level of success)
		problems (e.g., learning disabilities, peer relationships)

what goes on with a person's body and health. So I am going to ask you a number of questions that may not seem to you to be related to your reason for coming here, but they will help me make sure that I'm not missing anything." Then health-related, development, nutritional, substance-use, and exercise history can be gathered directly with greater ease. Table 1-5 summarizes the specific issues that the counselor reviews. The most common challenges that arise during the initial session with a new client have to do with medical or physical issues as well as substance use. Rather than attempt to deal with the intricacies of both of these areas cursorily in this context, they will be covered in depth in Chapters 2 and 3.

Throughout the process of gathering intake data, the clinician continually makes behavioral observations about the client, which will help round out the understanding of that person. The types of behavioral observations that can be made are listed in Table 1-6. Finally, the intake interview needs to result in an appreciation of the person's strengths. This issue can be addressed through directly questioning the client as well as through behavioral observation and inference. It is always interesting to hear how clients respond when asked what they perceive as their unique

strengths or abilities. Notably, the absence of an answer to this question may speak volumes. The therapist's focus on this issue also reaffirms his or her appreciation for the client as a whole human being, not just as a person who has problems. Strengths identified in the client can then be used for treatment planning and prognosis. Although there is no way to provide a complete listing of potential client strengths, some examples are provided in Table 1-7. This list will give the therapist an idea of the wide range of client characteristics that can be assessed in this regard; the list is not intended to limit exploration to these particular issues.

The gathering of all of these data is important because the clinician hopes to leave the initial interview with a relatively clear understanding of the client—not just of the presenting concern, but of the entire familial and social context in which this concern has arisen. Conceptualizing a client is like helping the client identify the core of the presenting concern and its relationship to the client's true self. It considers the motivations that fuel the presenting concern from the client's unique perspective and history. The therapist attempts to recognize and explain the client's reality as a means of deciding how best to help the

TABLE 1-5   Health History, Substance Use History, and Nutritional/Exercise History

CATEGORY	ISSUES TO BE ADDRESSED	SUBISSUES TO BE ADDRESSED
Health History	developmental issues	
		mother's pregnancy (i.e., in utero development and exposure)
		birth information
		developmental milestones
		previous mental health treatment
	prior therapy or counseling	
		prior psychological testing or assessment
		school assessments
		vocational assessments
	medical trauma	
		injuries, recent and past
		accidents, recent and past
		head injuries, recent and past
	physical health	
		current severe diagnosed illness
		past acute illness
		date, circumstances, and findings of last physical examination
		name of physician and other health care providers
		hospitalizations, recent and past
		current medical treatments other than medications
		current medications
Substance Use	prescription drug use	
		type(s)
		recency
		frequency
		amount
	over-counter drug use	
		type(s)
		recency
		frequency
		amount
	use of legal drugs: alcohol, tobacco products, caffeine	
		type(s)
		recency
		frequency
		amount
	use of illegal substances (e.g., marijuana, cocaine, amphetamines, hallucinogen, barbiturates, inhalants)	
		type(s)
		recency
		frequency
		amount

Continued

TABLE 1-5   Health History, Substance Use History, and Nutritional/Exercise History (continued)

CATEGORY	ISSUES TO BE ADDRESSED	SUBISSUES TO BE ADDRESSED
Substance Abuse	family history of substance use	
		medications (e.g., prescription and over-the-counter)
		legal drugs (e.g., alcohol, tobacco, caffeine)
		illegal substances (e.g., cocaine, inhalants, marijuana)
		issues related to adult children of substance users
Nutrition	daily food intake, exploring timing and quantities	
		breakfast
		lunch
		dinner
		snacks and desserts
	special diets	
		diets for physical illnesses (e.g., for diabetes or hypoglycemia, for heart or cardiovascular disease)
		vegetarian diets
		macrobiotic diets
	daily liquids intake, exploring timing and quantities	
		water
		juices
		soft drinks
		hot drinks
	daily exercise routine	
	awareness of nutrition and exercise needs	
	inappropriate use of food (e.g., under- or overeating; bingeing and purging)	
	inappropriate use of exercise	
	family attitudes about foods, liquids, and exercise	

client grow, change, and heal. Conceptualizations are not a straightforward process and are subject to revision as new information emerges about the client. In other words, although the clinician gives the initial conceptualization much thought and prepares it carefully based upon all the history that has been collected in the initial session, it generally can remain open to minor revision.

CONCEPTUALIZATION

Dealing with case conceptualization is clearly beyond the scope of this book. The complexity of this issue is underscored by the fact that entire books (Berman, 1997; Eells, 1997) have been written about this very topic. Developing a conceptualization that will be used to determine an individual treatment plan is a process, not a single static decision, and it is highly contextual and flexible. It must be clearly differentiated from a diagnosis; the terms *diagnosis* and *conceptualization* cannot really be used interchangeably, though many writers unfortunately continue to do so. A conceptualization is much broader than a diagnosis. A diagnosis considers a client's symptoms to arrive at a label that will be used to classify the client's behavior within a particular diagnostic category. A conceptualization, on the other hand, considers why, when, with whom, and how a client's pre-

TABLE 1-6   Behavioral Observations to Be Described by the Therapist

CATEGORY	ISSUES TO BE ADDRESSED
Appearance	dress
	grooming
	posture
	gestures and facial expressions
	manners and habits
	characteristic physical features
Cognitive Functioning	estimated intellectual level
	attention and concentration
	stream and clarity of thought and speech
	long- and short-term memory
	level of abstraction versus concreteness
	level of cognitive flexibility versus rigidity
	distorted thought content (e.g., fixed beliefs, obsessions, delusions)
	distorted thought processes (e.g., ruminations, loose or strange associations)
Affect and Mood	primary mood (i.e., subjective feelings)
	primary affect (i.e., expressed or observed feelings)
	range of affect
	appropriateness of affective expression
	congruence between affect and topic of conversation
	congruence between affect and mood
Sensorium/Perception	reality testing/appraisal
	perceptual disturbances (e.g., illusions, hallucinations)
	orientation to time, place, person, and self
	judgment
	alertness
Interpersonal Style	level of self-assurance displayed
	quality of interaction (e.g., hostile, challenging, trying to please, submissive, passive)
	ability to engage and self-disclose
	level of trust (i.e., ranging from paranoia to overly trusting)
	need for acceptance
Other Observations	changes in behavior over the course of the interview
	psychomotor movement
	speech patterns

senting concern developed; a conceptualization is not concerned with how best to label the client.

A conceptualization takes a look at three primary sets of factors: predisposing factors, precipitating factors, and perpetuating factors (Brems, 1999a). It then outlines the dynamics of the case, detailing intrapsychic, interpersonal matrix, and family-related dynamics that appear to drive the client's presenting concern and general existence. The conceptualization is not complete until it

TABLE 1-7   Examples of Client Strengths Identified by the Therapist

- good cognitive skills
- well educated
- good memory
- abstract reasoning skills
- no severe cognitive disturbances (such as delusions or ruminations)
- adequate support network
- stable intimate relationship
- some contact with family of origin
- loose network of acquaintances
- history of hard work and perseverance
- stable work history
- completion of chosen educational track
- follow-through on prior medical treatment regimen
- absence of significant childhood trauma
- no report of having been a victim
- no report having been a perpetrator
- no current health problems
- no history of substance use
- some evidence of interpersonal/social skills
- ability to express a range of affect
- absence of psychotic symptomatology
- hope about the possibility that therapy will be helpful
- self-motivation and interest in psychological issues

accounts for all problems, behaviors, cognitions, and affects presented by the client (Weiss, 1993). It considers the context for each presenting problem, integrating all apparently separate parts of the client into one cohesive and holistic network of events and experiences that can explain apparent inconsistencies or contradictions (Karoly, 1993). A good conceptualization is free of biases or stereotypes and keeps all attributions logical and rational (Olson, Jackson, & Nelson, 1997). In the words of Basch (1980), a therapist should not conceptualize "simply on the basis of the main complaint, nor should he [or she] center on a patient's symptom. The therapist should consider the context in which the complaint is made or in which the symptom occurs, for it is the context that often leads to an understanding

of what is going on with the patient and of what needs to be done for him" (p. 121).

The conceptualization is the logical outgrowth of the initial interview and guides the client's treatment from here on out. It becomes the basis of the client's treatment plan and helps the clinician understand the client from a unique and idiosyncratic perspective. The conceptualization reflects the clinician's theoretical orientation and as such can vary greatly from therapist to therapist.

REFERRAL

After therapists have conducted intakes, and even after they have made several special assessments (such as a substance use assessment or a mental status exam), they still commonly conclude that they are missing critical data. Data that are lacking, even after a thorough assessment by the original clinician, generally fall into one of two categories: (1) data that cannot be collected by the intake clinician because of limitations in training or expertise; or (2) data that should not be collected by the original clinician because of rapport, role, and/or boundary issues in the work with a client. With regard to the first category, the medical portion of the intake, for example, may have revealed some indication of memory disturbances that was only expanded upon, but not clarified, by the behavioral observations. In this case, the clinician must decide how to understand the client's memory problems and may need to conclude that she or he is unable to do so without additional data. Such additional data, however, may need to be collected by a neuropsychologist or a neurologist, that is, by professionals with skills not claimed by the intake clinician. Alternatively, with regard to the second category, the clinician may have indeed received vast training in neuropsychological assessment but may decide that it would be unwise to conduct this assessment personally, since such an action might intrude upon the therapeutic relationship. In either case, it is time for the initiation of a referral.

To clarify, referrals generally occur after an initial screening or assessment has been conducted by an intake clinician. A screening in this context refers to a quick intake that is focused and directive; assessment refers to an intake evaluation that is more comprehensive, where the clinician collects as many data as possible about the client. When screening or assessment results in questions the intake clinician is unable to answer, a referral is in order. Referrals can be made for additional assessment or for consultations. If additional assessment is the intention, the clinician refers the client to another provider who will conduct an assessment that goes beyond the original clinician's assessment. The primary contact is between the client and the additional provider (the referral target), with the referring

provider (the referral source) receiving a report, but not necessarily having verbal contact with the referral target. If the referral initiates a consultation, the primary contact may be between the referral target and the client or between the referral target and the referring clinician. Generally in such a case additional information is sought about the client that already exists (e.g., information from a psychiatrist who is already prescribing medications to a client). Common situations that result in referrals to other mental health, behavioral health, and health care specialists include questions about the following:

- cognitive and perceptual processes, such as evidenced in psychosis or organic brain disease (referral for neurology or neuropsychology)

- intellectual functioning and cognitive processing, not related to psychosis or brain disease, but related to day-to-day performance and academic achievement (referral for psychological evaluation with focus on cognitive testing [e.g., intelligence, achievement, and academic testing])

- coping ability and problem-solving style in specific circumstances (referral for psychological evaluation with focus on cognitive and personality variables [e.g., intelligence testing, objective personality assessment, and projective assessment])

- personality functioning and related intrapsychic and interpersonal dynamics within a context of community and culture (referral for psychological evaluation with focus on objective and projective assessment)

- the relationship between medical and psychological/emotional symptoms (referral for medical evaluation)

- medication regimens currently taken by the client (consultation with prescribing physician[s])

- psychiatric symptoms and their need for amelioration above and beyond psychotherapeutic or psychoeducational intervention (referral for psychiatric evaluation with assessment of need for psychotropic medication)

- the interface between psychological/emotional presenting concerns and perceptual deficits (referral for hearing and other perceptual skills [medical] screenings)

- communication difficulties beyond interpersonal or intrapsychic interfering factors (referral for speech and language evaluation)

- food intake or weight concerns that are integral to the psychological presentation and have potential physical implications (referral for nutrition evaluation by a physician or nutritionist)

- social and economic factors that interfere with healthy psychological adjustment and may have effects on medical well-being (referral for a social work evaluation)

All of these issues exceed the most usual presentation of clients for psychotherapy. However, all occur with sufficient frequency that therapists need to know how to deal with them and need to recognize them as issues that are best attended to through referral for ancillary services, not taken on by the therapists themselves. By not making referrals in these situations, therapists lay the foundation for potential mismanagement of their cases. Being conscientious about collecting a complete set of data from a client before finalizing conceptualizations and treatment plans is critical. Working with an incomplete or inaccurate set of data may lead to less-than-optimal interventions, at best slowing down the therapeutic process and at worst interfering with therapeutic progress. One example may be a client with severe depression of recent onset that appears to have almost completely immobilized the person: it interferes with proper physiological functioning (e.g., sleep, appetite, sexual interest) and results in social interference (e.g., missing work, self-isolation). A mental health clinician may be remiss by not referring such a client for a complete physical exam and possible psychotropic medication. The failure

to add antidepressant medication, at least for a while, to the treatment of a client with such a debilitating presentation of major depression may actually slow down or hinder therapeutic process. The clinician will be much more capable of therapeutic work if the medication has helped clear up some of the client's most interfering physiological manifestations of the depression. How to recognize the need for, and how to go about making, a referral will be the focus of Chapters 3 and 4.

DOCUMENTATION AND RECORD KEEPING

Writing case notes and reports is an essential clinical, ethical, and legal obligation of the care provider. Records must document and review the services that were provided and will guide treatment planning and implementation. In addition, they may be relevant for institutional purposes (e.g., for accreditation or review of a clinic), legal proceedings (e.g., malpractice suit against a therapist), and financial reasons (e.g., review of services for resource allocation). Records serve to protect the clinician from legal liability in case of lawsuits or ethical charges, and in some states they should be kept in accordance with state statutes regulating the content of mental health records (APA, 1993; NASW, 1993). Records must include a comprehensive, objective, consistent, retrievable, secure, and current range of information about the client, the clinician, and the services that were provided (Vandecreek & Knapp, 1997). Records are protected against unauthorized access and misuse, especially if a computer has been used for data storage and note writing (Swenson, 1997; Vandecreek & Knapp, 1997). A complete record must be kept for at least three years after the last contact with the client (unless the clinician practices in a state that has a law specifying a different time frame); a summary record must be kept for up to an additional 12 years (APA, 1993). Maintenance of records must be guaranteed to be confidential, safe, and done in such a way as to limit access (e.g., under lock and key, in a restricted area, in a computer with some form of encryption or a ROM-lock device). When the records can be disposed of, this must be accomplished in an equally confidential manner (e.g., the record may be shredded).

In keeping records, the clinician remembers that records can be subpoenaed, and hence documents carefully, weighing which material should be included. Points that need to be made are stated succinctly, without speculation or overly subjective interpretation. Notes must be clear and must honestly record what is known and what has transpired. They need to be respectful of the client, avoiding any type of judgment, prejudice, or hostility against the client. The clinician needs to consider carefully the implications the notes will have; the notes generally do not mention symptoms that are questionable or that have not been attended to through therapeutic intervention or referral. Records best conform to any clinic-specific rules, accreditation organization guidelines, or professional guidelines that are available to the clinician (Gutheil, 1980).

The intake interview is summarized in report format and the report becomes an important and permanent part of the client's mental health record. At a minimum, the intake report usually contains the following sections:

- identifying information
- history of presenting concern
- family history
- social history
- professional and academic history
- medical and related health history
- behavioral observations and brief mental status
- conceptualization
- diagnosis
- treatment recommendations
- clinician signature and date

If new data essential to information contained in the intake report emerge in sessions occurring after the intake interview, the best course of action is to prepare an Intake Report Addendum. All other information that needs to be recorded about the client during routine sessions is taken down in the form of progress notes. Progress notes are written after each contact with the client and include any contacts between sessions. The ideal progress note addresses client presentation, therapy or counseling process, clinician's interpretation, and plan for the next session(s) (see Brems, 1999a). Here presentation refers to client-related observations and information (such as objective, observable behavior, presentation, and mental status), any unusual behaviors noted in the client, new data relevant to the client's history, life changes related by the client, changes in condition or status of the client as compared to prior sessions, and the client's movement or progress toward treatment goals. Process refers to notes about content and themes of the session, interventions, their effectiveness, and the client's response, as well as here-and-now process as related to interactions the client had with individuals outside of therapy. Interpretations indicate the therapist's understanding of the process as it unfolded and reactions to and impressions about the session. The plan evaluates selected strategies and reveals the counselor's direction for the next session.

Progress notes emphasize client progress (i.e., changes and plans), as opposed to process notes, which tend to be slightly lengthier and more subjective. Perhaps in part due to an increased number of lawsuits involving counselors and therapists, progress notes have become the preferred mode of recording session occurrences. They are briefer (no more than a half to a full typed page), more objective, and less involved. The exception is the recording of sessions that involved a crisis. Such sessions are recorded in more detail and often are lengthy and specific. The special format of progress notes for crisis sessions is dealt with in the subsequent chapters according to the type of challenge the clinician and client have faced together.

SUMMARY AND CONCLUDING THOUGHTS

Meeting a new client presents the counselor with many interesting challenges, none of which tends to be problematic as long as the counselor is well prepared. This chapter has laid the foundation for a successful first contact with a client. Most importantly, clinicians need to be aware of the environmental and personal factors they bring to sessions, need to be clear about ethical and professional guidelines, and they need to have a set of structuring skills that guide their interview. They are cognizant of a clear data-gathering agenda and are sufficiently directive to collect all the necessary information to arrive at a comprehensive conceptualization of the client by the end of the first contact. At the same time, the counselor must be empathic and keeps in mind that forming rapport overrides data gathering. When a decision has to be made between either pursuing a certain line of questioning that greatly upsets the client or abandoning the quest for information in order to keep the client comfortable, the decision is easy. In the first session, the client needs to be kept relatively comfortable and needs to leave feeling better than before entering the treatment room. Although this is not always true for subsequent sessions, the initial session sets the tone for the client's perception of safety with the clinician, and hence a slightly different standard of comfort is followed.

The initial session rarely presents the counselor with overly burdensome challenges. Primarily, challenges that arise in the initial session have to do with decisions about which additional data may be essential to understanding the client most accurately, that is, to complete the conceptualization in the most descriptive and empathic manner. Sub-stance use and medical issues are two areas in which questions tend to arise. These topics will be discussed in detail in the following two chapters. With regard to substance use, Chapter 2 will demonstrate that much assessment and intervention can be done without immediate referral, although a referral usually will become necessary at some time. Chapter 3 will show that referral is always necessary, however, for medical and physical involvement.

Finally, sometimes a counselor leaves a session with a client knowing that several variables about the person remain unclear but not knowing how to go about filling in the clinical picture. In those instances, referral for psychological testing may be helpful in rounding out the client's conceptualization. Such assessment can also be useful when a client's intellectual potential is in question, when achievement appears to lag behind cognitive potential, or when vocational indecisiveness is part of the presenting concern. Chapter 4 outlines the clinical situations in which psychological testing may be indicated and describes how the clinician can make appropriate referrals for psychological assessment.

Special Note: The following informed consent was developed by this author based upon a sample provided by Drs. Bruce Bennett and Eric Harris of the APA during an APA- and AKPA-sponsored workshop on Risk Management in Psychology in Anchorage, Alaska in November 1995. The original sample (called an *Outpatient Service Contract*) can be requested from the APA Insurance Trust at 750 First Street NE, Suite 605, Washington, DC 20002-4242. The author's version of this consent form has been previously printed in Brems (1999a) and is reprinted here with permission.

Sample Informed Consent

Informed Consent

Psychological Services Center   Date:_____ _____
3211 Providence Drive   Name:_____
Anchorage, Alaska 99508

Welcome to the Psychological Services Center. This document contains important information about our center's professional services and business policies. Please read it carefully and note any questions you might have so you can discuss them with the clinician conducting your screening appointment. Once you sign this consent form, it will constitute an agreement between you, your therapist(s), and the Psychological Services Center (PSC).

Nature of Psychological Services

Psychotherapy is not easily described because it varies greatly depending on the therapist, the client, and the particular problems a client presents. There are often a variety of approaches that can be utilized to deal with the problem(s) that brought you to therapy. These services are generally unlike any services you may receive from a physician in that they require your active participation and cooperation.

Psychotherapy has both benefits and risks. Possible risks may include the experience of uncomfortable feelings (such as sadness, guilt, anxiety, anger, frustration, loneliness, or helplessness) or the recall of unpleasant events in your life. Potential benefits include significant reduction in feelings of distress, better relationships, better problem-solving and coping skills, and resolutions of specific problems. Given the nature of psychotherapy, it is difficult to predict what exactly will happen, but your therapist will do her or his best to make sure that you will be able to handle the risks and experience at least some of the benefits. However, psychotherapy remains an inexact science and no guarantees can be made regarding outcomes.

Procedures

Therapy usually starts with an evaluation. It is our practice at the PSC to conduct an evaluation that lasts from 2 to 4 sessions. This evaluation begins with a screening appointment and is followed up with an intake interview (that may last 1 to 3 sessions). During the evaluation, several decisions have to be made: the therapist has to decide if the PSC has the services needed to treat your presenting problem(s), you as the client have to decide if you are comfortable with the therapist that has been assigned to you, and both you and your assigned therapist have to decide on your goals for therapy and how to best achieve them.

In other words, by the end of the evaluation, your therapist will offer you initial impressions of what therapy will involve, should you decide to continue. Therapy generally involves a large commitment of time, money, and energy, so it is your right to be careful about the therapist you select. If you have questions about any of the PSC procedures or the therapist who was assigned to you, feel free to discuss these openly with the therapist. If you have doubts about the PSC or your assigned therapist, we will be happy to help you to make an appointment with another mental health professional.

If you decide to seek services at the PSC, your therapist will usually schedule one 50-minute session per week at a mutually agreed time (under some special circumstances sessions may be longer or more frequent). This appointment will be reserved for you on a regular basis and is considered a standing appointment (i.e., if you miss one week, you will still have the same appointment time next week). The overall length of psychotherapy (in weeks or months) is generally difficult to predict but is something your therapist will discuss with you when the initial treatment plan is shared with you after completion of the evaluation.

Continued

Sample Informed Consent (continued)

Fee-Related Issues

The PSC works on a sliding fee schedule that will be discussed with you during your screening appointment. Screening appointments always cost $____. Fees for therapy range from $____ to $____, depending on income and therapist. In addition to charging for weekly appointments, the PSC also charges special fees for other professional services you may require (such as telephone conversations which last longer than 10 minutes, meetings or consultations that you have requested with other professionals, etc.). In unusual circumstances, you may become involved in litigation wherein you request or require your therapist's (and her or his supervisor's) participation. You will be expected to pay for such professional time required even if your therapist is compelled to testify by another party. You will be expected to pay for each session at the time it is held, unless you and your therapist agreed otherwise. Payment schedules for other professional services will be agreed to at the time that these services are requested. In circumstances of unusual financial hardship, you may negotiate a fee adjustment or installment payment plan with your assigned therapist. Once your standing appointment hour is scheduled, you will be expected to pay for it (even if you missed it) unless you provide 24 hours' advance notice of cancellation.

To enable you and your therapist to set realistic treatment goals and priorities, it is important to evaluate what resources are available to pay for your treatment. If you have a health benefits policy, it will usually provide some coverage for mental health treatment when such treatment is provided by a licensed professional. Your therapist will provide you with whatever assistance possible to facilitate your receipt of the benefits to which you are entitled, including completing insurance forms as appropriate. However, you (*not* your insurance company) are responsible for full payment of the fee. If your therapist is a trainee, you will not be able to use your insurance benefits.

Carefully read the section in your insurance coverage booklet that describes mental health services and call your insurer if you have any questions. Your therapist will provide you with whatever information she or he has based on her or his experience and will be happy to try to help you understand the information you receive from your carrier. The escalation of the cost of health care has resulted in an increasing level of complexity about insurance benefits that often makes it difficult to determine exactly how much mental heath coverage is available. Managed health care plans such as HMOs and PPOs often require advance authorization before they will provide reimbursement for mental health services. These plans are often oriented towards a short-term treatment approach designed to resolve specific problems that are interfering with level of functioning. It may be necessary to seek additional approval after a certain number of sessions. Although a lot can be accomplished in short-term therapy, many clients feel that more services are necessary after the insurance benefits expire. Some managed care plans will not allow your therapist to provide reimbursed services to you once your benefits are no longer available. If this is the case, PSC staff will do their best to find another provider who will help you continue your psychotherapy.

Please be aware that most insurance agreements require you to authorize your therapist to provide a clinical diagnosis, and sometimes additional clinical information such as a treatment plans or summaries, or in rare cases, a copy of the entire record. This information will become part of the insurance company's files, and, in all likelihood, some of it will be computerized. All insurance companies claim to keep such information confidential, but once it is in their hands, your therapist has no control over what your insurer will do with the information. In some cases, the insurer may share the information with a national medical information data bank.

It is best to discuss all the information about your insurance coverage with your therapist, so you can decide what can be accomplished within the parameters of the benefits available to you and what will happen if the insurance benefits run out before you are ready to end treatment. It is important to remember that you always have the right to pay for psychological services yourself if you prefer to avoid involving your insurer.

Sample Informed Consent (continued)

Contact Hours

The PSC is open from 9 A.M. to 5 P.M. daily, but some evening and weekend appointments may be available. Your therapist is generally not available for telephone services but you can cancel and reschedule sessions with the assistance of the receptionist or through leaving messages on the confidential answering machine. If you need to reschedule an appointment, your therapist will make every effort to return your call on the same day, with the exception of calls made after hours or on weekends and holidays. If you are difficult to reach, please leave some times when you will be available. If you have an emergency but are unable to reach your therapist, call your family physician, emergency services at Southcentral Counseling Center, or the emergency room at the Providence or Charter Northstar Hospital. Please note that the PSC itself does not have emergency services or facilities.

Videotaping and Record-Keeping Procedures

Because the PSC is a training facility for therapists, all client sessions are videotaped. These tapes are made for supervision purposes only and are kept in a confidential locked area until the therapist has reviewed them with the supervisor. Once tapes have been reviewed with the supervisor, they are erased. Most tapes are erased within less than one week. In addition to videotapes, therapists also keep case notes. These notes are also kept under lock and key and are strictly confidential. The case notes are reviewed by the therapist's supervisor and then are filed permanently in a client's record. All records are locked and kept confidential. Information in the client records may be used for research and service planning purposes. Such use is entirely anonymous and no individual client data will ever be used. No client names are ever associated with data extracted from records for research purposes and all data will merely be presented in group data format, a format that preserves anonymity and never reveals individual client data.

Both law and the standards of the profession of psychology require that therapists keep treatment records. You are entitled to receive a copy of these records, unless your therapist believes that seeing them would be emotionally damaging to you. If this is the case, your therapist will be happy to provide your records to an appropriate mental health professional of your choice. Although you are entitled to receive a copy of your records if you wish to see them, your therapist may prefer to prepare an appropriate summary instead. Because client records are professional documents, they can be misinterpreted and can be upsetting. If you insist on seeing your records, it is best to review them with your therapist so that the two of you can discuss their content. Clients will be charged an appropriate fee for any preparation time that is required to comply with an informal request for record review. If you are under 18 years of age, please be aware that the law may provide your parents with the right to examine your treatment records. It is PSC policy to request an agreement from parents that they consent to give up access to your records. If they agree, your therapist will provide your parents only with general information on how your treatment is proceeding unless there is a high risk that you will seriously harm yourself or another person. In such instances, your therapist may be required by law to notify your parents of her or his concern. Parents of minors also can request to be provided with a summary of their child's treatment when it is complete. Before giving your parents any information, your therapist will discuss this matter with you and will do the best she or he can to resolve any objections you may have about what will be discussed. Please note that the PSC does not provide treatment to minors without their parents' consent.

Confidentiality

In general, the confidentiality of all communications between a client and a psychologist is protected by law, and your therapist can release information to others about your therapy only with your written permission. However, there are a number of exceptions:

Continued

Sample Informed Consent (continued)

- In most judicial proceedings, you have the right to prevent your therapist from providing any information about your treatment. However, in some circumstances (such as child custody proceedings and proceedings in which your emotional condition is an important element), a judge may require your therapist's testimony if the judge determines that resolution of the issues demands it.
- There are times when it may be helpful for other professionals to gain access to all or parts of your treatment records. Under such circumstances data can be released from your PSC record if you give your therapist written permission (in the form of a Release of Information) to do so. Such release cannot take place unless you consent in writing!
- There are some situations in which your therapist is legally required to take action to protect others from harm, even if such action requires revealing some information about your treatment: (1) If your therapist believes that a child, an elderly person, or a disabled person is being abused, she or he is required by law to file a report with the appropriate state agency; (2) If your therapist believes that you are threatening serious bodily harm to another person, she or he is required by law to take protective actions, which may include notifying the potential victim, notifying the police, or seeking appropriate hospitalization; (3) If you threaten to harm to yourself (e.g., suicide), your therapist is required to make all necessary arrangements to protect your safety, a process which may include seeking hospitalization for you, or contacting family members or others who can help provide protection.
- At the PSC, virtually all therapists are students in a clinical psychology program and therefore are not yet licensed to work independently. As such, all therapists in the PSC are working under the supervision of licensed psychologists who meet with the therapists on a weekly basis in a one-on-one meeting to review client cases. Your case will be discussed during these weekly meetings between your therapist and her or his supervisor. The supervisor is a part of the PSC staff and is bound by the same confidentiality laws as your therapist. The supervisor and therapist will also review therapy tapes as necessary to the therapist's education.
- At the PSC, all therapists participate in weekly staff meetings. These meetings are used as opportunities for consultation and each week a client case is presented. Your case may be one of the cases that will be presented during these meetings. The meetings are confidential and only PSC staff participate. Only first names of clients are used and all members of the PSC staff are bound by the same confidentiality laws as your therapist.

Signatures Verifying Agreement

Your signature below indicates that you have read the information in this document, that you have understood it, and that you agree to abide by its terms as long as you are a PSC client.

Client Signature and Date _____

Witness Signature and Date _____

2

The Challenge of Substance Use

> However disregulated the brain, addiction is not merely a neurochemical disease.
> However maladaptive, addiction is not merely a cognitive and behavioral disorder. Both
> the physiological and behavioral aspects of addiction are situated within a larger web of
> meaning—a meaning that is constituted by the culture and the particular individual's
> relationships with family, friends, lovers, and, if relevant, therapists.
>
> DAVID MARK AND JEFFREY FAUDE, *PSYCHOTHERAPY OF COCAINE ADDICTION*

General mental health practitioners usually have not spent a lot of time thinking or reading about substance use, and they often are not properly prepared to assess and treat substance-using clients. This lack of preparedness is an unfortunate reality, since substance use is one of the most common mental health problems in the United States, and there is a strong likelihood that substance-using individuals present for treatment in mental health settings (see Brems & Johnson, 1997 for a review). Every therapist and counselor needs to have some basic awareness of substance use issues and must be prepared to meet and identify the substance-using client. At a minimum, mental health care providers are knowledgeable about the types of commonly abused drugs, the conduct of thorough substance use assessment, special issues involved in making substance use diagnoses, and appropriate referral for treatment (Brems & Johnson, 1997; Chappel, 1993; Clement, Williams, & Water, 1993; Galanter, Egelko, Edwards, & Vergaray, 1994). This chapter is not about treatment of substance-using clients; it merely deals with identifying and

referring them for appropriate care. The assessment and diagnosis of substance use is a crucial and generic skill that prevents counselors from making incorrect general diagnoses. However, the treatment of substance use is a specialized area of intervention that requires additional supervised training and is beyond the scope of this book (Miller & Brown, 1997). Epidemiological data prove the prevalence of substance use in society in general and in mental health practice in particular, as well as its relationship to other mental illness and its grave consequences for the individual, the family, and the larger community. Such data underscore the need for therapists to learn proper assessment and diagnostic skills.

EPIDEMIOLOGICAL DATA

Epidemiological Catchment Area (ECA) studies funded by the National Institute of Mental Health (see Regier et al., 1990) found that 13.5% of the general population report a lifetime occurrence of alcohol abuse or dependence and

5.9–6.1% report drug abuse or dependence (Helzer & Przybeck, 1988; Regier et al., 1990). Conducted from 1991 to 1993, the National Comorbidity Survey (NCS) found similar rates with 14% of respondents reporting lifetime symptoms of alcohol abuse or dependence and 7.5% reporting lifetime drug dependence (Kessler et al., 1994; Warner, Kessler, Hughes, Anthony, & Nelson, 1995). This compares to ECA survey lifetime prevalence rates for the general population for phobias of 12.6%, simple phobias of 10.0%, major depression of 5.9%, agoraphobia of 5.2%, social phobia of 2.8%, and antisocial personality disorders of 2.6% (Regier et al., 1990). The NCS found rates for major depression of 17%, social phobias of 13%, simple phobias of 11%, and dysthymia of 6.4% (Kessler et al., 1994). These comparisons reveal that alcohol and other drug use and dependence rank among the greatest mental health risk disorders in the country.

Even more important for the mental health counselor, Primm (1992) indicates that the presence of a mental disorder almost triples the likelihood of the presence of a substance use disorder, as compared to the general population (2.7:1). Although overall coexistence rates for psychiatric and substance use disorders vary from 25–58% (Alexander, Craig, MacDonald, & Haugland, 1994; Miller, Belkin, & Gibson, 1994; Regier et al., 1990; Zimberg, 1993) to 55.9% (Lehman, Myers, Corty, & Thompson, 1994), they are consistently higher than pure substance use disorder rates (First & Gladis, 1993). The largest differences in the estimates of concurrent substance use disorder among mental health clients stem from different comorbidity rates for different diagnostic psychiatric subgroups. For example, Brown and Barlowe (1992) estimated a 15–25% comorbidity rate of substance use among clients presenting with symptoms of anxiety, whereas Newman and Gold (1992), in an inpatient eating disorders setting, revealed comorbidity of substance use and eating disorders of 39.2%. "Depending on specialization and setting, between one quarter and one half of clients treated by mental health care professionals for other medical or psychological problems evidence problems related to alcohol or other drug involvement" (Miller & Brown, 1997, p. 1269).

With regard to mental health practice, it is important to consider that clients presenting with emotional problems, who also have substance use disorder, tend to represent a higher social and personal cost and are more difficult to treat. These clients place a greater economic burden on their families (Clark, 1994), incur high productivity losses, have elevated associated health care costs, and stimulate high-cost comorbidity research (Riley, 1994). Comorbid clients represent close to two-thirds of all visits to general medical, specialty, volunteer, and self-help group services (Narrow et al., 1993). Alcoholic clients with comorbid mental health problems have greater symptom severity and seek more treatment; they report greater distress, more social problems, more trouble keeping jobs, and less satisfaction with family relationships (Dixon, McNary, & Lehman, 1995; Miller, 1995; Schmidt, 1992). Comorbidity also is associated with a more difficult course of treatment, greater rates of hospitalization, poor medication compliance, faster relapse, higher rates of criminal and suicidal behavior, and decreased treatment compliance (Cornelius et al., 1995; Hills, 1993; Ries, Mullen, & Cox, 1994; Weiss, Mirin, & Frances, 1992). In other words, comorbidity presents the counselor with great challenges in assessment, diagnosis, and treatment planning, especially if the mental health practitioner has not been sufficiently familiarized with issues of substance use (Miller, 1995).

TYPES OF DRUGS

The *Diagnostic and Statistical Manual of Mental Disorders*, Fourth Edition (DSM-IV) (American Psychiatric Association, 1994) provides diagnostic guidelines for 11 discrete categories of substances: alcohol, amphetamines, caffeine, cannabis, cocaine, hallucinogens, inhalants, nicotine, opiates, phencyclidine (PCP), and sedatives (including hypnotics and anxiolytics). For most of these substances, specific symptoms occur when the user is intoxicated or is deprived of the substance after prolonged use (withdrawal). The intoxication or withdrawal symptoms often enable a mental health care provider to diagnose abuse of a substance even if the client does not admit to it. Thus, familiarity with the intoxication and withdrawal symptoms of commonly abused drugs is helpful for all counselors and therapists. Given the general acceptance of the DSM-IV as the organizing principle for diagnoses in the mental health field, these 11 groups of substances will be reviewed briefly to give the therapist some idea about their nature and effects. Many of the definitions, intoxication symptoms and signs, and withdrawal symptoms presented below are based on information contained in the DSM-IV. Many of the substances included in this listing are perfectly legal, easily and widely obtainable, and at times even specifically prescribed for use (e.g., sedatives, some opiates, and, of course, alcohol, caffeine, and nicotine). The legality and easy availability is exactly what can make some of these substances dangerous and is why their use can be difficult to identify as problematic or pathological. Counselors need to be aware that even using legal drugs can represent a diagnosable disorder. Two excellent layperson's references that are helpful for therapists not specializing in substance abuse treatment are Schuckit's (1998) *Educating Yourself about Alcohol and Drugs*, which also includes a thorough list of street names for the most commonly used drugs, and Kuhn, Swartzwelder, and Wilson's (1998) *Buzzed: The Straight Facts about the Most Used and Abused Drugs from Alcohol to Ecstasy*.

Alcohol

The figures provided in the epidemiological data section have already revealed that as a legal substance, alcohol is the most commonly abused drug in the United States. Its use is a regular and widespread practice that has become a large com-

ponent of social, and even religious, life (Bukstein, 1995). The amount of alcohol in alcoholic drinks and other substances (e.g., mouthwash, cough syrup) varies widely. One standard beer contains about 3–6% alcohol; a standard glass of wine, 5–15%; and a shot of liquor, 40–50% (some forms of liquor may be even more concentrated: for example, Austrian rum can contain up to 90% alcohol). The effects of alcohol can be felt within 10 minutes of consumption and a standard drink will have effects for up to one hour. Absorption rates vary somewhat depending on whether food is consumed with the alcohol. Eating decreases the rate of absorption, especially if high-protein foods are consumed.

Alcohol is a central nervous system (CNS) depressant that has sedating and tiring effects and specific intoxication and withdrawal symptoms. Tolerance develops with chronic use. Alcohol intoxication symptoms include maladaptive behavioral and psychological changes (such as inappropriate behavior, mood lability, and poor judgment), along with a variety of signs (such as slurred speech, incoordination, unsteady gait, nystagmus [i.e., shifty eyes that cannot focus], attention and memory impairment, stupor, and even coma). Alcohol withdrawal symptoms include autonomic hyperactivity (e.g., sweating, increased pulse rate), hand tremor, insomnia, nausea, vomiting, hallucinations or illusions, agitation, anxiety, and even seizures. Withdrawal symptoms generally occur within 12 hours of last ingestion.

Alcohol has been considered a gateway drug and commonly leads to or involves the use of other drugs (especially cannabis, cocaine, and nicotine). Alcohol is sometimes used to self-medicate withdrawal or intoxication symptoms of other drugs (e.g., cocaine). Alcohol use appears to increase the risk for violence, accidents, and suicide. Long-term use has many negative medical consequences, including damage to the liver, heart, and central nervous system (Kuhn, Swartzwelder, & Wilson, 1998). Long-term consequences, along with general information about alcohol use, are summarized in Table 2-1.

TABLE 2-1   Alcohol—A CNS Depressant

GENERAL INFORMATION

Street Names	Hooch, booze, liquor, devil's brew
Examples	Beer, wine, whiskey, gin, vodka
Primary Effects	Relaxation, calmness, pleasure, talkativeness, tiredness, sleepiness, inability to think clearly
Cost/Availability	Cheap and readily available
Legal Status	Completely legal in U.S. for adults
Prevalence	Ubiquitous social use; 7–14% problem use
Gender Ratio of Problem Use	5 men: 1 woman

USE PATTERNS AND SYMPTOMS

Route of Administration	Oral via drink
Single Dose/Absorption Rate	12 oz. beer; 1 oz. hard liquor; 15–20 mg/dl (one dose) per hour
Effect Onset and Duration by Dose	Within 10 minutes; for up to one hour
Traces Detectable	Over one hour in blood
Typical Pattern of Use	Episodic (bingeing) or regular (daily to multiple daily)
Intoxication Pattern	Inappropriate behavior, mood lability, poor judgment
Intoxication Symptoms	Slurred speech, incoordination, unsteady gait, nystagmus, attention and memory problems, stupor
Withdrawal Symptoms	Autonomic hyperactivity, hand tremor, insomnia, nausea, hallucinations, illusions, agitation, anxiety, seizures

RISKS/DANGERS

Addiction Potential	Moderate
Overdose Risk	Low; higher if drinking very fast or combining with other drugs
Lethal or Toxic Dose	Depends on ETOH concentration; blood level of 300–400 mg/dl
Signs and Causes of Overdose	Inhibition of respiration and pulse, difficult to arouse, unconsciousness
Dangerous Interactions and Contraindications	Antibiotic, anticoagulant, antidepressant, antipsychotic, antihistamine, antiseizure, heart, pain relieving, and sedative medications
Long-Term Medical Risks	Liver disease, heart disease, gastrointestinal problems, brain/CNS impairment, peripheral neuropathy, malnutrition
Long-Term Psychological Risk	Social isolation, suicide, impaired learning, depression
Other Risks	Blackouts, accidents, injury

Given the general social acceptance of alcohol consumption in a wide variety of social and even professional settings, judging clinically whether a person's behavior surrounding alcohol constitutes a disorder may be difficult. In fact, the DSM-IV reports that 90% of all adults have had some experience with alcohol and 60% have had one or more adverse life events due to alcohol consumption (American Psychiatric Association, 1994). The continuum of alcohol use ranges from appropriate social use that does not represent a problem to the over-consumption of alcohol that greatly interferes with functioning. Generally speaking, consumers who fall on either end of this continuum do not present a diagnostic puzzle. However, consumers who fall in the middle range of use can challenge a clinician's decision about diagnosis and treatment plans (Kitchens, 1994).

Amphetamines

Representing a class of stimulant drugs that cause feelings of power, alertness, and stimulation, amphetamines include substances such as amphetamine, dextroamphetamine, and methamphetamine. Most are taken orally or intravenously, with the exception of methamphetamine which may be snorted (i.e., taken nasally) or (if in its purest form, called "ice") smoked. Some amphetamines can be obtained legally through prescription (e.g., medications for obesity, weight loss, narcolepsy, or attention deficit disorder with names such as Ritalin or Cylert); most are obtained illegally. The onset of effects varies by method of ingestion. Oral administration results in a 30-minute delay of onset of effects, which last about two to four hours, whereas injection results in immediate yet shorter effects. Injection also results in more intense effects and this phenomenon has been called a "rush." This is often followed by a "crash," a rapid dissipation of effects that leaves the person craving another rush and hence reinforces the cycle of use. The two most typical patterns of use for amphetamines are episodic and daily. The episodic (binge) pattern is more commonly tied to injection use. Amphetamine metabolites can be detected in urine for up to three days. Tolerance develops rapidly.

Amphetamines have specific intoxication and withdrawal sytmptoms. Amphetamine intoxication may recognized through symptoms such as maladaptive behavioral or psychological changes (including affective blunting, interpersonal sensitivity, hypervigilance, and anxiety), and physiological signs (including tachycardia [i.e., fast heart action] and bradycardia [i.e., slowed heart action], dilation of the pupils, changed blood pressure, perspiration or chills, nausea, vomiting, weight loss, psychomotor changes, muscular weakness, confusion, seizures, and even coma).

Withdrawal symptoms include fatigue, unpleasant dreams, insomnia or hypersomnia, increased appetite, and psychomotor changes. Many unpleasant mood changes can occur after long-term use. Extreme reactions to high doses of amphetamines can lead to visual and tactile hallucinations, which the user usually considers to be caused by the substance use. Overdose risk is present with high doses of amphetamines as they may induce cardiac arrest, hyperthermia, or seizures. Details about the drug are summarized in Table 2-2.

Caffeine

This substance stimulates the central nervous system, resulting in alertness, increased ability to concentrate, and even euphoria, but it also causes nervousness and agitation, especially with increasing doses. Caffeine is of course perfectly legal, highly available, and widely used by the general public. What many consumers fail to realize is that caffeine is not only present in its most well-known forms (coffee and caffeinated soft drinks) but also is hidden in many other commonly consumed substances such as prescription and over-the-counter medications, candy bars, and some teas. An average 8 oz. cup of brewed coffee has anywhere from 75–150 mg of caffeine; stimulant medications that are sold over the counter can have as much as 100–200 mg per tablet; weight loss medications may contain 75–200 mg per tablet; an average candy bar contains approximately 5 mg of caffeine.

Caffeine is likely to create tolerance in frequent users. Caffeine can induce a specific intoxication syndrome that is more likely among infrequent users who have not developed tolerance. Given caffeine's long half-life of 2–6 hours, its symptoms can last 4–16 hours. Intoxication symptoms include restlessness, nervousness, excitement, insomnia, flushed face, diuresis, gastrointestinal disturbance, muscle twitching, rambling flow of thought, cardiac arrythmia, agitation, and periods of inexhaustibility. Although they are not formally

TABLE 2-2   Amphetamines—CNS Stimulants

GENERAL INFORMATION

Street Names	Speed, uppers, poppers, bennies, ice, pep pills
Examples	Amphetamine, methamphetamine, dextroamphetamine, methylphenidate
Primary Effects	Euphoria, sense of power, alertness, energy, decreased need for sleep, decreased fatigue
Cost/Availability	Cheap to moderate; some readily available over the counter (OTC) or with prescription
Legal Status	Some illegal; some legal with prescription; some OTC
Prevalence	Moderate: 2% problem use
Gender Ratio of Problem Use	3–4 men : 1 woman

USE PATTERNS AND SYMPTOMS

Route of Administration	Oral, injection, or snorting
Single Dose/Absorption Rate	Varies with substance and route of administration; 50 mg on average
Effect Onset and Duration by Dose	30 minutes and lasting 2–4 hours; unless injected, which produces rush and crash
Traces Detectable	1–3 days in urine
Typical Pattern of Use	Episodic (bingeing) or regular (daily to multiple daily)
Intoxication Pattern	Affective blunting, hypervigilance, interpersonal sensitivity
Intoxication Symptoms	Change in heart rate, blood pressure, and body temperature; pupillary dilation, nausea, agitation, weakness, confusion, seizures
Withdrawal Symptoms	Fatigue, unpleasant dreams, poor sleep, increased appetite, agitation

RISKS/DANGERS

Addiction Potential	High
Overdose Risk	High, especially with high doses
Lethal or Toxic Dose	Depends on tolerance, level of exertion, and type of drug; 50 to 60 mg for chronic users; 200 mg for naïve users; worse if combined with other drugs
Signs and Causes of Overdose	Cardiac arrest, hyperthermia, seizures
Dangerous Interactions and Contraindications	MAO inhibitor type antidepressants, OTC decongestants
Long-Term Medical Risks	Hyperthermia, malnutrition, sexual dysfunction, sleep disorder
Long-Term Psychological Risk	Visual and tactile hallucinations, paranoia, depression, aggression
Other Risks	Legal problems, interpersonal violence

recognized as a diagnosable disorder, caffeine withdrawal symptoms do occur; they include headaches, drowsiness, anxiety, depression, nausea, and vomiting. Caffeine has negative medical consequences with long-term use. Specifically, gastrointestinal distress and heart problems have been reported. Table 2-3 presents a summary of information about caffeine.

Cannabis

This group of perception-distorting substances is derived from the cannabis plant, which grows wild around the world. This class of drugs includes products such as marijuana (the dried leaves, tops, and stems of the plant, which are rolled into cigarettes), hashish (the dried resinous

TABLE 2-3   Caffeine—A CNS Stimulant

GENERAL INFORMATION

Street Names	None known
Examples	Coffee, tea, chocolate, OTC and prescription medications (e.g., weight loss)
Primary Effects	Alertness, ability to concentrate, euphoria
Cost/Availability	Cheap and highly available
Legal Status	Completely legal in U.S. for all ages
Prevalence	Ubiquitous use
Gender Ratio of Problem Use	Male > female

USE PATTERNS AND SYMPTOMS

Route of Administration	Oral via drink, food, or pills
Single Dose/Absorption Rate	20–150 mg (8 oz. coffee contains 75 mg); average daily intake in U.S. is 200 mg; half-life of 2–6 hours
Effect Onset and Duration by Dose	Immediate onset; effects last 1–4 hours per dose
Traces Detectable	Detectable in urine
Typical Pattern of Use	Regular daily to multiple daily use
Intoxication Pattern	No consistent pattern identified; diagnosed via symptoms
Intoxication Symptoms	Restlessness, agitation, insomnia, nervousness, excitement
Withdrawal Symptoms	Headaches, drowsiness, anxiety, depression, nausea, vomiting (not formally recognized in DSM-IV)

RISKS/DANGERS

Addiction Potential	Moderate
Overdose Risk	Rare, but possible (especially for young children)
Lethal or Toxic Dose	For children: 35 mg /kg bodyweight; For adults: more than 10 g/day
Signs and Causes of Overdose	Nausea, vomiting, tremors, confusion, irregular heart rate, respiratory failure, seizures
Dangerous Interactions and Contraindications	Medications to raise blood pressure, MAO inhibitor type antidepressants, other stimulants
Long-Term Medical Risks	Gastrointestinal distress, cardiac problems, respiratory failure
Long-Term Psychological Risk	Anxiety, agitation, nervousness, sleep disturbance
Other Risks	None known

substance that is exuded from the tops and leaves), or hashish oil (a distillate of hashish). Most commonly, all forms of cannabis are smoked, though they may be taken orally in other ways (e.g., mixed with food or drink). Metabolites of cannabis can be detected in the body (stored in fatty tissues and released into the blood stream) for up to one month. There is high variability in the level of the active ingredient, delta-9-tetrahydrocannabinol (THC), in the various forms and batches of cannabis, which makes the onset and duration of its effects somewhat unpredictable. Effects from this substance tend to be felt within 30 minutes of ingestion and last for 3–4 hours, but may also continue to recur for as long as 12–24 hours as more THC is released from fatty tissue. The level of THC concentration can be approximately 1% in low grade marijuana, 4–8% in higher grade marijuana, 7–14% in hashish, and up to 50% in hashish oil.

Although the DSM-IV claims that concentrations of THC have increased in cannabis products in recent years, other resources claim that this statement is based on confounded research (Kuhn, Swartzwelder, & Wilson, 1998).

Although cannabis intoxication has been formally recognized, no cannabis withdrawal disorder has been identified. Intoxication symptoms generally begin with euphoria, grandiosity, and inappropriate laughter. They include maladaptive behavioral or psychological changes (such as impaired judgment, poor motor coordination, and slowed reaction time) along with a number of physiological signs (conjunctival infection, increased appetite, dry mouth, and tachycardia). With long-term use, however, the effects may be sedating or lethargy inducing. No specific withdrawal syndrome has been identified although some symptoms may develop. Adverse reactions, such as hallucinations, delusions, paranoia, and anxiety, are possible with very high doses of highly concentrated cannabis products. Symptoms associated with cannabis use, in addition to general information about cannabis, are provided in Table 2-4.

Cannabis has been found helpful in the alleviation of symptoms associated with chemotherapy for cancer (by reducing vomiting) and AIDS (by increasing appetite and weight). The legalization of cannabis for both general and medical use has been attempted repeatedly in the absence of any evidence of an addictive nature. In fact, in recent elections its medical use was legalized in several U.S. states (including Alaska, California, and Oregon).

Cocaine

Cocaine is a CNS stimulant that is derived from the coca plant. It can take many forms, including coca leaves, coca paste, cocaine hydrochloride, and cocaine alkaloid. Native populations in Central and South America, where the coca plant grows, most commonly use the coca leaves and coca paste, whereas the latter two preparations are more commonly used in other parts of the

world. Cocaine hydrochloride, which is usually referred to by the generic label cocaine, is generally snorted as a powder or dissolved in water and injected. Cocaine alkaloid, known as crack in the United States, vaporizes easily and is therefore most commonly inhaled. Cocaine is often mixed with other drugs, especially stimulants, PCP, or heroin. The latter mixture is called speedball and is injected with fast and intense effects. Inhalation of cocaine results in faster onset and shorter duration of effects than most other forms of cocaine. Crack has the fastest onset and delivers a briefer euphoria than regular cocaine. Its effects are essentially immediate and last for only minutes. The intensely pleasurable euphoria produced by all forms of cocaine is quickly followed by intense dysphoria and sometimes paranoia. Patterns of use tend to be either daily or episodic (bingeing). Cocaine metabolites can be detected in urine for 1–3 days after a single dose and for up to 7–12 days with regular daily use. Tolerance inevitably occurs with regular use.

Cocaine can induce intoxication and withdrawal disorders. Intoxication symptoms include symptoms of maladaptive behavior and psychological changes (such as interpersonal sensitivity, hypervigilance, changes in sociability, and affective blunting) as well as physiological signs (such as tachycardia, bradycardia, dilation of the pupils, changes in blood pressure, perspiration or chills, nausea, vomiting, weight loss, psychomotor changes, muscular weakness, respiratory depression, confusion, dyskinesia, and even coma). Withdrawal symptoms include dysphoria and fatigue, unpleasant dreams, changes in sleep pattern, increased appetite, and psychomotor changes (agitation or retardation). However, withdrawal symptoms tend to be transitory and are seen only with high dose or frequency of use.

Long-term consequences of cocaine use include complications such as mood symptoms, paranoia, weight loss, and malnutrition (because cocaine is an appetite suppressant). Injection use exposes users to high risk for infectious and sexually transmitted diseases. Specifically, high rates of tuberculosis, hepatitis, chlamydia, and

TABLE 2-4   Cannabinoids—Mildly Perception Distorting Substances

GENERAL INFORMATION

Street Names	Pot, Mary Jane, grass, reefer, weed, joint
Examples	Marijuana, hashish, sinsemilla, hash oil
Primary Effects	Relaxation, mood elevation, sedation, tranquility, drowsiness
Cost/Availability	Moderate and widely available
Legal Status	Mostly illegal (some legal medical use)
Prevalence	33% "trial" use; 4% problem use
Gender Ratio of Problem Use	Male > female

USE PATTERNS AND SYMPTOMS

Route of Administration	Smoking and other oral routes (e.g., via food)
Single Dose/Absorption Rate	High variability depending on THC concentration (e.g., low grade marijuana: 1%; hash oil: 50%)
Effect Onset and Duration by Dose	Within 30 minutes; lasting an average of 3–4 hours, with some effects lasting for up to 24 hours or longer
Traces Detectable	THC stored up to one month in fatty tissue and released into blood
Typical Pattern of Use	Regular daily to multiple daily use
Intoxication Pattern	Impaired judgment and motor control, social withdrawal, anxiety
Intoxication Symptoms	Increased appetite, dry mouth, tachycardia, conjunctival infection, euphoria, grandiosity
Withdrawal Symptoms	With high doses only; not recognized in DSM-IV; irritability, restlessness

RISKS/DANGERS

Addiction Potential	Low
Overdose Risk	Very low to none
Lethal or Toxic Dose	Not known
Signs and Causes of Overdose	Very high dose can cause anxiety
Dangerous Interactions and Contraindications	May interact with blood pressure medications; can cause heart problems if combined with cocaine
Long-Term Medical Risks	Increased heart rate, bronchitis, asthma; possibly may impair immune function
Long-Term Psychological Risk	Paranoia, anxiety, delusions, hallucinations (rare)
Other Risks	Accidents due to impaired judgment, slowed reaction time, and poor coordination

HIV/AIDS have been reported. Snorting carries risk for nasal and sinus problems; smoking is associated with respiratory ailments. A summary of data on cocaine use is provided in Table 2-5.

Hallucinogens

This group of drugs includes LSD, morning glory seeds, mescaline, and psilocybin, used primarily to induce alterations in sensation, emotions, and perception; in some cases these drugs are used for their stimulating effects. Not included are PCP and cannabis, although they also may have hallucinogenic effects; these are discussed separately because their symptoms differ significantly from this class of substances. The effects of hallucinogens include detachment, mood swings, altered sense of time and space, religious or mystical experiences, and illusions

TABLE 2-5   Cocaine—A CNS Stimulant

GENERAL INFORMATION

Street Names	Crack, snow, coke, gold dust, rock
Examples	Coca leaves, coca paste, cocaine hydrochloride, cocaine alkaloid
Primary Effects	Intense euphoria, sense of powerfulness, decreased need for sleep or food, alertness, increased energy
Cost/Availability	Expensive but available
Legal Status	Illegal
Prevalence	20% "trial" use; possibly as high as 3–5% problem use
Gender Ratio of Problem Use	Equal numbers of men and women

USE PATTERNS AND SYMPTOMS

Route of Administration	Inhaling, injecting, or snorting
Single Dose/Absorption Rate	Varies with substance and route of administration; average half-life is 1 hour
Effect Onset and Duration by Dose	Within minutes and of brief duration; injection use creates rush and crash
Traces Detectable	1–3 days in urine after single dose; 7–10 days with habitual use
Typical Pattern of Use	Episodic (bingeing) or regular (daily to multiple daily)
Intoxication Pattern	Affective blunting, hypervigilance, changed sociability, interpersonal sensitivity
Intoxication Symptoms	Change in heart rate, blood pressure, and motor activity; pupillary dilation, nausea, vomiting
Withdrawal Symptoms	Symptoms transitory and only with high doses: dysphoria, fatigue, bad dreams; change in sleep, appetite, and motor activity

RISKS/DANGERS

Addiction Potential	Very high, especially with crack
Overdose Risk	Very high
Lethal or Toxic Dose	Single average dose can result in death
Signs and Causes of Overdose	Seizures, cardiac arrest, stroke, respiratory failure
Dangerous Interactions and Contraindications	Heart medications, caffeine, buspirone, and other substances that increase likelihood of seizure
Long-Term Medical Risks	Weight loss, malnutrition; respiratory, cardiac, nasal, and sinus problems
Long-Term Psychological Risk	Paranoia, mood disturbance, sleep disturbance, nightmares
Other Risks	Injection use: infectious disease; behavior patterns increase risk for sexually transmitted diseases

or hallucinations. The most common forms of ingestion are smoking and oral routes, though injection is chosen by some. The onset and persistence of effects varies greatly across this large group of heterogeneous substances and also is influenced by the level of active ingredients contained in the drug. As is true in most cases, injection results in faster effects that are of shorter duration. Most hallucinogens have half-lives that result in slow onset and long duration of symptoms, although there are some notable exceptions (e.g., DMT, effects of which are felt within 10 minutes, peak within 30, and cease within 60). The most common pattern of use appears to be episodic, rather than daily. Tolerance develops quickly for the euphoric and psychedelic effects of hallucinogens.

Hallucinogens produce specific intoxication symptoms, including maladaptive behavioral or psychological changes (such as paranoia, depression, and ideas of reference), perceptual changes despite full alertness (e.g., intensification of perceptions, depersonalization, derealizations, illusions, hallucinations), and physiological signs (such as dilation of the pupils, tachycardia, sweating, palpitations, blurred vision, tremors, and incoordination). Despite their name, hallucinogens rarely cause outright hallucinations; instead, they result in heightened perception with magical thinking (e.g., thinking one has the ability to fly). Further, if hallucinations do occur, the user is generally aware that they are drug induced. Intoxication increases risk for dangerous behavior (e.g., jumping out of a window due to the drug-induced belief that one can fly). Also associated with hallucinogens is hallucinogen persisting perception disorder, or flashbacks. This refers to the reexperience of the perceptual disturbances noted during intoxication despite not having ingested the drug. Common triggers for flashbacks are dark environments, fatigue, stress, or ingestion of other drugs. Flashbacks may occur for several weeks up to five years postuse. Other adverse effects can involve the development of panic attacks. Although overdose risk is generally low, some substances in this class can be lethal in a single dose (e.g., belladonna drugs, such as atropine, can stimulate the heart rate and body temperature to lethal levels). Additionally, hallucinogens are often adulterated with other substances, which increases the risk of adverse, even lethal, effects. No withdrawal disorder or symptoms are noted, though craving has been reported by some. Table 2-6 presents a summary of information on hallucinogens.

Inhalants

Inhalants represent a dangerous and varied class of abused substances that are widely and legally available. They include primarily aliphatic and aromatic hydrocarbons (e.g., gasoline, glue, spray paint, paint thinners), and secondarily, halogenated compounds (e.g., cleaner, correction fluids, spray can propellants) that are inhaled (sniffed) for their intoxicating effects. Inhalation may be accomplished by holding a fluid-soaked cloth in front of mouth and nose or breathing in the fluid vapors from a container. Some users heat the fluid to accelerate vaporization of the inhaled substance. Inhalants create rapid onset of effects (within seconds lasting a few minutes up to an hour), such as immediate giddiness and delayed confusion. Inhalants have severe physical and psychological effects. They are extremely damaging to the brain and can cause respiratory tract, lung, heart, kidney, and liver problems, as well as confusion, suicidality, and depression. Effects vary widely with substances and are not yet fully known. Inhalants appear to be used more commonly in rural or economically depressed areas where access to other substances may be limited (American Psychiatric Association, 1994; Schuckit, 1998). Tolerance can develop with heavy use.

Only a specific intoxication disorder has been identified for inhalant use. Intoxication symptoms include maladaptive behavioral or psychological changes (such as belligerence and assault) along with physiological signs (such as dizziness, nystagmus, incoordination, slurred speech, unsteady gait, lethargy, depressed reflexes, psychomotor retardation, tremor, muscle weakness, blurred vision, stupor, euphoria, and coma). For some, intoxication also includes hallucinations or other perceptual disturbances. A withdrawal syndrome has not (yet) been identified formally, although symptoms have been reported 24–48 hours after use, lasting for up to five days. These include sleep disturbance, nausea, illusions, and diaphoresis (increased sweating).

In addition to the previously described ill effects on the user's medical and psychological state, a multitude of other risks is associated with inhalant use. Users are at high risk for injuries, accidents, and death (see Kuhn, Swartzwelder, & Wilson, 1998, p. 121, for a description of "sudden sniffing death") due to suicide, burns from flammable substances, cardiac arrhythmia, and oxygen

TABLE 2-6   Hallucinogens—Perception Distorting Substances

GENERAL INFORMATION

Street Names	Acid, mescal, mushrooms, boomers, buttons, peyote
Examples	Ergot, lysergic acid diethylamide (LSD), phenylalkylamines, alkaloids, MDMA
Primary Effects	Increased sensory perception, restlessness, stimulation
Cost/Availability	Varies greatly by drug
Legal Status	Illegal
Prevalence	8–26% "trial" use; .3% problem use
Gender Ratio of Problem Use	3 men : 1 woman

USE PATTERNS AND SYMPTOMS

Route of Administration	Injecting, smoking and other oral routes
Single Dose/Absorption Rate	Vary with drug (e.g., Parlodel 10–15 mg); long half-life increases length of effects
Effect Onset and Duration by Dose	Varies, but average of 2–8 hours; e.g., LSD: 30–60 minutes onset; up to 12 hours duration
Traces Detectable	Varies greatly
Typical Pattern of Use	Episodic (bingeing)
Intoxication Pattern	Depression, paranoia, ideas of reference, poor judgment
Intoxication Symptoms	Tremors, changes in heart rate, incoordination; change in perception (hallucinations, depersonalization, etc.)
Withdrawal Symptoms	No withdrawal syndrome recognized despite reports of craving

RISKS/DANGERS

Addiction Potential	Low to moderate
Overdose Risk	Low to moderate for most; high for some
Lethal or Toxic Dose	Single dose for some (e.g., belladonna)
Signs and Causes of Overdose	Increased body temperature and blood pressure, respiratory failure, seizures
Dangerous Interactions and Contraindications	Often cut with contaminants that increase risk of medical or lethal side effects; dangerous with many medications and stimulants
Long-Term Medical Risks	Few long-term physical risks or effects
Long-Term Psychological Risk	Panic attacks, flashbacks, psychosis
Other Risks	Injury or death of self or other due to distorted perceptions

deprivation. Of deaths associated with inhalant use, 26% are caused by accidents and 28% by suicides (Kuhn, Swartzwelder, & Wilson, 1998). The most common pattern of use is episodic (bingeing) by groups of adolescents. Prolonged solitary use is generally associated with greater dependence. Table 2-7 details use patterns and other information about inhalant use. It is notable and frightening that 15% of high-school students indicate having used inhalants at least once in their lifetime (Kuhn, Swartzwelder, & Wilson, 1998). Most commonly inhalants are a secondary drug of choice.

Nicotine

Nicotine is a stimulant drug that enhances attention, concentration, and possibly memory; for some users it also has calming or antianxiety effects. It is used by some to induce a relaxation response in stressful situations. Of course it is a legal and widely available drug of choice and is obtained through the smoking of cigars, cigarettes, and pipes, as well as through the use of chewing tobacco or snuff and prescription medications containing nicotine (e.g., nicotine gum

TABLE 2-7   Inhalants—Multi-Effect Substances

GENERAL INFORMATION

Street Names	Bolt, bullet, rush
Examples	Gasoline, glue, spray paint, paint thinner, correction fluid
Primary Effects	Giddiness, loss of inhibition, confusion; many other variable effects by type of drug
Cost/Availability	Inexpensive and readily available
Legal Status	Legal
Prevalence	15% "trial" use; < 1% problem use
Gender Ratio of Problem Use	4–5 men : 1 woman

USE PATTERNS AND SYMPTOMS

Route of Administration	Inhalation
Single Dose/Absorption Rate	Highly variable half-lives and doses
Effect Onset and Duration by Dose	Immediate to within minutes; lasting for up to 1 hour
Traces Detectable	Not known; variable by substance
Typical Pattern of Use	Episodic use in group settings or singular regular use
Intoxication Pattern	Belligerence, assaultiveness, poor judgment
Intoxication Symptoms	Dizziness, incoordination, slurred speech, unsteady gait, lethargy, tremors, blurred vision, nystagmus, and others
Withdrawal Symptoms	Sleep disturbance, nausea, illusions, diaphoresis (not recognized in DSM-IV)

RISKS/DANGERS

Addiction Potential	High
Overdose Risk	High, especially for first time use
Lethal or Toxic Dose	Single dose or "session" can result in death
Signs and Causes of Overdose	Cardiac arrhythmia, oxygen deprivation, muscle Incoordination
Dangerous Interactions and Contraindications	Sedatives (opiates, barbiturates, benzodiazepines)
Long-Term Medical Risks	Damage to brain, liver, kidney, heart, CNS, lungs, and respiratory tract
Long-Term Psychological Risk	Hallucinations, suicide, confusion, other perceptual disturbances
Other Risks	Burns due to flammable nature of substance, accidents

and nicotine patch). Nicotine crosses the blood-brain barrier and has quick central nervous system and behavioral arousal effects. Its effects are noticeable within five minutes or less and last up to half an hour. Nicotine metabolites can be detected in urine for several days. Use of nicotine-containing substances among adolescents has been identified as a strong correlate of later use of alcohol, cocaine, and marijuana (U.S. Department of Health and Human Services, 1988), and it has been labeled a gateway drug (Schuckit, 1998). Tolerance develops with continued use.

Although no intoxication disorder has been specified, a withdrawal syndrome can be formally diagnosed. Withdrawal symptoms include dysphoric mood, insomnia, irritability, frustration, anger, anxiety, difficulty concentrating, restlessness, decreased heart rate, increased appetite, and weight gain. Withdrawal symptoms are related to the route of ingestion and the nicotine concentration of the preferred nicotine-containing substance. Severe craving between episodes of use perpetuates consumption. It has been speculated that nicotine is a common means of self-medicating other mental disorders because 55–90% (depending upon source) of persons with other mental disorders use nicotine, as compared to 30% among the general population (American Psychiatric Association, 1994). Long-term medical risks, such as pulmonary disease, lung cancer, heart disease, and skin wrinkling or coloring, are associated with the use of nicotine-containing substances. Further, nicotine increases the metabolism of many other drugs. In other words, while using nicotine, individuals may require a larger dose of other medications because nicotine causes the body to more quickly break down these substances. Discontinuation of nicotine use may then result in overdoses of such medications since they are no longer broken down at the same quick rate and are allowed to accumulate in the body. Table 2-8 collects the data concerning nicotine use.

Opiates

Opiates are analgesic drugs—that is, through their action on the central nervous system they induce painkilling effects. They include legal as well as illegal substances that are used to reduce pain or inspire relaxation and euphoria (though first-time use may cause dysphoria instead). Opiates such as morphine (a natural opioid) and codeine (a synthetic form) can be legally obtained through prescription as analgesics, anesthetics, antidiarrheal agents, or cough suppressants. The prescribed opioids are generally taken orally. The most commonly used illegal opioid is heroin, a semisynthetic. Heroin is usually injected, though it may also be smoked or snorted. Onset and duration of effects depend on the route of administration. Injection results in a rush, a sudden and intense onset of symptoms, which may persist for up to 3–6 hours. Other opiates have much longer persistence, creating effects for up to 12 or 24 hours (e.g., methadone) or even longer. Metabolites of opiates can be detected in urine for up to 36 hours. Tolerance develops with prolonged use.

Both intoxication and withdrawal disorders can be diagnosed for opioid use. Intoxication symptoms include maladaptive behavioral or psychological changes (such as euphoria followed by apathy), pupillary constriction (dilation if overdose induces anoxia), as well as several other physiological signs (such as drowsiness, coma, slurred speech, and attention and memory impairment). Opioid withdrawal is indicated by symptoms including dysphoric mood, nausea, vomiting, muscle aches, dilation of pupils, lacrimation, rhinorrhea, diarrhea, yawning, fever, and insomnia. These symptoms tend to become more severe with increased tolerance and time period of use.

Opiate use carries a number of high risks. It increases the likelihood for depressive and sleep disorders, can cause liver damage (usually due to contaminants), and creates problems with body temperature regulation. Injection carries the usual risks of increased likelihood of infectious disease, including hepatitis, sexually transmitted disease, tuberculosis, and HIV/AIDS. Risks associated with opiate use, and other general information about the drug, can be found in Table 2-9. Of all substances of abuse, opiates (especially if injected) pose the highest risk for overdose. This is partly due to the variable level of purity, which ranges significantly from batch to batch, making it difficult to predict how much of the drug is actually ingested. In general, opiates available today are purer than they were two decades ago (Kuhn, Swartzwelder, & Wilson, 1998). The American Psychiatric Association (1994) reports that 1 out of 100 untreated opiate users die from use-related problems, including overdose, injuries, accidents, or medical complications. For example, in one sample of heroin users, 25% ultimately died for reasons directly related to their opiate use (from suicides, accidents, infectious disease, and killing by police) (Schuckit, 1998).

Phencyclidine (PCP)

Originally developed as anesthetics, these synthetically produced substances became street drugs in the 1960s due to their perception-altering effects. PCP-type drugs are taken orally, in-

TABLE 2-8   Nicotine—A CNS Stimulant

GENERAL INFORMATION

Street Names	Snuff, death sticks, coffin nails
Examples	Cigars, cigarettes, chewing tobacco, nicotine gum and patch
Primary Effects	Increased attention and concentration, calming and antianxiety effects; may enhance memory
Cost/Availability	Moderate and readily available
Legal Status	Legal for adults
Prevalence	55% lifetime; 30% current use
Gender Ratio of Problem Use	Almost equal in U.S. (males > females)

USE PATTERNS AND SYMPTOMS

Route of Administration	Smoking, other oral routes, skin absorption
Single Dose/Absorption Rate	1 mg nicotine per cigarette; 3–5 mg in snuff chewed for 30 minutes; 20 minute half-life (smoked)
Effect Onset and Duration by Dose	Within 5 minutes and lasting up to 30 minutes
Traces Detectable	In urine for several days
Typical Pattern of Use	Regular multiple daily use
Intoxication Pattern	None recognized in DSM-IV
Intoxication Symptoms	None recognized in DSM-IV
Withdrawal Symptoms	Weight gain, dysphoria, insomnia, irritability, frustration, anger, anxiety, decreased concentration and heart rate

RISKS/DANGERS

Addiction Potential	High
Overdose Risk	Extremely low
Lethal or Toxic Dose	Possibly extremely high multiple doses
Signs and Causes of Overdose	Tremors, convulsions, respiratory failure due to muscle paralysis
Dangerous Interactions and Contraindications	With cocaine can cause lethal heart attack; heart medications; increases metabolism of several medications
Long-Term Medical Risks	Pulmonary and respiratory disease (e.g., emphysema, lung cancer), heart disease, skin wrinkling and staining
Long-Term Psychological Risk	Depression (cause and effect not established)
Other Risks	Gateway drug for alcohol and cannabis

jected, or smoked. Although not the only drug included in this classification, PCP (which has many street names, including angel dust, hog, and peace pill) is the most commonly used one. It is a white crystalline powder (hence the name angel dust) that dissolves easily in water. PCP's effects can last for several days and include detachment and depersonalization that affect functioning for extended periods of time. PCP use is complicated by the reality that it produces widely variable effects at the same time. "Put simply, taking PCP can produce a state similar to getting drunk, taking an amphetamine, and taking a hallucinogen simultaneously" (Kuhn, Swartzwelder, & Wilson, 1998, p. 93). Metabolites of PCP can be detected in urine for weeks after prolonged or high-dose use. Multiple daily use of this widely available, easily obtained drug is the

TABLE 2-9   Opiates—Analgesic CNS Active Substances

GENERAL INFORMATION

Street Names	Smack, snow, horse, Miss Emma, Chinese molasses
Examples	Dilaudid, Percodan, Darvon, morphine, codeine, heroin, opium
Primary Effects	Euphoria, relaxation, pleasure followed by dreamy state; reduced pain perception
Cost/Availability	Variable price, readily obtainable
Legal Status	Some illegal; some legal with prescription; some OTC
Prevalence	6% "trial" use; .7% problem use
Gender Ratio of Problem Use	3–4 men : 1 woman

USE PATTERNS AND SYMPTOMS

Route of Administration	Injecting, smoking, snorting, and other oral routes
Single Dose/Absorption Rate	Varies with drug and route; morphine: 60 mg (lasts 4–6 hrs); codeine, 200 mg (lasts 4–6 hrs); Fentanyl .1 mg (lasts up to 2 hrs)
Effect Onset and Duration by Dose	Varies with drug; from within 2 minutes to 30 minutes; lasting a few minutes to several hours
Traces Detectable	In urine up to 36 hours
Typical Pattern of Use	Regular multiple daily use
Intoxication Pattern	Euphoria followed by apathy, impaired judgment, change in psychomotor activity
Intoxication Symptoms	Pupillary changes, drowsiness, slurred speech, coma, impaired attention and memory
Withdrawal Symptoms	Dysphoria, nausea, fever, yawning, vomiting, muscle aches, lacrimation, diarrhea, rhinorrhea, insomnia

RISKS/DANGERS

Addiction Potential	High
Overdose Risk	Very high
Lethal or Toxic Dose	Single dose, depending on purity and contaminants
Signs and Causes of Overdose	Labored breathing, decreased heart rate, tremors, pinpoint pupils, coma, fluid-filled lungs shutting down oxygen supply
Dangerous Interactions and Contraindications	Contamination increases medical risks; drugs that slow breathing (e.g., alcohol, barbiturates, benzodiazepines)
Long-Term Medical Risks	Liver disease, problem with body temperature regulation, constipation
Long-Term Psychological Risk	Depressive disorders, sleep disturbances, nervousness and anxiety associated with craving
Other Risks	High death rate due to medical complications and accidents; legal problems

most common pattern of use. No tolerance has been demonstrated.

Intoxication includes the following symptoms: maladaptive behavioral or psychological changes (such as belligerence and assault) and physiological symptoms (such as nystagmus, hypertension, tachycardia, numbness, decreased responsiveness to pain, ataxia, dysarthria, muscle rigidity, seizures, hyperacusis, and coma). At times hallucinations and/or illusions may also occur, hence its occasional classification as a hallucinogen. No withdrawal syndrome has been identified.

PCP use can be especially dangerous because the drug often contains contaminants, some of which can be life threatening. Another risk is presented by the serious symptoms of PCP overdoses that can kill (via seizures, coma, or respiratory arrest). Common medical complications include hyperthermia, hypertension, and seizures. Injection use carries the usual increased risk for infectious diseases such as hepatitis, tuberculosis, sexually transmitted disease, and HIV/AIDS. PCP users appear to be at high risk for violence, injury, and accidents due lack of insight, poor judgment, disorganized thinking, and numbing to pain induced by the drug. Table 2-10 outlines additional details on PCP use.

Sedatives

This class of CNS depressants includes sedatives, anxiolytics, and hypnotic drugs such as benzodiazepines, barbiturates, and hypnotics. They can be obtained easily and legally in the form of sleeping medications and antianxiety medications (anxiolytics). Most are taken orally. Onset and persistence of sedative effects vary widely by type of drug and level of active ingredient. In low doses these substances merely result in mild sedation; in high doses they produce a heightened sense of well-being, possibly to the degree of euphoria. One danger associated with sedative use is the rapid development of tolerance that results in larger and larger doses of the drug being ingested. It is not uncommon for users to ingest up to 20 times the amount of prescribed substance. This excess consumption can lead to overdose risk that can be life threatening for some forms of sedatives, if only because of the increased risk for accidents that users experience while under the influence of these drugs. Half-lives for this class of drugs vary widely, with some being fast acting and more addictive and others being longer acting and having delayed onset of withdrawal symptoms (e.g., diazepam). Generally speaking, however, high levels of physiological dependence lead to quick tolerance and withdrawal as it becomes harder and harder for users to reach desired effects. For illegal forms of sedatives, the most common pattern of use starts with intermittent use that leads to daily use. For the legal forms, use begins with daily prescribed ingestion, increasing to multiple (excessive) daily doses over time.

Intoxication and withdrawal disorders have been identified. Intoxication symptoms include maladaptive behavioral and psychological changes (such as inappropriate sexual behavior, affect lability, impaired judgment and functioning, and aggression), and physiological signs (such as slurred speech, incoordination, unsteady gait, nystagmus, attention and memory problems, stupor, and coma). Withdrawal symptoms include autonomic hyperactivity, hand tremor, insomnia, nausea, vomiting, hallucinations and illusions, psychomotor agitation, anxiety, and grand mal seizures. Unexpected side effects that occur with some regularity include anxiety, nightmares, hostility, and rage (Kuhn, Swartzwelder, & Wilson, 1998). Intense and frequent intoxication can lead to depression and suicide; memory impairments similar to alcoholic blackouts have also been reported.

Sedatives may be commonly used to deal with side effects of stimulant drugs such as cocaine or amphetamines or to boost effects of methadone. Accidental overdoses can occur especially among individuals with high levels of tolerance, leading to potentially lethal respiratory depression and hypotension. Lethality is also increased as sedatives are combined with other substances. Injection use carries with it the usual risks for increased likelihood of infectious disease. Long-term medical risks include liver damage and loss of coordination. Table 2-11 outlines such risks and includes general information about the use of sedatives.

DIAGNOSIS OF SUBSTANCE USE

The easiest and most straightforward approach to diagnosis of substance use is the nosology presented by the DSM-IV. It outlines two types of substance-related disorders: substance use disorders and substance-induced disorders. Substance use disorders are further broken down into substance abuse and substance dependence. Both abuse and dependence apply to 9 of the 11

TABLE 2-10   Phencyclidine—Perception Altering Stimulant Substances

GENERAL INFORMATION

Street Names	Angel dust, super acid, whack, comos, hog, tranq, peace pill
Examples	PCP, sernylan, ketamines, thiopene analog (TCP)
Primary Effects	Euphoria, stimulation, decreased inhibition and pain perception
Cost/Availability	Moderate cost; easy to obtain
Legal Status	Illegal
Prevalence	13% "trial" use; <1% problem use
Gender Ratio of Problem Use	3 men : 1 woman

USE PATTERNS AND SYMPTOMS

Route of Administration	Injecting, smoking, snorting, and other oral routes
Single Dose/Absorption Rate	10 mg; long half-life
Effect Onset and Duration by Dose	Within 15–30 minutes; peaks at 2 hours; lasts 4–6 hours depending on dose
Traces Detectable	In urine for weeks with high use
Typical Pattern of Use	Regular daily use
Intoxication Pattern	Belligerence, poor judgment, disorganized thought, assaultiveness, insensitivity to pain
Intoxication Symptoms	Nystagmus, hypertension, tachycardia, numbness, ataxia, dysarthria, muscle rigidity
Withdrawal Symptoms	None recognized in DSM-IV, although some craving has been reported

RISKS/DANGERS

Addiction Potential	Low
Overdose Risk	High
Lethal or Toxic Dose	One large dose or several single doses in a short period of time
Signs and Causes of Overdose	Respiratory depression, high body temperature and blood pressure, coma, seizures
Dangerous Interactions and Contraindications	Often mixed with other drugs; often contaminated with other substances which increases medical risks
Long-Term Medical Risks	Hyperthermia, hypertension, seizures, coma
Long-Term Psychological Risk	Psychotic-like symptoms, aggression, confusion, agitation, violence
Other Risks	High risk for accidents due to numbness to pain and poor judgment and insight

substances mentioned earlier in this chapter, with the two exceptions being caffeine and nicotine. No substance use disorder (neither abuse nor dependence) can be diagnosed for the consumption of caffeine; only substance dependence can be diagnosed for nicotine.

Abuse is diagnosed if the client evidences "a maladaptive pattern of substance use leading to clinically significant impairment of distress" (American Psychiatric Association, 1994, p. 182). Clinically significant impairment is defined

as the presence of at least one of the following four symptoms experienced over the past 12 months:

- recurrent use that results in a failure to fulfill a major life role (e.g., work, school, or home)

- recurrent use in situations in which drug use is physically dangerous (e.g., driving)

- recurrent use-related legal problems (e.g., citations for driving under the influence)

TABLE 2-11   Sedatives—CNS Depressants

GENERAL INFORMATION

Street Names	Roofies, easy lay, downers, tranqs, sleepers, yellow jackets
Examples	Benzodiazepines, carbamates, methaqualone, barbiturates, hypnotics
Primary Effects	Reduction of anxiety, lightheadedness, relaxation, mellow feelings, drowsiness, incoordination, slurred speech, impaired learning and memory
Cost/Availability	Variable price according to legal status; widely available
Legal Status	Some illegal; some legal with prescription; some OTC
Prevalence	Almost ubiquitous occasional use; 1% problem use
Gender Ratio of Problem Use	Almost equal (women > men)

USE PATTERNS AND SYMPTOMS

Route of Administration	Oral
Single Dose/Absorption Rate	Dose varies across drugs—e.g., diazepam: 5 mg; chloralhydrate: 50 mg/kg; half-lives vary—e.g., Triazolam: 3 hrs.; Flurazepam: 100 hrs.
Effect Onset and Duration by Dose	Gradual onsets with long duration (on average; high variability with drug)
Traces Detectable	In blood or urine for up to one week
Typical Pattern of Use	Intermittent or regular daily to multiple daily use
Intoxication Pattern	Inappropriate sexual or aggressive behavior, mood lability, impaired judgment
Intoxication Symptoms	Slurred speech, incoordination, unsteady gait, nystagmus, impaired attention and memory, coma, stupor
Withdrawal Symptoms	Autonomic hyperactivity, tremor, insomnia, nausea, agitation, anxiety, hallucinations, delirium, vomiting

RISKS/DANGERS

Addiction Potential	High
Overdose Risk	Low in isolation; high in certain combinations
Lethal or Toxic Dose	High risk with tolerance or combination with other drugs
Signs and Causes of Overdose	Respiratory depression, hypotension, slurred speech, loss of reflexes, unarousable sleep, nausea, memory impairment
Dangerous Interactions and Contraindications	Alcohol, opiates, anesthetics, inhalants; may speed the metabolism of several other medications
Long-Term Medical Risks	Irritation of mucosa, lightheadedness, liver damage, vertigo, incoordination
Long-Term Psychological Risk	Depression, suicide, irritability, tiredness
Other Risks	Accidents especially while driving or operating machinery; risk for sexual assault

■ continued use despite persistent negative social or interpersonal effects (e.g., fights with partner)

Clients can only be diagnosed with abuse if they have not received a prior or current diagnosis of dependence on the same substance.

Dependence is defined as "a maladaptive pattern of substance use leading to clinically significant impairment of distress" (American Psychiatric Association, 1994, p. 181); however, in this case, clinically significant impairment is defined through the presence of at least three symptoms of a list of seven, again occurring over the span of a 12-month period. The seven possible symptoms show a greater severity of use than those listed in the abuse category, and they point the therapist toward a pattern of use that has resulted in continuously increased use. These symptoms include:

- tolerance (evidenced either through need for increased amounts of the substance to reach the desired effect or through diminished effect with ingestion of the same amount over time)

- withdrawal (met either through the presence of withdrawal symptoms or through the use of the substance to prevent such symptoms)

- use of the substance in larger amounts or for longer time periods than originally intended

- persistent desire for the substance or unsuccessful attempt to cut down on its use

- large time investment to procure the substance, to use it, or to recover from its use

- reduction of time spent on other activities to make time for substance-use-related activity

- continued use despite awareness of the substance's negative effect on major life areas

Some substance use treatment professionals have criticized the classification of substance use disorders by the DSM-IV as simplistic and too straightforward. They have argued that substance use cannot be forced into two arbitrary categories of abuse or dependence but rather represents a continuum that ranges from nonuse to dependence. Lewis, Dana, and Blevins (1988) suggest that the continuum of use is best conceptualized as ranging from no use to moderate use with no associated problems; to heavy use with no associated life problems; to heavy use with associated moderate problems; to heavy use with serious problems; and finally to dependence with serious life and health problems. Similarly, researchers within the National Institute on Drug Abuse believe that the current DSM system is not sufficiently specialized to take into consideration all components of substance use and dependence to make a reliable functional diagnosis (Blaine, Horton, & Towle, 1995). They too suggest that that a "sharper definition and criteria for abuse, hazardous use, or mild problems"

(p. 13) is necessary in the future to provide clearer conceptual definitions and diagnostic criteria. They support a model presented by Skinner (1990) that takes in to consideration consumption and related problems; this model suggests that dependence should be diagnosed only when heavy consumption and severe problems converge and that abuse needs to be graded further to be able to differentiate between levels of use and complications. Certainly level of use does range widely, and it is sometimes difficult to make a judgment regarding proper diagnosis, however, the differentiation regarding abuse versus dependence appears to be valid. The continuum of use must still be taken into consideration when deciding upon a course of treatment with a client; however, it does not necessarily dictate diagnosis.

The second type of substance-related disorders specified in the DSM-IV are substance-induced disorders. There are three types of substance-induced disorders: intoxication, withdrawal, and other substance-induced mental disorders (e.g., delirium, amnestic disorder, mood disorder, psychotic disorder). Intoxication is marked by the "development of a reversible substance-specific syndrome due to the recent ingestion of (or exposure to) a substance . . . [and] clinically maladaptive behavioral or psychological changes that are due to the effect of the substance on the central nervous system and develop during or shortly after use of the substance" (American Psychiatric Association, 1994, p. 184). The specific syndrome or changes vary from substance to substance. Withdrawal is diagnosed if the following two conditions are met: "development of a substance-specific syndrome due to the cessation of (or reduction in) substance use that has been heavy and prolonged [and a] substance-specific syndrome [that] causes clinically significant distress or impairment in social, occupational, or other important areas of functioning" (American Psychiatric Association, 1994, p. 185). As with intoxication, actual manifestation of withdrawal symptoms depends on the drug of choice. Not all substances result in withdrawal syndromes (e.g., cannabis); and some have very specific withdrawal symptoms that indicate a disorder in and of itself (e.g., hallucinogen persisting perception disorder [flashbacks]).

The final DSM-IV category of substance-induced disorders varies widely in terms of manifestation (or associated symptomatology), both across substances and even within specific substances. Essentially, this category refers to any other mental disorder the symptoms of which can be traced to the use of a substance (e.g., alcohol intoxication delirium; amphetamine-induced mood disorder; inhalant-induced persisting dementia). Being able to recognize the symptoms of what may be a variety of mental disorders (ranging from organic brain/cognitive disorders to mood disorders, psychotic disorders, and so forth) as a substance-induced disorder is a critical skill for all therapists. It is for this reason that a thorough substance use assessment is crucial if a client gives any indication of regular drug use. Not assessing a client's substance use and its potential effects on her or his life (and behavior, cognition, and affect) may mean misdiagnosing clients as depressed, psychotic, or otherwise impaired, when in reality they are experiencing these symptoms as a result of their drug use. Clearly, the treatment implications differ greatly depending on whether certain symptoms are manifestations of a pure psychiatric disorder or a substance-induced disorder.

ASSESSMENT OF SUBSTANCE USE

Given that occurrence of substance use is the rule rather than the exception in mental health settings (see Kessler, 1994), it becomes important for counselors and therapists to make reliable and valid diagnoses. Since correct diagnosis is directly related to counselor skill and client truthfulness, competent substance use assessment clearly has its place in the repertoire of skills therapists must bring to the intake assessment of each and every client. Substance use assessment must be con-

ducted matter-of-factly, not with an accusatory or condescending tone (Boylan, Malley, & Scott, 1995). Most substance-using clients arrive with a certain amount of guilt about their habit; lecturing or using moralistic tones will only serve to make these clients more defensive and withdrawn (Perkinson, 1997). A number of issues need to be attended to in order to assure the success of substance use assessment.

Dealing with Intoxicated Clients

As previously outlined, intoxication symptoms vary greatly by substance ingested. To reiterate quickly, clients may present with slurred speech, incoordination, or unsteady gait if under the influence of CNS depressants (e.g., alcohol, sedatives, opiates, inhalants); with agitation, talkativeness, vigilance, or grandiosity if under the influence of CNS stimulants (e.g., cocaine, amphetamines); with euphoria and altered state of time if under the influence of cannabis; with hallucinations if under the influence of hallucinogens or PCP; or with giddiness or confusion if under the influence of inhalants (Perkinson, 1997). More subtle presentations are possible as well: the client may have no obvious symptoms but may be clearly impaired. Sometimes the smell of alcohol or inhalant fumes may clue in the clinician that the client is not in a normal alert state of consciousness.

It is a necessary precondition for therapy that both client and therapist are of clear mind and can process information optimally and efficiently. If a client arrives for the initial interview or a subsequent session under the influence of alcohol or other drugs, it is best to reschedule the session. A client who is under the influence of chemicals will not have the cognitive wherewithal to benefit from treatment. Nor will this person be able to provide reliable information that will help the clinician understand and conceptualize the case. The clinician explains this to the client and the appointment is rescheduled. Explanations are best given very succinctly and nonjudgmentally. Arguing with an intoxicated client is a losing proposition. The person may not be entirely rational, and even the best therapeutic explanation may be lost and certainly forgotten as soon as the client leaves. Thus, it is more important to act than to talk. Some intoxicated clients may be quite loquacious. The therapist best avoids getting drawn into a prolonged conversation with such a client but rather turns toward arranging safe transportation for the client.

Some clinicians include a statement in their informed consent or discuss in the screening or intake session that they will refuse services to clients who are clearly under the influence of a mind-altering substance. If this has been done, then the client merely is reminded at the time of the event that no services can be provided given the person's state of intoxication. The therapist who wisely has chosen to reschedule an impaired client's session has some responsibility about the client's behavior outside of the therapy room after the client leaves the premises. Given this responsibility, it is crucial to explore how an impaired client transported him- or herself to the session. If he or she drove under the influence and plans to leave by car as well, the therapist will need to intervene. Specifically, such a client now represents a potential danger to self and others. The therapist will need to persuade the client to make alternative arrangements for transportation, such as phoning a friend or family member to pick up the client or ordering a taxi that will take the client home. If clients refuse such intervention, they need to be warned that, given the therapist's legal and ethical obligations to protect and warn, the clinician will notify the police of the client's intent to drive under the influence. Clearly, this situation has the potential to become a major interference in the therapeutic relationship but the therapist must react responsibly and within the parameters of the law.

Assessment Validity and Reliability

It is the job of the therapist to create an easy and honest flow of information during the assessment session that will allow for reliable and valid diag-

nosis and treatment planning. The assessment will most likely be an extension of the intake interview (i.e., a clinical interview) and will incorporate a range of specialized questions that explore the client's substance use history. It is also possible to expand assessment through the use of collateral informants, biochemical analyses (Carey & Teitelbaum, 1996), or questionnaires (Kitchens, 1994). The first two procedures are rarely practical in outpatient mental health work; the third can be useful, and many substance use questionnaires have been developed to aid the clinician. Because most of these instruments have psychometric properties that leave something to be desired, the cautious therapist does not rely exclusively on their administration but uses them as supplements to the clinical interview.

Whether they administer questionnaires, use structured interviews, or rely on both, therapists need to be aware that substance-abusing or -dependent clients may not be the most reliable and forthcoming historians. Meyer and Deitsch (1996) have outlined several specific behavioral observations clinicians can make to help them decide whether a substance use assessment, or any assessment, for that matter, is valid. Namely, these authors indicate that a valid substance use assessment is possible if the client:

- is not currently using a substance
- does not perceive any objective or interpersonal coercion for treatment (i.e., is neither court referred nor urged by a significant other)
- is certain about confidentiality and is trusting of the counselor
- is willing to have self-report data checked against more objective reports (e.g., urinalyses or family reports)
- is generally compliant
- is not antisocial
- has intrinsic motivation for treatment

If any of the aforementioned points are not met, information gleaned from questionnaires and interviews may not be reliable or truthful.

Comorbidity

In both mental health and substance use settings, misdiagnoses and missed diagnoses are common problems for comorbid patients (El-Guebaly, 1990). Misdiagnosis of cooccurrence results when a clinician fails to differentiate between symptoms and syndromes (Ziedonis, 1992). For example, a client may evidence occasional use of alcohol without having a diagnosable substance use disorder. However, lack of familiarity with substance use diagnosis or inadequate attention to assessment may result in an inappropriate diagnostic label. Misdiagnosis can also occur when substance use or its sequelae result in symptoms that appear psychiatric in nature. Many substance use disorders can have symptoms that mimic psychopathology, particularly affective disorders (Mirin, Weiss, Michael, & Griffin, 1988), anxiety disorders (Schuckit & Monteiro, 1988), and psychoses (Zweben, Smith, & Stewart, 1991). For example, toxic symptoms of prolonged alcohol, stimulant, inhalant, and hallucinogen use may result in psychotic-looking (e.g., paranoid) states that can be misdiagnosed as psychosis. Acute and prolonged withdrawal symptoms from several drugs can mimic affective disorders. Sustained alcohol or opiate intoxication can result in depression and irritability secondary to CNS depressant functions (Schuckit & Monteiro, 1988). Likewise, protracted withdrawal from cocaine can result in symptoms identical to a major depression after as many as 12 weeks of sobriety (Horton, 1995). Substance-induced organic brain symptoms are sometimes misunderstood as psychotic symptoms. For example, alcohol hallucinosis may be misdiagnosed as schizophrenia; amphetamine-induced delusional disorders may be misunderstood as paranoid schizophrenia because both auditory hallucinations and paranoid ideation are part of the picture; and methadone withdrawal may result in an organic affective syndrome that is often confused with major depression (Mirin et al., 1988; Zweben, Smith, & Stewart, 1991).

In other words, a number of symptoms have to be evaluated carefully to determine whether they are manifestations of use, intoxication, acute withdrawal, protracted withdrawal, or organic/neurological consequences of drug addiction as opposed to an underlying or additional, independent psychiatric disorder that is not substance induced (Mirin et al., 1988; Ries, Mullen, & Cox, 1994). Such careful separation of drug-induced disorders from independent psychiatric disorders (Good, 1993) is necessary for reliable and valid diagnosis of cooccurrence. This is best achieved by delaying testing and assessment for a time period sufficient to allow for substance-related and -induced symptoms to subside. Although consensus exists in the literature that such a delay is essential, suggestions for the length of the interval of delay range from 4 weeks to 3 months (c.f., American Psychiatric Association, 1994; Ziedonis, 1992). The decision is perhaps best made with a given client in mind because, more than anything, the drug that is being used will determine the appropriate waiting period for final diagnosis (e.g., protracted cocaine abuse may require a three-month waiting period, whereas an individual acutely abusing hallucinogens may be successfully diagnosed after two weeks).

Another issue to consider when assessing cooccurrence is primacy and/or relationships of disorders. Two primary possibilities exist in this regard. First, one disorder may be the cause of the other. If this is the case, it is important to determine which disorder is primary (i.e., did a psychiatric disorder cause substance abuse, or did substance abuse result in psychiatric disorder?). Second, there may be an interaction or independent coexistence of substance use disorder and another mental illness (Miller, 1995). In this case, the clinician must consider that a substance use disorder may mimic a psychiatric disorder, that a substance use disorder may mask other psychiatric symptoms, or that psychiatric disorder may mimic behaviors more commonly associated with substance use disorder (Ries, Mullen, & Cox, 1994).

To address the issue of primacy of disorder (Chiauzzi, 1994; Miller, 1995), a thorough family drug and psychiatric history (Kessler, 1995), as well as information on onset of substance use and psychiatric symptoms, needs to be collected. These data can then be used to make a determination about the most likely scenario of the coexistence of the two psychiatric disorders. First and Gladis (1993) suggest using the following criteria to decide about primacy of substance use disorder rather than mental illness:

- substance use symptoms preceded psychiatric symptoms in terms of lifetime onset
- family history of substance use symptoms is more frequent and more pronounced than family history of psychiatric symptoms
- abstinence resolved the psychiatric symptoms

Primacy of the psychiatric diagnosis is similarly determined by meeting the following three criteria:

- psychiatric symptoms precede substance use symptoms in terms of lifetime onset
- family history of psychiatric symptoms is more frequent and more pronounced than family history of substance use
- resolution of psychiatric symptoms resolves substance use symptoms

A case of independent coexistence of substance use and psychiatric disorder is considered if the client meets the following criteria:

- joint onset of substance use and psychiatric symptoms
- family history of both or neither substance use and psychiatric disorder
- resolution of one set of symptoms without change in the symptomatic picture of the other

To facilitate the process of evaluating primacy and independence, a time line of symptoms can be developed to determine which symptoms (psychiatric or substance use related) appeared first. This time line can be provided by the client but may be more reliable if corroborated

through interviews with persons who have been familiar with the client over the course of several years (such as family members, long-term friends, or care providers). Optimally a longitudinal approach to diagnosis would be taken, tracking the client through direct observation over an expanded period of time (El-Guebaly, 1990); however, since such long-term tracking requires the collaboration of treatment staff who encounter the patient in one or several settings over the course of months or even years, it rarely occurs.

A time line is not only useful to the assessment of primacy but may also answer related questions. Specifically, it may assist the counselor in determining whether an acute or chronic substance use disorder caused the development of the other psychiatric disorder, whether it merely provoked the reemergence of psychiatric symptoms of an earlier existing disorder, or whether the substance use disorder worsened the severity of the psychiatric disorder. Similarly, the same relationships can be explored to assess whether the psychiatric disorder led to substance use (e.g., as a means of self-medicating), provoked the reemergence of prior substance use, or worsened the symptoms of existing substance use (Ries, Mullen, & Cox, 1994). Finally, family psychological, psychiatric, and substance use history is taken. Although this process is started with the client, it may need corroboration from family members or others familiar with the family of the patient.

Related to the issue of primacy, but worth mentioning separately, is the issue of substance use as a means of self-medicating a preexisting psychiatric illness. Some clients presenting with psychiatric and substance use symptoms may be easily identified as having a primary psychiatric diagnosis if the clinician can determine that the substance use serves to reduce undesired emotional or behavioral sequelae of the psychiatric disorder. For example, patients with attention-deficit hyperactivity disorder are at "increased risk for abuse of stimulant drugs, including cocaine. In such patients, stimulants both reduce anxiety and help focus attention" (Mirin et al., 1988, p. 153). Similarly, manic episodes of a bipolar client may be dealt with through increased alcohol intake (Schuckit & Monteiro, 1988), and opiates may be used to reduce psychotic symptoms (Zweben, Smith, & Stewart, 1991). The degree to which this self-medication process has led to a substance abuse disorder will vary; however, in terms of resolving the primacy issue, an understanding of the possible self-medicating function of the substance use behavior may prove helpful.

Type and Detail of Collected Information

A range of questions needs to be asked of the potentially substance-using client for appropriate diagnosis, referral, and treatment planning. These questions elicit detailed information about substance use (e.g., frequency, recency, intensity, duration, age of onset), associated behaviors and problems (e.g., blackouts, intoxication symptoms, withdrawal symptoms), and effects on the client's life in general (e.g., impact on work history, intimate relationships, social interactions). Table 2-12 summarizes these areas of functioning.

It is important to explore all substances that may have been used independently and not to place them all in a single category of abused drugs (Hood & Johnson, 1997). Vastly different patterns of use are possible across different drugs and research shows with great consistency that the trend of substance use is one toward polysubstance abuse (i.e., the use of more than one substance at a time or in a lifetime) (Brems & Namyniuk, 1999). Exploring each substance with regard to frequency of use (how many times per day, week, month, year), recency of use (e.g., within one day, one week, one month of the intake interview), intensity of use (i.e., how much at each incidence of use), variability (i.e., changes in pattern of use; e.g., alcoholic binges), and duration (i.e., for how long each time) is critical. It is also helpful to determine how long a particular substance has been used, that is, to inquire about age of onset for occasional and

then regular use. Once all substances have been explored, it is also helpful to ask the client to identify a drug of choice. This may not always be self-evident from frequency and intensity of use. A client may identify cocaine as drug of choice despite having used alcohol more frequently, recently, and intensively. Such a response may suggest that the client uses alcohol due to lack of availability or higher cost of cocaine. The importance of identifying the client's drug of choice, of course, rests in the subsequent tailoring of treatment. In this example, treating the client in an alcohol treatment program would miss the point and merely would result in relapse with regard to cocaine use.

Questions regarding intoxication and withdrawal symptoms are extremely important because these are the symptoms that may help the therapist distinguish abuse from dependence, as well as diagnose additional substance-induced disorders. It is helpful for the therapist to have knowledge of common substance intoxication and withdrawal symptoms in order to be able to ask pointed questions about symptoms concerning the drugs the client identified as drugs of choice or frequently used drugs. In the absence of this knowledge, the therapist can ask questions about generic common intoxication and withdrawal symptoms; Tables 2-1 through 2-11 include relatively comprehensive listings of possible intoxication and withdrawal symptoms for the drugs covered in this chapter. By comparing the list obtained from the client to the lists for each individual substance, the therapist can later determine the proper diagnosis.

The circumstances surrounding drug use are also important to explore because they may help the counselor identify situational triggers for use (Carey & Teitelbaum, 1996; Hood & Johnson, 1997). Whether a client uses alone or with others can be valuable information, especially if a pattern develops (such a pattern may consist of isolated use or use with a certain group of friends). The presence of friends who do not use the substance has good prognostic implications, whereas the loss of all nonusing friends has negative implications.

TABLE 2-12 Substance-Use-Related Issues to Be Addressed Through Assessment

CATEGORY OF QUESTIONING	INFORMATION NEEDED
Type and Pattern of Substance Use*	age of onset recency duration frequency intensity time of day and place of use variability in pattern of use drug tolerance use to prevent withdrawal symptoms drug of choice
Common Intoxication and Withdrawal Symptoms	attention and/or memory impairment blackouts blurred vision changes in appetite and/or weight disorientation and/or decreased alertness flashbacks heart arrhythmias (e.g., tachycardia, bradycardia) incoordination and/or unsteady gait insomnia or hypersomnia nausea and/or vomiting nystagmus (rhythmic movements or tremors of the eyes) palpitations (unusually rapid or irregular beating of the heart) perceptual disturbances perspiration or chills psychomotor agitation or retardation pupillary dilation or constriction seizures slurred speech tremors or twitching
Investment in Drug Use	methods of securing of money to pay for drugs criminal behavior related to drug use or procurement time spent using, hung-over, intoxicated, or withdrawing time spent on drug procurement, use, and after-effects
Attempts at Quitting	periods of sobriety number of prior attempts at quitting or decreasing amounts inability to control use willingness to give up activities or responsibilities to use absence of desire to quit

TABLE 2-12 Substance-Use-Related Issues to Be Addressed Through Assessment (continued)

CATEGORY OF QUESTIONING	INFORMATION NEEDED
Circumstances Around Use	situational triggers for use
	mood antecedents and consequences of use
	expectations about use
	friends and family who use
	friends and family who do not use
	peer or family pressure for or against use
	use alone or with someone else
Potential Affected Areas of Functioning	family relationships
	social relationship
	employment history
	academic history
	coping skills
	decision-making and problem-solving ability
	physical health
	financial situation
	criminal record
Common Associated Psychiatric Symptoms	mood disorder (including suicidal ideation and action)
	psychotic disorder
	cognitive disorder
	anxiety disorder
	impulse control disorder

*(Repeat this line of questions for each specific drug, addressing all items for each individual drug.)

Expectations about the consequences of alcohol and other drug use can illuminate the reinforcing factors that maintain substance use (Carey & Teitelbaum, 1996), as in a case where the client expects greater social ease and camaraderie through use. It is also important to explore whether the client has unsuccessfully attempted to quit and whether he or she has experienced periods of nonuse. A client who has attempted to quit many times without success and without assistance may have some motivation for change, whereas a client who has never thought of quitting may still not be inclined to do so. A client with no extended periods of sobriety may have a worse prognosis than one who has been able to abstain for one reason or another (e.g., women who abstain during pregnancy).

The issue of illegal activity is an important one that often has direct implications for the severity of the use. A client who has had repeated criminal convictions due to use-related behavior (e.g., DUIs, disorderly conduct) may be more severely addicted than one who has never experienced such repercussions. Similarly, the amount of time invested in the substance use habit may provide some insight into severity of use. A client whose entire day is preoccupied with drug procurement, use, and recovery is less likely to be treatable on an outpatient basis than is a client with a mild habit using only on weekends.

Several areas of information ancillary to substance use can often illuminate the important differentiation between use and abuse. Commonly associated difficulties can emerge in a multitude of areas, including family relationships, social contexts, employment and career settings, and academic settings (Boylan, Malley, & Scott, 1995; Meyer & Deitsch, 1996). Also often affected are the client's financial situation and physical health, as well as involvement with law enforcement. Assessing how vastly the client's network of relationships (both intimate and superficial) has been affected by drug use provides great insight with regard to the level of addiction and the possible level of motivation for treatment. Often family relationships are severely affected when clients turn away from nonusing parents or spouses. Even more complicated may be the client who uses with his or her spouse or intimate partner, for in these situations drugs can serve as a major cementing feature of the relationship. Violence within the intimate relationship deserves exploration because it, too, is often related to drug use. In terms of social relationships, the clinician explores whether there are drugs in the client's social network; that is, if all of his or her friends use or if there are any left who do not. With regard to employment history, it is important to find out if clients have used on the job, are going to work intoxicated or withdrawing, and if their ability to work has been affected by the use. This is of particular importance, in terms of physical safety, if a client's work involves operation of heavy machinery or otherwise requires attention and concentration to assure safety.

Finally, it is important to find out if the client's financial situation has suffered due to drug use, if there have been changes in physical health, and if there has been involvement with the criminal justice system. Any mental health symptoms revealed by the client must be put in the context of the substance use in order to make accurate diagnoses of mental illness per se versus substance-induced mental illness. Perkinson (1997) provides a thorough structured interview outline for the assessment of substance use for those counselors who would like to review a sample of specific phrasing of questions.

Use of Supplemental Questionnaires

Once the clinical interview is complete, the care provider can decide whether to administer additional questionnaires to elicit more information or corroborate findings. A number of instruments have been developed for this purpose, though none is entirely satisfactory as far as validity and reliability are concerned. The most commonly used instruments are the Michigan Alcohol Screening Test (MAST), the Addiction Severity Index (ASI), and the Substance Abuse Subtle Screening Inventory (SASSI). Others exist (e.g., subscales of the MMPI-2), and the interested reader is referred to Allen and Columbus (1995) and Litten and Allen (1994) for more information. The simplest instrument that has been developed to assess problematic drinking is the CAGE Questionnaire (Ewing, 1984), which consists of four easily asked questions. Affirmative answers to the four questions alert the counselor to potential drinking problems:

- Has the client attempted to *C*ut down on drinking?
- Has the client been *A*nnoyed because others have criticized the client's drinking?
- Has the client felt *G*uilty about drinking?
- Has the client used drinking as an *E*ye opener in the mornings?

Michigan Alcohol Screening Test.   The MAST (Selzer, 1971) was developed as a "rapid and effective screening for lifetime alcohol-related problems and alcoholism" (Allen & Columbus, 1995, p. 386) for a variety of populations. It consists of 25 brief items that are self-administered in approximately 10 minutes and responded to on a true-false basis. Scoring is accomplished after reverse-scoring 4 of the 25 items and assigning weighted scores. These weighted scores are then summed; the sum represents a total score reflecting severity of alcohol-related problems. A total score of "five points or more would place the subject in an 'alcoholic' category. Four points would be suggestive of alcoholism, three points or less would indicate the subject was not alcoholic" (Allen & Columbus, 1995, p. 391). Psychometric studies (e.g., Hedlund & Vieweg, 1984) report adequate internal consistency but marginal construct validity. Nevertheless, the MAST has been recommended as a screening tool by the National Institute on Alcohol Abuse and Alcoholism (see Connors, 1995).

Addiction Severity Index. The ASI (McLellan et al., 1985) was developed to measure seven key problem areas associated with drug use. The ASI rates each problem area with regard to severity, providing information that will assist with treatment planning. The ASI is a structured interview that can be completed in about 45 minutes (including scoring time) and can be used in conjunction with a clinical interview as previously described. It has reportedly excellent interrater reliability and test-retest reliability, as well as acceptable concurrent validity. It can be administered repeatedly throughout treatment to assess progress. The seven areas tapped by the ASI are drug and alcohol use, medical status, employment status, illegal activity, family and social relations, and psychiatric conditions over a lifetime and in the last 30 days. Each area is assessed via 20 to 30 questions, the answers to which are scored objectively and subjectively. Thus, the ASI is more comprehensive than the MAST but also more difficult and time-consuming to administer. However, training aids for interested users are available from the National Institute on Drug Abuse (NIDA). The ASI is rated by the NIDA as one of the most widely used drug screening instruments for treatment and research purposes (NIDA, 1994).

Substance Abuse Subtle Screening Inventory. The SASSI (Miller, 1985) was developed as a subtle measure of substance use disorder to classify respondents as chemical abusers or nonabusers. The SASSI screens test takers without using obvious drug related items. For this reason the SASSI is particularly useful in non-substance use treatment settings (Department of Health and Human Services, 1993). This feature requires that the SASSI be used in conjunction with a clinical interview, once a categorization of substance abuse or dependence has been made. The SASSI consists of 52 true-false items and can be scored to provide subscales that render a substance use profile. Administration time averages 10 minutes. The subscales of the SASSI measure the following: use of alcohol, use of other drugs, more obvious attributes of substance use, subtle attributes of substance use, defensiveness in test-taking behavior, current risk for substance abuse, and validity. Six of the scales are used in a formula to provide a dichotomous determination of whether the respondent is substance dependent. The current scale has good internal consistency and validity (Miller, 1985).

REFERRAL AND TREATMENT PLANNING

Regardless of how difficult a clinician may perceive the substance use assessment to be, only once all data have been collected and after it has been determined that a substance use disorder exists, does the real challenge begin. The first question to arise after the mental health counselor has decided to make a substance use diagnosis is whether to attempt to treat a client with this diagnosis. As indicated from the outset, the quest of this chapter is to make sure that no mental health care provider misdiagnoses or fails to identify clients who have substance use related problems; this chapter does not address the complexity of substance use treatment. The question of who is the best health care provider for substance-using clients cannot be answered entirely satisfactorily (Miller & Brown, 1997). One thing is for certain, however: a therapist who has no experience with substance use treatment best serves the substance-using client by making a referral to a care provider who does. Many specific treatment options exist that are better equipped to deal with the special dynamics inherent to working with substance-using clients. Such a referral does not preclude the referring provider from working with the client concurrently, with a counseling or therapy focus. It merely acknowledges the existence of an overarching problem that also needs to be addressed and worked on simultaneously by an expert in the substance use treatment field. Some clinicians, on the other hand, prefer a straight referral, ending their work with any client who actively uses substances in a diagnosable manner.

The referral decisions that have to be made are not always clear-cut. Most important to the mental health care provider is the question of whether substance-using clients are best treated on an outpatient basis or whether inpatient treatment is necessary. Certainly, if there is an anticipated period of withdrawal after severe and prolonged use, detoxification may need to take place in a controlled medical environment. However, even after detoxification, because the client continues to be vulnerable to relapse, inpatient treatment may remain necessary. A consistent argument has been made that clients who are actively using substances or who are still under the influence are best not seen in outpatient treatment. A client who is severely addicted and using probably needs to be referred for inpatient evaluation and treatment. A few specific guidelines about inpatient versus outpatient treatment will be provided in the following paragraphs.

If a client appears able to benefit from outpatient treatment, either because inpatient treatment has been completed or because the severity of the substance use was judged not to warrant inpatient services, auxiliary services also need to be discussed as part of the referral. Common recommendations include referrals to physicians, who will monitor and assess physical health; to support groups (e.g., AA or NA) to help the client establish a nonusing social network; and perhaps to a psychoeducational program of groups conducted by service providers specializing in substance use treatment. The latter is particularly indicated if the client still manifests significant denial or a profound inability to stop use (Hood & Johnson, 1997). Substance use, even once stopped, generally implies that there will be a number of related presenting concerns that need to be addressed in the treatment plan. These tend to include legal problems, relationship problems, educational and career interferences, the need for social skills training and social support development, physical health and fitness problems, the need for alternative coping strategies to deal with stress, the development of leisure skills, the treatment of psychological symptoms (ranging from symptoms to disorders), and relapse prevention (Namyniuk, Brems, & Clarson, 1997; Nystul, 1993). Additionally, substance use does not occur in a vacuum, and there is significant consensus in the literature that it always has implications for family members. Hence, treatment planning with a substance-using individual needs to acknowledge that the client's family may need therapeutic support of one form another as well (Namyniuk, Brems, & Clarson, 1997).

Referral Criteria

The American Society of Addiction Medicine (ASAM) has provided a set of guidelines for assessment and decision making regarding treatment selection for substance-using clients (ASAM, 1996). These guidelines specify that the client needs to be evaluated along six dimensions to evaluate treatment needs (these overlap with the assessment criteria outlined earlier):

- intoxication and withdrawal complications: risks associated with both; support network to assist with outpatient detoxification

- biomedical conditions and complications: acute or chronic medical conditions that might interact with the client's substance use habit and might affect treatment

- emotional and behavioral conditions and complications: acute or chronic psychiatric conditions or emotional variables that might interact with the client's substance use habit and might affect treatment

- treatment acceptance versus resistance: level of self-motivation and anticipated compliance; perception of coercion into treatment on the part of the client
- relapse potential: awareness of relapse triggers; risk for continued use without treatment; awareness of treatment needs and coping resources
- recovery circumstances: living environment variables that hinder or facilitate recovery; legal, vocational, or criminal justice complications or motivators for recovery

The answers to these six considerations will help the counselor make a referral plan that is commensurate with the client's level of need by determining level of risk. The higher the risk implied in the answers to the six domains, the more restrictive the setting to which the client will be referred; the lower the risk, the greater the likelihood that nonrestrictive, outpatient approaches will be most appropriate. The levels of care that can be considered range from early intervention, outpatient treatment, intensive outpatient treatment/partial hospitalization, residential treatment/inpatient services, to medically managed intensive inpatient services (Perkinson, 1997).

Early intervention is appropriate to low-risk clients with minimal symptoms and no current diagnosable disorder. It involves education about risk of substance use and can be delivered in a variety of settings, including in the context of psychotherapy or counseling. As such, this level of intervention may be possible at the level of the assessing mental health care provider, as long as that professional has some basic knowledge about substance use. Outpatient treatment is best provided by clinicians who specialize in substance use treatment. It can consist of individual sessions and/or group therapy and is similar in approach to mental health counseling, though more frequent contacts with a provider may be arranged. Clients appropriate for this level of intervention have a substance use diagnosis but have minimal intoxication or withdrawal symptoms and ancillary medical or psychiatric problems. They have intrinsic motivation for cooperation with treatment and continued out-of-control use does not pose a threat. These clients generally have a supportive environment that will assist, rather than hinder, recovery.

Intensive outpatient or partial hospitalization programs involve the client in structured day and evening programs that usually require at least nine hours of participation per week. Such settings have the resources to meet clients' medical and psychiatric needs and often can provide access to detoxification facilities and medical staff. Clients who are appropriate for these types of programs experience some mild and easily managed intoxication or withdrawal symptoms due to their substance use disorder; they can be maintained without 24-hour per day supervision given their lower relapse potential. These programs offer a supportive environment that will assist clients in their quest for treatment. These clients are motivated for treatment and are willing to cooperate and seek assistance as needed.

Clients with high relapse potential and severe intoxication or withdrawal symptoms are best referred for residential or inpatient services. These clients tend to have fluctuating commitments to and motivation for treatment and may present with medical or psychiatric conditions that complicate their treatment. They may also have medical conditions related to withdrawal that require close medical supervision. These clients still have questionable ability to remain abstinent, and their living environment may not be entirely supportive of sobriety. There is some variability in the level of care that can be offered in inpatient settings. Some of these settings are fairly flexible, open-door facilities that allow the client the freedom to come and go as needed for job or family obligations; other settings are highly restrictive, allowing absences only with specially approved passes or while under supervision. Inpatient programs involve many hours of treatment per day and generally have structured group and individual treatment programs. These facilities can manage most medical and psychiatric issues that emerge, including availability of detoxification facilities. Clients with severe medical disorders secondary to or interacting with their substance use may need additional medical supervision that can be provided only in specially equipped facilities. Such clients are in need for referrals to medically managed intensive inpatient treatment facilities. A few suggested guidelines to assist in the referral decision-making process are provided in Table 2-13. This table is not definitive; it merely provides suggestions that can be used alongside proper clinical judgment to arrive at the best possible plan of action given what is known about the client. This table cannot take the place of proper and intensive assessment and in-depth knowledge of all risk factors. It only differentiates between inpatient versus outpatient treatment decisions, since the need for more subtle nuances of care is too difficult to quantify.

Referral Process

Clearly, mental health care providers need to have some familiarity with the types of substance use treatment facilities available in their communities for appropriate referral to take place. At a minimum, it is helpful for the clinician to call a specialized substance use professional in the community (by looking in the phone book) to talk about who provides outpatient services and where inpatient services are rendered. A visit to a facility in the community would also be helpful because being able to describe the setting to which one refers a client can be extremely helpful to allay the client's fears and anxieties. Having a relationship with a substance use treatment center in the local community is helpful because such facilities often have an intake counselor on call who can help with referral and assessment decisions on the spot. Knowing these care providers can help the clinician during times when treatment and referral decisions are difficult and urgent to make. The clinician may be

able to initiate the call to the facility in the client's presence, allowing the client to make phone contact with the next care provider. Such interfacing is also helpful if the counselor remains uncertain about the level of intervention that is necessary for the client. Having a relationship with a treatment facility in that case allows for a quick phone consultation that can help clarify these issues.

Another essential component of the referral process is the preparation of the client. Given the reality that the client sought services from a mental heath care provider, it is likely that some level of denial is present about the need for specialized services for the use of a substance. The client may be resistant to a referral for substance use treatment services and may not want to acknowledge or accept that a substance use problem exists. A first step in making the referral, therefore, is the assessment of the client's level of awareness and acceptance of an existing substance use problem. This is best accomplished by questioning the client about the connection between the presenting problems and the client's pattern of substance use. Optimally, when thus queried, the client admits to a relationship between the two and recognizes that the presenting problems cannot be entirely independent of the substance use. Unfortunately, not all clients will react this favorably to the clinician's suggestion that the two issues are related. If the client denies a relationship between the substance use problem and the presenting concerns, it becomes the first therapeutic task of the clinician to point out the connections and help the client recognize them.

How does a clinician go about making the connection for the client who is very much in denial? This process is similar to the process family members go through when they are attempting to get another member to recognize a substance use problem. The process begins by telling the client that the substance use is creating a range of problems in his or her life and that these problems, in all likelihood, cannot be resolved without addressing the issue of substance use. The more specific the clinician can be in the se-

TABLE 2-13   Suggested Guidelines for Inpatient Versus Outpatient Referral Decisions

ASAM* VARIABLE	OUTPATIENT TREATMENT OPTION	INPATIENT TREATMENT OPTION
Intoxication or Withdrawal Symptoms	■ No acute intoxication ■ Minimal risk for severe withdrawal symptoms	■ Risk of severe withdrawal ■ Past treatment entry failure after detoxification
Medical Condition or Complications	■ Any existing medical conditions are stable and do not require hospitalization ■ Any existing medical conditions not directly affected by substance use	■ Concurrent medical illness that requires medical attention ■ Medical consequences of substance use that may be a threat to future health ■ Medical condition that may be aggravated by continued substance use
Emotional or Psychological Conditions or Complications	■ No risk of harm to self or others ■ Mental faculties sufficiently intact to comprehend treatment interventions ■ No primary comorbid psychiatric disorder ■ Emotional or psychological problems interfere with abstinence	■ Some risk of harm to self or other ■ Emotional, psychological, or behavioral symptoms that impair daily living skills and coping abilities ■ Coexisting personality disorder that results in unpredictable dysfunctional behavior ■ Violence during intoxication
Treatment Acceptance versus Resistance	■ Willingness to cooperate with treatment schedules ■ Some level of motivation for treatment because of recognition of a problem with substance use	■ Fluctuating willingness for cooperation ■ Inability to perceive a problem with current pattern of substance use
Relapse Potential	■ Ability to remain abstinent between appointments with sufficient regular supports	■ Inability to control use if the substance is present ■ Inability to curb use without constant supervision ■ Other symptoms require this level of care
Recovery Circumstances	■ Environment is supportive of recovery ■ Willingness and anticipated ability to create a supportive environment if it is absent currently	■ Environment not conducive to recovery ■ Dangerous environment ■ Occupation in which continued use would pose threat to self or others

*American Society of Addiction Medicine

lection of life concerns that are clearly influenced by the client's substance use, the more convincing the argument. A general statement about substance use having negative effects on relationships, for example, is not likely to be of great assistance. Clients will merely deny that this has happened in their life. However, being able to refer back to very specific concerns or patterns in a specific relationship that was discussed during the assessment interview and clearly connecting these specific problems with the client's substance use can have a great impact. Connecting the substance use to various problems in the client's life is an essential means of attempting to create internal motivation in the client for seeking assistance for the substance use pattern.

Once the connections have been made, and regardless of the level of acceptance by the client, in the next step of the referral process, the clinician explains that an expert in substance use treatment will need to be involved for optimal treatment outcome. The level and type of involvement will be determined by the preceding criteria for inpatient versus outpatient treatment; the clinician will also make personal choices about whether to conduct counseling or therapy with clients who are using substances and who are concurrently treated for their substance use. Concurrent treatment requires good coordination of care and regular communication among all involved care providers (with appropriate releases of information, of course). Once the client has accepted the need for a referral, clinician and client will work together to make the referral happen. The specifics of this work will depend on the level of intervention needed by the client. Most importantly, however, the clinician works collaboratively with the client until at least some contacts have been established that are deemed crucial to the proper treatment of the client. In other words, the client is not just sent home with some names and numbers, never to be seen again by the referring clinician. Instead, some time will be taken in session to make contacts with collaborating professionals. These contacts can range from making calls to an outpatient counselor, who will be the substance use expert in the outpatient treatment of the client, to calls to a residential treatment facility for immediate transfer of the client for inpatient treatment. For optimal results, the clinician needs to communicate caring and show concern and should not demonstrate an eagerness to get rid of the client.

If, even after the clinician's most caring attempts, a client decides to refuse referral for substance use treatment, whether such treatment was to take place collaboratively with or instead of mental health treatment, the mental health care provider must decide what to do with the client.

There are at least two broad choices in such instances. First, the clinician can choose to refuse services to the client. In such a case, the counselor would explain to the client that appropriate treatment is not possible without the involvement of substance use professionals, given the counselor's personal level of expertise in the area. The client is then given resources (e.g., names of inpatient and outpatient providers) for a referral and is strongly encouraged to make these contacts. If the client absolutely refuses to acknowledge the importance of substance use, but definitely needs treatment, the counselor who has decided not to see this client may need to refer the client to another mental health care provider (i.e., *not* a substance use treatment professional per se) who is known to the counselor (but not necessarily to the client) to have some expertise in substance use treatment. The client may well accept such a mental health referral, even after having refused substance use-related referrals.

Second, the mental health care provider can decide to see the client even without the support of a substance use professional, given that otherwise the client in all likelihood will not seek any services at all. The argument in this case is that the mental health care provider, inexperienced with substance use issues, can still help the client with the presenting concern, even though substance use issues may not be dealt with optimally. If this is the choice, the clinician has some responsibility to explain to the client the limitations of the work that they can do together. It will be important for the clinician to let the client know that he or she does not believe this is the best treatment of choice and that the issue of substance use will always be prominent, even in this provider's work with the client. The counselor who chooses this course of action will then need to become educated about substance use treatment, at least in a rudimentary form, and will always consider how the mental health issues that are being worked on may be affected by the substance use and vice versa. The clinician will also periodically revisit the issue with the client, attempting to make the appropriate referrals as mental health treatment progresses. Seeking consultation with a professional experienced in the area of substance use treatment is tantamount to success and ethical treatment delivery in these cases.

DOCUMENTATION AND RECORD KEEPING

Documentation of a client's substance use pattern can be accomplished either as part of the intake report (if the data are available then), as an intake report addendum (if it emerges soon thereafter), or as part of a progress note. Regardless of where this documentation is placed, it needs to be very thorough, reflecting that the clinician asked all information necessary to make an accurate diagnosis. Hence, if recorded in the form of a progress note, this note will be longer than usual.

If the client is referred to a substance use treatment provider, a clear description of the referral process in the progress note is important. Any consultation with other professionals to enhance the decision-making process is also best put in writing. Such progress note entries are then supported through appropriate ROIs, the originals of which are kept in the client's chart (copies are mailed to the other provider[s]). If the client was deemed in need of a referral but refused it, this information is carefully documented. The original clinician verbally outlines for the client, and then documents in a progress note in the client's chart, all limitations about the treatment that will be provided to the client, given the client's refusal to involve substance use treatment experts. This documentation keeps the clinician safe from later lawsuits that could otherwise claim inappropriate treatment planning.

Occasionally, when referrals are initiated, the professional to whom the client was referred (referral target) requests information above and beyond that contained in a standard screening summary, intake report, or progress note. In such instances, the referring clinician (referral source) prepares a special referral report. Whenever this is the case, the original of this document is retained in the client's chart, and the copy is mailed to the requesting provider. If any records are received from the referral target, these documents become a permanent part of the client's chart. However, these documents generally cannot be released by the receiving clinician. Instead, if another provider needs them, they should be requested from the practitioner who originally prepared them.

SUMMARY AND CONCLUDING THOUGHTS

Controversy continues to exist about the ultimate treatment goal for substance-using clients. Nystul (1993) and Nugent (1994) argue that it is not clear whether no use or controlled use should be the goal; others, however, believe that only no use can be a proper treatment goal (e.g., American Psychiatric Association, 1994; Meyer & Deitsch, 1996). This controversy will not likely be put to rest any time soon, and until that time each clinician will have to make a personal decision that fits with her or his conceptualization of substance use and moral values. Regardless of where one stands on the issue of complete versus controlled abstinence, counselors and therapists need to be knowledgeable about making accurate substance use diagnoses. Only a thorough assessment can help the clinician decide whether a client has a real problem with substance use that needs to be treated by a specialized care provider. Once the determination has been made that the client has a diagnosable substance use disorder, the referral process is initiated. By that time the clinician already has a clear idea as to the level of treatment that will be necessary. Appropriate referral is then made for inpatient or outpatient services that can either be delivered collaboratively or exclusively.

3

The Challenge of Medical Involvement

We have neglected our emotional reality, and the source of our self-nourishment: our bodies.

STANLEY KELEMAN

As indicated in Chapter 1, medical and physical concerns that arise in the initial session often leave the counselor feeling somewhat puzzled or uncertain about the conceptualization of the client's case. Whenever medical issues arise, the most prudent course of action is to seek consultation with or make a referral to a medical practitioner. A medical practitioner may be defined as any physician with an M.D. (traditional medical education), N.D. (naturopathic medical education), D.C. (chiropractic medical education), or D.O. (osteopathic medical education); a certified nutritionist; a licensed nurse practitioner with psychiatric specialization; or a similarly qualified and specialized health care provider, depending on circumstances and referral needs. It is best to choose a medical practitioner who has specialized in the area of physical functioning that has prompted the referral or consultation. For example, a psychotropic medication referral is best directed to a psychiatrist (M.D.) who specializes in the age group or disorder with which the client presents. A referral due to a suspicion of an underlying medical disorder that presents with psy-

chological symptoms is best directed to a physician who has a reputation for sensitivity to psychological issues. Consultations due to medications a client is already taking need, of course, to be initiated with the prescribing physician. These are not hard and fast rules, just guidelines that tend to work best. Getting to know medical practitioners in one's community is the basis of all successful referrals. It can be downright dangerous to make referrals without knowing the target provider (more about this later). To reiterate from Chapter 1, medical (or related) referrals are indicated in all of the following situations (with the preferred provider indicated in parentheses):

- when information is needed about the relationship between medical and psychological/emotional symptoms (nontraditional physician)

- when information is needed about medication regimens currently taken by the client (prescribing physician)

- when information is needed about psychiatric symptoms and their need for amelioration above and beyond psychotherapeutic or psychoeducational intervention (psychiatrist)

- when information is needed about food intake or weight concerns that are integral to the psychological presentation and have potential physical implications (nutritionist or nontraditional physician)

- when information is needed about the interface between psychological/emotional presenting concerns and perceptual deficits (physician for initiation of hearing, vision, or other perceptual skills screenings)

Failure to refer can have tragic consequences in cases when the physical health of the client is in question or when physical factors may play a potential role in the client's presentation. For example, the possible physical complications of a client who presents with anorexia nervosa require the monitoring of physiological consequences by a physician. A mental health care professional who does not involve a medical doctor in such a case—especially if the case is severe, chronic, and protracted—risks the client's death from physical complications secondary to the psychological disorder. This is simply one of many examples that drive home the importance of knowledgeable and timely referrals. Throughout, this chapter will highlight situations that require the involvement of medical specialists in the care of clients presenting for psychotherapy. These situations reiterate that referral and collaboration are useful and necessary processes that do not imply lack of skill on the part of the referring clinician. This chapter also provides a framework for recognizing the need for referrals and a schema for referrals that can be used by individual clinicians flexibly, creatively, and idiosyncratically, as required by the special needs of each client.

Making proper medical referrals can reduce a client's pain and suffering and can, on rare occasions, be a matter of life and death. Not only do clinicians refer clients to medical providers because they need psychotropic medications, but they also make referrals for a variety of other reasons:

- because the client presents with a psychological disorder that has physiological consequences or implications (e.g., bulimia which may result in a variety of physical concerns, ranging from esophageal ulcers to even more severe conditions)

- because the client presents with psychological symptoms that can have physical causes and because the client has not had a recent physical exam (e.g., panic attacks or depression may need differential diagnostic procedures that rule out physical disorder)

- because the client presents with emotional or psychological reactions that can be secondary to medications but that have not been discussed with a physician; or because he or she may present with physical symptoms that may or may not be part of a psychological disorder (e.g., hyperventilation, sleep disturbances, rashes)

Some psychologists have taken the matter of attention to physical processes in their clients even further by engaging in advanced training to enable them to conduct physical examinations of patients (see Folen, Kellar, James, Porter, & Peterson, 1998). These professionals advocate for a more active medical interface in psychology.

The importance of an interface between medicine and psychology is clarified by some figures provided by Tomb (1995) who indicates that 50–80% of patients treated in medical clinics actually have a diagnosable psychiatric disorder; 60% of patients treated by general medical practitioners actually need mental health care; and 50% of patients in psychiatric clinics have undiagnosed medical illnesses. Further, over half of patient visits to primary care physicians are related to psychosocial problems, although they are presented to the care provider in the form of physical complaints (Wickramasekera, Davies, & Davies, 1996). On the other hand, Klonoff and Landrine (1997) report evidence that a significant percentage of psychiatric patients suffer from an undiagnosed medical illness, with estimates ranging from 41–83% depending on setting and symptoms.

These numbers clearly emphasize that psychological providers must be aware of clients' physical needs and that medical providers must be conscious of the patients' psychological or emotional states. When providers fail to address clients' total needs, they not only compromise client care, but they become vulnerable to lawsuits. In fact, lawsuits are increasing against psychological providers who failed to investigate possible physical diagnoses that would have explained clients' symptoms (Klonoff & Landrine, 1997). It behooves each and every mental health care provider to be knowledgeable about medical referral issues and to have a collaborative relationship with a medical provider in the community. When referrals are made to physicians (or other medical providers, such as physician's assistants, nutritionists, nurse practitioners, etc.), the clinician has the responsibility to coordinate and facilitate this process. The referral process will be described in detail in the context of a psychotropic medication referral in the following section. Although the process may differ somewhat based on the actual reason for referral, this prototype will suffice to give the reader a clear concept of how to interface with a medical professional.

PSYCHOLOGICAL SYMPTOMS AND DISORDERS REQUIRING PSYCHOTROPIC MEDICATION REFERRAL

There are a number of psychological disorders that are better managed when concurrently treated with medications; among the most obvious are schizophrenia, severe major depression, severe anxiety, and panic disorders (Cormier & Cormier, 1999). Although the individual needs of the client are considered in every case, Seligman (1996) proposes some general rules for when psychological providers need to make medication referrals: she suggests that providers always make a medication referral in cases of psychosis, bipolar disorder, Tourette's syndrome, pervasive develop-

mental disorder, and obsessive compulsive disorder; providers strongly need to consider a medication referral in cases of major depression, attention-deficit hyperactivity disorder, eating disorders, cyclothymia, and panic disorder; they generally do not need to consider a psychotropic medication referral if the presentation is limited to adjustment disorders, learning problems, oppositional defiant disorders, and other mild diagnostic categories (see DSM-IV for other categories of disorders not listed here). Freimuth (1996) adds that medication referrals should be strongly considered any time symptoms appear recalcitrant to change induced by psychotherapy alone; she cautions, however, that medication referrals must never be initiated as a quick fix that takes the place of psychological therapeutic interventions. Gitlin (1997) suggests that pharmacological referrals are indicated any time there are complaints that focus on or include bodily feelings or sensations (e.g., insomnia, fatigue), changes in cognitive capacity (e.g., memory disturbances, changes in concentration and attention), impairment in reality testing, medical symptoms (e.g., headaches, stomachaches), nonresponse to psychotherapy, family history of psychiatric illness or medical illness with psychological symptoms, and feelings of suicidality. He also points out, however, that "pharmacotherapy never precludes other methods of treatment . . . [because] there may be a variety of different, valid therapeutic approaches [and that] psychotherapy is likely to enhance the treatment and help patients in ways not measured in research studies" (p. 7). In other words, a medication referral does not imply that psychotherapy was, is, or will be unsuccessful; it merely acknowledges that humans are more complex than *just* mind or *just* body and need to be treated in a manner that reflects this complexity.

Some psychotherapists are hesitant to make medication referrals. They fear that medications undermine the therapeutic process because they suggest dependency, helplessness, and passivity on the part of the client (Freimuth, 1996). Others are hesitant to refer for psychotropic medication because they fear that clients will falsely perceive the medication as a magic cure that can eliminate symptoms simply and effortlessly (Sansone & Shaffer, 1997). These clinicians believe either that reduction of symptoms via medication reduces motivation for psychotherapeutic treatment or that reduction of symptoms that are amenable to psychotropic medication will result in symptom substitution (Gitlin, 1997). Although these arguments have some merit, in some cases clients are so impaired by severe psychological symptoms that they are incapable of participating in psychotherapy until some symptom reduction via medication has been achieved. Hence, these arguments are not feasible in all cases and are considered on a case-by-case basis. Finally, some clients may interpret a referral for medication as a suggestion that they are weak, incapable of taking care of their own problems, or that their disorder is extremely and overwhelmingly severe (Sansone & Shaffer, 1997). Some clinicians use this as reason to reject a psychotropic medication referral. However, if this is the only reason providers reject a medication referral, they must ask themselves whether they have made an appropriate treatment decision or whether this is a countertransferential issue. Would the same provider also hesitate to use other interventions (e.g., the preparation of a relaxation tape for the client) because of the client's concern that he or she does not have complete personal control over this technique?

Medication referrals can serve a useful purpose and can be helpful to clients (a summary of diagnostic categories for which psychotropic medication referrals are helpful is provided in Table 3-1). They must be weighed carefully in terms of their possible advantages (e.g., symptom resolution or amelioration) and disadvantages (e.g., medication side effects, decrease in motivation for nonmedical treatment). If psychotropic medications are rejected, this rejection needs to be based on sound clinical judgment about the potential negative consequences of a medical referral and, of course, its lack of necessity. If a referral is made, it must be done so based on equally careful consideration of the potential

usefulness of psychotropic medication for the client. Having access to respected and trusted medical providers greatly facilities the psychotherapist's decision about whether to make the referral.

Collaborating with Other Professionals

The most common model of collaboration around psychotropic medications that a nonmedical mental health care provider will encounter is the split treatment model. In this model, the client is cared for by a medical provider, who manages the medication, and a therapist or counselor, who manages the psychotherapeutic intervention; that is, the model integrates "two professionals performing different therapeutic functions" (Gitlin, 1997, p. 327). Because some people believe it is the source of a variety of case management problems, this model has received some criticism. For example, Freimuth (1996) points out that the split treatment model can foster splitting on the part of the client (where the client plays one professional against the other), can result in poor communication between providers, and can lead to a lack of knowledge about the totality of the client by both professionals, with each assuming responsibility only for a certain aspect of the client. Gitlin (1997) agrees and adds that the split treatment model can interfere with the client's ability to integrate various parts of the self (most notably the client draws boundaries between the biological or physical and the psychological or emotional aspects of the self) and can lead to medicating for symptom relief only. However, these criticisms can be overcome if both providers are willing to keep lines of communication open, respect what each contributes to the client's treatment, and reach out to each other directly, without putting the client in the middle (Sansone & Shaffer, 1997). If the involved providers make these efforts, the advantages of the split treatment model can outweigh its potential drawbacks. Specifically, counseling or psychotherapy can focus on psychological issues and

TABLE 3-1   Diagnostic Categories Requiring Consideration of Psychotropic Medication Referral

DIAGNOSTIC CATEGORY	LEVEL OF NEED FOR MEDICATION REFERRAL	SAMPLE REFERENCE*
Anorexia Nervosa	Low to Medium	Hoffman & Halmi (1993)
Bipolar Disorder	High	Post (1993)
Bulimia Nervosa	Medium	Zwaan & Mitchell (1993)
Delirium	Low to High	Katz (1993)
Delusional Disorder	Low to Medium	Breier (1993)
Dementia	Low to High	Loebel, Dager, & Kitchell (1993)
Dissociative Identity Disorder	Medium to High	Davis (1993)
Generalized Anxiety Disorder	Low to Medium	Cowley (1993)
Major Depression	High to Medium	Tollefson (1993)
Obsessive Compulsive Disorder	High	Pigott, Grady, & Rubenstein (1993)
Panic Disorder	Medium to High	Sheehan & Ashok (1993)
Pervasive Development Disorder	Medium to High	Seligman (1996)
Posttraumatic Stress Disorder	Medium	Hammond, Scurfield, & Risse (1993)
Psychogenic Amnesia	Low	Davis (1993)
Psychogenic Fugue	Low	Davis (1993)
Schizophrenia	High	Potkin, Albers, & Richmond (1993)
Sexual Dysfunction	Low	Heiman (1993)
Sleep Disorders		
Narcolepsy	High	Gillin (1993)
Sleep Apnea	High	Gillin (1993)
Parasomnias	Medium	Giles & Buysse (1993)
Substance Use Disorders	Low to Medium	Seligman (1996)

*The information in this table is based on a wide range of resources; for each disorder identified as potentially needing psychotropic medication referral, a single reference is provided in the Table; three helpful overall references are Dunner (1993), Gitlin (1997), and Klein and Rowland (1996).

can clarify that medication intervention is not a magical cure but merely an ancillary aid for the management of the most severe physical symptoms; psychotherapy can also remain focused on psychotherapy, without having to be redirected to medication issues, such as titrating dosages, monitoring side effects, and checking physical consequences of medication. The pharmacological care provider, on the other hand, can concentrate on tracking and assessing medical issues and can make sure that these aspects of the client's symptom management do not interfere with the therapeutic relationship. In other words, from the mental health care provider's perspective, this model keeps the psychotherapeutic and prescribing roles separate, allowing for unimpeded unfolding of the therapeutic relationship. Such a model can work to the client's advantage as long as both providers put effort toward communicating with one another.

Such effort may mean that the counselor must learn how to communicate with a medical doctor, learning to speak in diagnostic language that is concise and uses DSM-IV jargon for disorders and symptoms (Diamond, 1998). The pharmacotherapist must learn how to communicate with a psychological provider and must be willing to concede that medications alone will not serve to heal, change, or improve the client's well-being. Both providers must use the client's well-being as the guiding principle in their work with the client and with each other; they must respect one another and communicate directly with each other. Both must be aware of the possibility of splitting, especially when treating personality disordered clients, and each must be capable of revising treatment plans based on input from the other provider (Sansone & Shaffer, 1997; Seligman, 1996). Neither must be threatened by the involvement of the other, and both must be able to recognize the concrete advantages of the other's involvement in the management of a given client. Both need to retain awareness of the fact that the other provider is a busy professional and that interactions and communications need to be brief, concise, specific, and direct (Gitlin, 1997). Finally, Gitlin (1997) suggests that both care providers must be willing to call the other whenever necessary, and in particular when

- clinical conditions change acutely
- the client becomes significantly suicidal
- a major change in medications is about to be made
- specific clinical questions arise that affect both the physical and psychological realm of the client's being
- medication side effects are noted
- splitting is occurring (e.g., the client devalues one provider in the presence of the other)

If such collaboration is possible, the split treatment model can work to clients' advantage, because two professionals are collaborating on their behalf. It must be noted that proper releases of information (ROIs; see Chapter 1) must be signed by the client to allow the two providers to communicate freely. If a client refuses to allow the two providers to interact, the client may impede treatment by setting up a situation in which splitting can occur, causing providers to work separately, not collaboratively. If the client refuses to sign an ROI, each provider needs to be careful not to interfere with the other's treatment—a difficult prospect indeed in the absence of direct information.

The Referral Process

The counselor or psychotherapist initiates the referral process by discussing with the client the belief that psychotropic medication may be helpful to the management of the client's presenting symptoms, providing the client with information about psychotropic medications and the process of securing them. If the client agrees that a trial of psychotropic medication may be helpful, the clinician initiates the next step of the referral process by preparing the client for what to expect from a medication referral. This information will vary somewhat depending on the providers who are involved in the process. It is important for a mental health care provider without a medical background to have established a consistent collaborative relationship with a medical provider in their community. Such familiarity will make preparation and collaboration much easier for all parties involved. A clinician chooses a pharmacotherapist in the same manner he or she chooses any other referral target for a client. Namely, the therapist needs to consider the following traits and characteristics of the potential collaborator (see Gitlin, 1997):

- general clinical competence (credential and interpersonal skills)
- acceptance of and respect for psychotherapy or counseling by the medical provider
- psychopharmacological competence (i.e., training and experience with psychotropic medications)
- comfort with a split treatment model that is collaborative and interactive
- ability to communicate with the psychotherapist or counselor
- ability to communicate with clients
- willingness to educate clients and to explain necessary psychotropic medication-related

procedures and issues (e.g., titration, side effects) (Ganz, 1988)

- ability to be open with clients while providing information at a pace and in language that can be understood and digested by the client (Diamond, 1998)

Once providers have chosen a pharmacotherapist, they can easily prepare the client for the psychotropic medication visit, since they know what to expect from the medical provider. Most commonly, a pharmacotherapist will meet with the client anywhere from one to three times for a diagnostic interview to determine the necessity and potential helpfulness of psychotropic medication. Additionally, some diagnostic lab work may be necessary, the results of which may assist in making choices about types of medications and dosages. Each visit with the medical provider may last about 15–30 minutes, a reality that needs to be pointed out to the psychotherapy client who is used to 50-minute sessions. The client also is made aware (and needs to sign proper ROIs) that the pharmacotherapist and psychotherapist or counselor will discuss the client and will consult from time to time. A good pharmacotherapist will give feedback to the client after the evaluation is complete and before the medication is dispensed. This information also needs to be shared with the mental health care provider who should not have to rely on second-hand information from the client, who may or may not understand everything the pharmacotherapist conveys. The client needs to be aware that both providers will retain regular contact to monitor the impact of treatment and to coordinate efforts.

Clearly, knowing when to refer for psychotropic medication is only one aspect of a mental health clinician's responsibility; mental health care providers also need to be knowledgeable about many other aspects of the referral process so that they can help clients who are in need of psychotropic medication. Additionally, therapists or counselors need to know enough about psychotropic medications to be aware of what these substances can do for clients, both how they can be helpful and harmful. Because the mental health care provider generally has more contact with an individual client than does a pharmacotherapist, the mental health care provider needs to be knowledgeable about psychotropic medications to assist with the monitoring of such medications in a number of contexts (Adler & Griffith, 1991; Diamond, 1998; Seligman, 1996). Therapists or counselors can be helpful in monitoring or recognizing

- treatment compliance
- medication side effects
- desired effects (i.e., effectiveness of the medication)
- new circumstances that may require changes in medication (e.g., pregnancy)
- behavior that could affect the effectiveness or effects of the medications (e.g., use of other drugs and alcohol)
- danger signals that require immediate medical attention

In all of these instances, the clinician can be helpful in terms of making sure that the client communicates with the medical provider, thus facilitating the proper dosing and management of the psychotropic medication regimen. Of course, monitoring cannot be the sole responsibility of the nonmedical care provider; the prescribing physician is really the one who tracks the client's adjustment to medication and assesses its effectiveness formally and through medical tests, if indicated (Fast & Preskorn, 1993). However, nonmedical care providers can be of substantial use to the drug monitoring process given their more frequent contact with clients. It goes without saying that the nonmedical mental health care provider is never tempted to give medical advice to the client, no matter how closely this opinion agrees with the opinion of the client's medical provider. If the therapist has concerns about the medication management in the case of an individual client, the best course of action is to contact the medical provider directly.

As previously noted, maintaining a positive and collaborative relationship with the medical professional is central to the optimal medication management of the client.

Critical Information about Psychotropic Medication

The referral process and subsequent collaborative efforts among client, pharmacotherapist, and counselor or psychotherapist will be greatly facilitated by a clinician who is knowledgeable about the basics of psychotropic medication. These basics include the types of medications available and their potential side effects. There are essentially four primary and two secondary (or less important) groups of psychotropic medications. The primary groups are:

- antidepressants
- anxiolytics
- antipsychotic (or neuroleptic medications)
- mood-stabilizers (or antimania medications)

The secondary groups are:

- hypnotics/sedatives (e.g., sleep aids)
- stimulants (e.g., medications for hyperactivity)

This chapter will not cover these two secondary groups (refer to Klein & Rowland, 1996, for information). Each of the groups of psychotropic medications has several classes of drugs within it, and each class of medications has specific desirable and undesirable (or side) effects. In other words, all psychotropic medications are impure, in that they do not only have a single biological effect (i.e., the desired effect of reducing or eliminating targeted symptoms) but also other physiological consequences (i.e., side effects). Side effects are a reality of all psychotropic medications that must be discussed with a client, preferably by the medical provider, but also by the counselor. "Withholding information about side effects or alternative treatments is paternalistic and disrespectful; moreover, it interferes with the kind of long-term relationship building that promotes effective treatment and helps the client make responsible decisions about . . . medication" (Diamond, 1998, p. 6). Common side effects are listed in Table 3-2 (and numbered for use in Tables 3-3 to 3-6). The danger of side effects should not rule out the use of psychotropic medication but certainly should be considered in weighing risk-benefit ratios for each individual client. When discussing side effects with clients, it may be important to stress "(1) the difference between likely and unlikely side effects, (2) the need to distinguish between uncomfortable and dangerous side effects, (3) the accommodation to side effects that commonly occurs, and (4) ways to diminish side effects" (Gitlin, 1997, p. 219). Only then will the client be able to make a sound decision based on all information available, and only then can the clinician successfully assist with the monitoring of psychotropic medication, without overreacting to minor or nondangerous side effects and without underreacting to those effects that are potentially life threatening. Side effects vary greatly from medication to medication and from client to client, and they also vary across time within a given medication or client. Thus, although a number of side effects are listed in Tables 3-3 to 3-6, they may not always be present during each client's use. Finally, clinicians need to be aware of side effects to recognize when complaints of new symptoms may represent possible manifestations of side effects of psychotropic medications as opposed to true new psychological or physical symptoms. Such knowledge will help the therapist explain and normalize the symptomatology for the client. For example, for a client recently put on certain medications for depression, changes in sexual desire and/or performance could be explained by the medication, and such explanation may have a reassuring effect on the client.

In addition to side effects, for each class of medications, a number of variables have to be considered by the medical provider. Many of these variables are also important information for the psychotherapist; some are less relevant to the non-medical practitioner (e.g., mode of action).

TABLE 3-2   Common Psychotropic Medication Side Effects

SYMPTOM/SYMPTOM COMPLEX*	DEFINITION AND/OR EXAMPLE
1. headaches	
2. gastrointestinal symptoms	nausea, bloating, abdominal discomfort, diarrhea, etc.
3. sexual dysfunctions	delayed orgasm, decreased sexual desire, sustained erection, menstrual difficulties
4. psychomotor effects	restlessness, pacing, difficulty sleeping, etc.
5. orthostatic hypertension	drop in blood pressure upon changing positions, results in dizziness and lightheadedness
6. weight gain	
7. anticholinergic effects	dry mouth, blurred vision, rapid heart beat, constipation, urinary retention, impairment in recall, confusion
8. cardiac symptoms	arrhythmias, cardiac conduction abnormalities
9. sedation and akinesia	feelings of being slowed down, lack of vigor or energy
10. fatal overdose potential	
11. edema	excess accumulation of fluids in connective tissue
12. ataxia	inability to coordinate voluntary muscle movement
13. addiction and/or withdrawal	
14. lethargy, apathy, drowsiness	
15. hair loss	
16. hypertension	increased blood pressure
17. cognitive impairment	concentration/attention difficulty, memory disturbance
18. tremor	trembling or shaking
19. metallic taste	
20. hematologic risks	depression of bone marrow production; depression in blood cell production
21. seizures	
22. dystonic reactions	abrupt spasms of voluntary muscles (if the spasm occurs in the larynx, this side effect becomes a medical emergency)
23. akathisia	uncomfortable restlessness; fidgetiness
24. Parkinsonism	pill-rolling tremor, stiffness, shuffling gait, masked facies (expressionless face), drooling, difficulty initiating or ending motion
25. tardive dyskinesia	involuntary, repetitive movement (see note below)
26. neuroleptic malignant syndrome	CNS dysfunction, autoimmune dysfunction, hyperthermia, muscular rigidity (see note below)

*The numbers next to each side effect are for reference in Tables 3-3 to 3-6. Tardive dyskinesia and neuroleptic malignant syndrome are defined in detail in the narrative section on Antipsychotics.

Following is a brief definition of the variables that are important for the non-medical provider and client to understand. Missing from this discussion of variables is the issue of mode of action for each class of medications; to do justice to such a discussion would go beyond the scope of the current chapter and the interested reader is referred to primary sources for this information (e.g., Diamond, 1998; Klein & Rowland, 1996). First, it is important to know each medication's predominant clinical indication (i.e., the symptoms for which it is most effective); second, therapists need to be aware that all psychotropic medications have trade names (the names under which a given drug is marketed) and generic names (the label that relates to or reflects the chemical composition of the drug); third, it is helpful for clinicians to know the average effective dosage for each medication, along with likely initial dosage (if this differs from average dosage); fourth, it is useful for counselors to be aware of the half-life of a given medication (the time required for half of the drug to be cleared from a client's body), because half-life has implications for the time it takes for a medication to reach steady state (the stable or therapeutically effective level of the drug achieved in the client's body with regular administration; usually equal to five times the half-life of a medication); fifth, the clinician needs to know the most common side effects of a given drug; and sixth, clinicians have to be cognizant of any possible contraindications (i.e., foods, actions, medical conditions, or concurrently taken medications that may interact with the psychotropic medication to result in an undesirable [heightened, lessened, changed] effect of the medication in the client's body).

Antidepressants. Simply put, there are two types of antidepressants: inhibitors of monoamine oxidase and inhibitors of monoamine neurotransmitter uptake (Richelson, 1993), also referred to as MAOIs and non-MAOIs respectively (Sansone & Shaffer, 1997). The non-MAOIs are often subdivided further based upon the structure of the compound (e.g., tricyclic, tetracyclic) (Klein & Rowland, 1996) or the function of the compound (e.g., serotinergic [selective serotonin reuptake inhibitors or SSRIs] versus nonserotinergic) (Quitkin & Taylor, 1996; Sansone & Shaffer, 1997) and are labeled as either first- or second-generation non-MAOIs (Sansone & Shaffer, 1997). Antidepressants are useful for severe acute depressions with physiological changes in functioning (e.g., appetite, weight, sleep). They are also used (though less successfully so) with more chronic (or long-standing) depressions or dysthymias, some anxiety disorders, panic disorder, and obsessive-compulsive disorder (Richelson, 1993; Sansone & Shaffer, 1997); success has also

been reported for bulimia, chronic pain, and reduction of craving in cocaine withdrawal or smoking cessation (Diamond, 1998; Gitlin, 1997) Generally speaking, antidepressants have long half-lives and take several weeks to reach steady state (though some of the newer generation antidepressants may result in somewhat faster initial symptom relief). Therefore, medication trials with antidepressants usually need to last at least six weeks to determine their effectiveness properly (Quitkin & Taylor, 1996). Antidepressants, on average, are administered for at least four months post-symptom-resolution, typically on a tapering schedule (Richelson, 1993). More information about antidepressants is provided in Table 3-3. For additional detail, the reader can refer to the most recent volume of the *Physician's Desk Reference* or to Diamond (1998).

Anxiolytics. There are four types of anxiolytics: benzodiazepines, azaspirones, serotonin reuptake blockers, and antihistamines (Wingerson & Roy-Byrne, 1993). Each grouping has several sub-types of medications, each of which varies somewhat with regard to specific clinical indication, side effects, and so forth. Most generally speaking, anxiolytics are indicated in cases of severe acute and chronic anxiety, with some additional usefulness for performance anxiety, panic attacks, withdrawal symptoms from depressants (e.g., alcohol or barbiturates), and mood instability (Sansone & Shaffer, 1997). Occasional use has also been reported for anticonvulsant purposes (Diamond, 1998). Finally, use of benzodiazepines is indicated for clients who are acutely out of control either due to substance intoxication or psychosis. In such cases, benzodiazepines tend to be a safer choice than antipsychotics, especially if the substance of intoxication is unknown. Combinations of benzodiazepines with antipsychotics for out-of-control clients can lower the dose of the latter resulting in fewer side effects for the client (Diamond, 1998).

Careful use is indicated with benzodiazepines: they are addictive, can cause sedation, are dangerous with patients who suffer from liver

TABLE 3-3   **Antidepressants: Important Characteristics and Variables**

TRADE NAMES	GENERIC NAME	MEAN DOSAGE (INITIAL DOSAGE)	SIDE EFFECTS	CONTRAINDICATIONS/ DANGERS
NON-MAOIS: TRICYCLIC ANTIDEPRESSANTS				
Elavil, Endep, Ami-tril, Etrafon, Limbitrol	amitriptyline	100–300 mg (25–50)	5,6,7,8,9,10	avoid use after MI;* many drug interactions (including with alcohol)
Ascendin	amoxapine	150–400 mg (50–100)		avoid use after MI; many drug interactions
Anafranil	clomipramine	100–250 mg (25–50)	5,6,7,8,9	heart disease, Raynaud's, diabetes, asthma; interacts with CNS active drugs
Norpramin, Pertofrane	desipramine	100–300 mg (25–50)	5–10	avoid use after MI; many drug interactions (including with alcohol and tobacco)
Sinequan, Adapin	doxepin	100–300 mg (25–50)	5–10	glaucoma, urinary retention; many drug interactions (including alcohol and tobacco)
Tofranil, Janimine, SK-Pramine	imipramine	100–300 mg (25–50)	5,7,9,10	avoid use after MI; many drug interactions, including with alcohol
Aventyl, Pamelor	nortriptyline	50–150 mg (10–25)	6–10	avoid use after MI; many drug interactions
Vivactil	protriptyline	15–60 mg (10)	5,7,8,10	avoid use after MI; many drug interactions, including alcohol
Surmontil	trimipramine	100–300 mg (25–50)	4,7,8,9,12	avoid use after MI; many drug interactions, including decongestants
NON-MAOIS: NEW GENERATION ANTIDEPRESSANTS				
Wellbutrin	bupropion	300–550 mg (100)	3,11,12	avoid use with seizure patients, bulimics, or anorexics; no concomitant use of MAOIs
Ludiomil	maprotiline	100–225 mg (25–50)	5,7,9	avoid use in combination with other antidepressants
Desyrel	trazodone	50–600 mg (50)	3,9	no concomitant MAOI use; interacts with digoxin and phenytoin
Serzone	nefazodone	300–600 mg (100)	3,7	no concomitant use of other antidepressants, esp. MAOIs; many drug interactions

TABLE 3-3   Antidepressants: Important Characteristics and Variables (continued)

TRADE NAMES	GENERIC NAME	MEAN DOSAGE (INITIAL DOSAGE)	SIDE EFFECTS	CONTRAINDICATIONS/ DANGERS
NON-MAOIS: NEW GENERATION ANTIDEPRESSANTS (CONTINUED)				
Remeron	mitrazapine	15–45 mg (15)	6,7,9	agranulocytosis (i.e., rapid cessation of production of white blood cells)
NON-MAOIS: NEW GENERATION ANTIDEPRESSANTS—SSRIS‡				
Prozac	fluoxetine	20–40 mg (20)	1,2,4	not within 5 weeks of MAOI use; interacts with tryptophan (in foods); also as Anafranil
Effexor	venlafaxine	75–150 mg (varies)	1,2,3,4,5,16	many drug interactions
Luvox	fluvoxamine	100–300 mg (50–100)	1,2,3,4,9	many drug interactions
Zoloft	sertraline	50–200 mg (50)	1,2,3,4	no concomitant MAOI use; many drug interactions, including alcohol
Paxil	paroxetine	10–50 mg (20)	1,2,3,4	no concomitant MAOI use; many drug interactions, including alcohol
MAOIS				
Marplan	isocarboxazid	10–20 mg (10)	5,6,7,9	impairment of renal or liver functioning; avoid with other antidepressants; many drug and food interactions†
Nardil	phenelzine	60–90 mg (15)	5,6,7,9	liver disease; many drug interactions (including alcohol); food interactions†
Parnate	tranylcypromine	30–60 mg (30)	5,6,7,9	liver disease; many drug interactions (including antihistamines); food interactions†

*myocardial infarction (heart attack)

†All MAOIs interact with foods containing tryptophan and tyramine (aged mature cheeses; smoked, aged, or pickled fish and meats; meats extracts [bouillon]; yeast; red wine; and fava beans) and with certain medications (ephedrine, phenylephrine, phenylpropanolamine, amphetamines, cocaine, and tricyclic antidepressants).

‡All SSRIs (selective serotonin reuptake inhibitors) interfere with the metabolism of many other medications, causing potentially toxic or lethal serum levels; they also cause severe withdrawal symptoms not noted with other antidepressants if stopped abruptly, characterized by dizziness, headache, nausea, vivid dreams, sleep disturbances, paresthesias, and irritability.

disease, and can result in withdrawal symptoms when use ends (Sellers, 1996; Wingerson & Roy-Byrne, 1993). Half-lives of antianxiety medications tend to be short, making anxiolytics fast-acting, ranging from minutes or hours for the quickest to 7–10 days for the slowest (aza-spirones). Information about anxiolytics is provided in Table 3-4. For additional detail, the reader can refer to the most recent edition of the *Physician's Desk Reference*, to Klein and Rowland (1996), or to Wingerson and Roy-Byrne (1993).

Antipsychotics. There are a minimum of seven classes of antipsychotic (neuroleptic) medications: benzisoxazoles, phenothiazines, thioxanthenes, butyrophenones, dihydroindolones, dibenzodiazepines, and dibenzodiazepines; all of which reportedly have a wide therapeutic index, are relatively safe, and are nonaddictive (Schultz & Sajatovic, 1993). Approximately equivalent in efficacy (though at often greatly disparate dosages), these medications all have slightly different actions. They differ foremost with regard to adverse effects (Klein & Rowland, 1996) and selection of class of medication should be carefully based on each client's individual needs and presentation (e.g., considering level of agitation because some classes are more sedating than others) (Schulz & Sajatovic, 1993). Antipsychotics are most commonly prescribed for hallucinations and delusions that are part of a schizophrenic condition, a psychotic depression, or an organic mental disorder. However, they have also enjoyed some use for severely violent or aggressive behavior, overwhelming anxiety states, manic behavior with psychotic features, and obsessive-compulsive disorders that are resistant to other interventions (Diamond, 1998; Sansone & Shaffer, 1997). Time to effectiveness can be as brief as 24–48 hours to first symptom relief; however, marked improvement often requires several weeks to months, though as "a general rule, 50% of the ultimate improvement is likely to be seen in the first 3–4 weeks" (Kane, 1996, p. 115). Dosages are generally started low to retain final dosage at the lowest possible level, and some recommend prescrib-

ing augmenting agents to boost the medication's effectiveness (Schultz & Saratovic, 1993). A differentiation must be made between maintenance and acute dosages of neuroleptics, with acute dosages being tapered to maintenance levels as soon as possible (Diamond, 1998). The most important aspect of neuroleptic management is the monitoring and treating of side effects which can range from mild to severe.

The most dangerous side effects caused by psychotropic medications are those that can occur with clients who take neuroleptic medications. Specifically, there are two symptom complexes that require early recognition and immediate intervention: tardive dyskinesia (TD) and neuroleptic malignant syndrome (NMS). TD is one of four possible extrapyramidal side effects (see Table 3-2) that occurs in about 15–20% of all patients who are placed on neuroleptics for a long-term course of treatment and up to 70% of chronic patients in high risk groups (e.g., elderly, diabetics, patients with a concurrent bipolar disorder) (Casey, 1993; Post, 1993). TD involves involuntary and repetitive purposeless movements (e.g., chewing, tongue movements, lip smacking, eye blinking, head movements, neck movements, lip movements) that can be irreversible if no immediate intervention takes place. If such symptoms are detected, the client needs to be referred back to the medical provider immediately for medical management that can include the reduction or adjustment of the medication causing the TD or the use of other medications to control it. Medical intervention is also important to distinguish TD from medical illnesses, such as Tourette's, late-stage Alzheimer's, Huntington's, or Wilson's disease (Casey, 1993).

NMS is an "acute disorder of thermal regulation and neuromotor control, carrying a reported mortality rate of 21%" (Davis, Janicak, & Khan, 1993, p. 170). It is due to this high mortality rate that even nonmedical providers must learn to recognize the symptoms of NMS in their clients who are taking neuroleptics. NMS can occur, not only in response to neuroleptics, but also in response to antidepressants that have a

TABLE 3-4   Anxiolytics: Important Characteristics and Variables

TRADE NAMES	GENERIC NAME	MEAN DOSAGE	SIDE EFFECTS	SPEED OF EFFECT
BENZODIAZEPINES*				
Xanax, Alplax, Tafil, Xanor, etc.	alprazolam	1–4 mg	5,9,12,13,17	rapid
Librium, Lipoxide	chlordiazepoxide	15–40 mg	5,9,12,13,17	intermediate
Clonopin	clonazepam	.3–3 mg	3,5,9,12,13,17	intermediate–slow
Tranxene	clorazepate	15–60 mg	4,5,9,12,13,17	slow
Valium, Vivol, Diazelmuls	diazepam	5–40 mg	4,5,9,12,13,17	intermediate
Dalmane	flurazepam	15–30 mg	5,9,12,13,17	ultraslow
Ativan, Lorax, Almazine, Alzapam	lorazepam	1–6 mg	5,9,12,13,17	rapid
Serax, Zapex	oxazepam	45–120 mg	1,5,9,12,13	rapid
Restoril	temazepam	15–30 mg	5,9,12,13,17	rapid
Halcion	triazolam	.125–.5 mg	5,9,12,13,17	rapid
Paxipam	halazepam	60–160 mg	5,9,12,13,14,17	slow
Centrax	praxepam	20–60 mg	2,5,9,12,13,14,17	slow
AZASPIRONES				
BuSpar	buspirone	30–60 mg	1,2,4,5	avoid use with MAOIs; nonaddictive (safe for use with addicted clients)
SEROTONIN REUPTAKE BLOCKERS (ALSO SEE ANTIDEPRESSANTS)				
Anafranil	clomipramine	110–250 mg	5,6,7,8,9	bronchospastic problems, congestive heart failure, diabetes, Raynaud's
Prozac	fluoxetine	20 mg	1,2,4	as Anafranil
ANTIHISTAMINES				
Vistaril, Atarax	hydroxyzine	20–100 mg	5,7,9	nonaddictive (safe for use with addicted clients)

*For all benzodiazepines, contraindication is present for patients with glaucoma, liver disease, and abuse potential; further, cross-tolerance exists between benzodiazepines and alcohol and other drugs; caution is necessary as overdose potential exists if benzodiazepines are mixed with other drugs/substances.

neuroleptic metabolite (e.g., Ascendin). NMS involves muscular rigidity, altered consciousness, clouded sensorium, fluctuating blood pressure, tachypnea (increased rate of respiration), other respiratory distress, motor disturbances, and diaphoresis (profuse/excessive sweating). Although it can occur any time, NMS usually develops within two weeks of initiation of neuroleptics. It occurs in up to 2.4% of patients, but more frequently among high-risk groups (i.e., patients with organic mental disorder, agitation, dehydration). It develops quickly and can last 7–14 days (or longer if depot [time-released] drugs are used). Treatment requires medical intervention that involves the discontinuation of the triggering medication, reevaluation of psychotropic medication needs and sensitivities, supportive measures to treat the symptoms of NMS (e.g., cooling blankets, ice packs, oxygen, hydration), and/or medication for the NMS (Davis, Janicak, & Khan, 1993). Referral to a medical provider is also important to make accurate differential diagnosis, which can involve heat stroke, encephalitis, tetanus, and lethal catatonia.

Additional information about antipsychotics is provided in Table 3-5. For more detail about antipsychotic medication, the reader can refer to the most recent edition of the *Physician's Desk Reference* for each specific drug or to Diamond (1998).

Antimania Medication. The most commonly prescribed and first successful antimania (or mood stabilizing) medication developed is lithium, which has now been joined by a few anticonvulsants (carbamazepine, valproic acid and valproic acid/sodium valproate combinations) for the treatment and prevention of bipolar disorder (Bowden, 1996). A few other drugs are also being used experimentally but have not yet proven their effectiveness (e.g., clonidine, lamotrigine) (Diamond, 1998; Post, 1993). Although bipolar disorder is clearly the most common indication for antimania drugs, they have also been used to a lesser extent, but successfully in the treatment of aggression, impulse-control difficulties, and rage reactions (Sansone & Shaffer, 1997). Use of lithium requires regular medical monitoring because blood levels of the drug must be checked to evaluate its effectiveness. Dosages are generally kept to a minimum because side effects increase with the amount of drug used. Other medications can be used to augment the effectiveness of lithium to keep dosages optimally low. Half-life for lithium is 24 hours and levels are drawn every two days until a stable and therapeutic blood level has been achieved; then levels are drawn every two to four months. Half-lives for anticonvulsants are shorter (10–15 hours) and these medications also require regular blood level checks. Additional information about antimania medications is provided in Table 3-6. For more detail about these medications, the reader can refer to the most recent edition of the *Physician's Desk Reference* for each specific drug or to Klein and Rowland (1996).

PSYCHOLOGICAL DISORDERS AND SYMPTOMS REQUIRING MEDICAL EVALUATION AND DIAGNOSIS

A variety of psychological disorders or symptoms require medical input, not necessarily for psychotropic medication referral (though this may be the case, as outlined above), but rather for purposes of differential diagnosis, evaluation of potential physical consequences or complications of the psychological symptoms involved in the disorder, or coexistence of physical illness (Dunner, 1993). Both medical and psychological providers often fail to consider the possibility that clients or patients presenting for one reason (e.g., medical concerns or psychological problems) may actually also or instead suffer from a condition in the realm of the other provider (e.g., Wickramasekera, Davies, & Davies, 1996). In other words, medical doctors often fail to make proper psychological referrals when patients could benefit from psychotherapy (Dagadakis, 1993), overlooking possible psychiatric disorders as the proper diagnosis (e.g., Roy-Byrne, 1996), and mental health care providers often fail to make referrals to medical doctors when clients may have been in need of medical evaluation (Tomb, 1995), resulting in treatment failure (e.g., Hornig-Rohan & Amsterdam, 1994; Klonoff & Landrine, 1997). These oversights could be easily prevented if both types of providers would consider the importance of the other group and would be more open to working jointly or collaboratively,

TABLE 3-5   Antipsychotics: Important Characteristics and Variables

TRADE NAMES	GENERIC NAME	MEAN DOSAGE	SIDE EFFECTS*	METHODS OF ADMINISTRATION
PHENOTHIAZINES				
Thorazine	chlorpromazine	100–600 mg	3,5,20,22–26	capsules, syrup, tablets, concentrate, parenteral,† suppository
Mellaril	thioridazine	50–300 mg	3,22–26	concentrate, suspension, tablets
Prolixin, Permital	fluphenazine	1–40 mg	3,20,22–26	elixir, parenteral, tablets, decanoate,† concentrate
Trilafon	perphenazine	8–16 mg	22–26	tablets, concentrate, parenteral
Compazine	prochlorperazine	30–150 mg	5,15,22–26	tablets, capsules, syrup, suppositories, parenteral
Serentil	mesoridazine besylate	200–400 mg	22–26	tablets, concentrate, parenteral
Stelazine	trifluoperazine	3–30 mg	4,18,21–26	concentrate, parenteral, tablets
Vesprin	triflupromazine	50–150	5,20,22–26	parenteral
BUTYROPHENONES				
Haldol	haloperidol	1–40 mg	3,4,22–26	concentrate, decanoate, parenteral, tablets
THIOXANTHENES				
Navane	thiothixene	5–60 mg	3,5,8,22–26	concentrate, parenteral, capsules, solution
Taractan	chlorprothixine	75–200 mg	7,8,22–26	tablets, concentrate, parenteral
DIBENZODIAZEPINES				
Clozaril§	clozapine	300–600 mg	3,6,8,9,20–26	tablets
Loxitane	loxapine	60–160	5,7,9,22–26	capsules, parenteral, concentrate
BENZISOXAZOLE DERIVATIVES				
Risperdal	risperidone	2–6 mg	3,5,6,22–26	tablets
DIHYDROINDOLONES				
Moban	molindone HCL	20–225 mg	4,5,14,22–26	tablets, concentrate
OTHER				
Zyprexa	olanzapine	10–20 mg	2,3,5,6,7,23	tablets
Seroquel	quetiapine	300–400 mg	5,6,7,9	tablets

*All antipsychotics are contraindicated for patients with cardiovascular disease, chronic respiratory disease, epilepsy, Parkinson's disease, glaucoma, and myasthenia gravis; caution is necessary with drug-combining as antipsychotics have many drug interactions, including with alcohol and nicotine.

†by intravenous or intramuscular injection

‡depot drugs, which may require slightly different dosing

§Due to a danger of agranulocytosis, regular white blood cell counts are required to reduce the risk of loss of ability to fight infection.

TABLE 3-6   Antimania Medications: Important Characteristics and Variables

TRADE NAMES	GENERIC NAME	MEAN DOSAGE/ BLOOD LEVEL	SIDE EFFECTS	CONTRAINDICATIONS/ DANGERS
Lithonate, Litholoid, Eskalith CR	lithium carbonate, lithium citrate	600–1800 mg .6–1.2 mEq/L*	2,6,15,18,19	psoriasis, renal disease; many drug interactions; risk of birth defects
Tegretol	carbamazepine	>100 mg 6–12 mg/ml†	2,9,14,18	bone marrow depression; discontinue MAOIs first; many drug interactions; risk of birth defects; reduced effectiveness of oral contraceptives
Depakote, Depakene	divalproex sodium, valproic acid	1000–1500 mg 45–125 mg/ml†	2,6,14,15,18	hepatic disease or dysfunction; many drug interactions; risk for birth defects
NEW DRUGS UNDER INVESTIGATION				
Lamictal	lamotrigine	100–200 mg	1,2,14	not yet known; potentially fatal rash
Neurotin	gabapentin	900–1800 mg	1,2,9,14,18	not yet known
Isoptin, Calan	verapimil	80–480 mg	1,2,5	neurotoxic if combined with carbamazepine or lithium; liver problems, arrhythmias

*milliequivalent per liter

†milligram per milliliter

perhaps through a split treatment model similar to the one previously suggested for psychotropic medication referrals.

Mental health care providers need to be aware that clients can benefit from medical intervention and evaluation above and beyond psychotropic medication. Some clients will present with psychological symptoms that are not necessarily exclusively in the realm of psychiatric (or psychological) disorder, but perhaps are better explained and treated through medical intervention (i.e., differential diagnosis); others present with psychological disorders that have strong implications for their physical well-being; and yet others present with psychological symptoms that are highly correlated with additional medical illness (a condition referred to as medical comorbidity). Recognition of such disorders and patterns on the part of the mental health care provider is critical to the optimal treatment of

persons presenting for counseling or psychotherapy (Jorge & Robinson, 1993).

To be prepared for referrals beyond those for psychopharmacological evaluation, a clinician must attend to a few additional concerns that can facilitate communication with physicians. First, it is recommended that nonmedical therapists identify not only a pharmacotherapist, but also a few medical specialists with whom they can easily collaborate. Klonoff and Landrine (1997) recommend that the optimal list of specialists would include an endocrinologist, neurologist, gynecologist, and internist. When referring to a medical provider, the counselor needs to be prepared to offer hypotheses about what might be going on with the client, suggesting concrete ideas of differential diagnoses based on physical data. To do so, the clinician needs to learn some basic medical jargon that will facilitate communication and respect. Further, medical providers will take the nonmedical referral source more seriously if this individual refers to the client by using the label *patient* (Klonoff & Landrine, 1997). Obtaining basic preliminary physical data from the client greatly facilitates the referral and increases the likelihood of correct medical diagnosis. Such preliminary physical data can be gleaned through the general intake interview (see Chapters 1 and 2) and may consist, for example, of information such as weight patterns, sleep patterns, changes in physical functioning, substance use, as well as specific physical symptoms and their context. The more physical data mental health care providers can offer, the more seriously the physician will take them and the referral (Klonoff & Landrine, 1997). Presenting this wealth of medical information in the most concise and brief manner, as opposed to embedding it in a lengthy psychosocial history, will reap the greatest benefit (Diamond, 1998). Finally, in collaborating with a medical provider, it is important for the therapist never to pretend to understand information when in reality the therapist does not. Asking questions to be well informed about clients' medical conditions and about any medical tests they may face is essential to optimal client-therapist interaction. If clinicians do not understand what their clients will encounter, they cannot be of any help to them in terms of preparing for or processing medical issues (Klonoff & Landrine, 1997).

Psychological Disorders and Symptoms That Require Medical Differential Diagnosis

Many psychological symptoms can either be part of a psychiatric disorder (those listed in DSM-IV) or a medical disorder, and at times they can be part of both. Psychiatric and medical disorders can coexist, greatly complicating the intervention required for a given client. Not making appropriate medical referrals when symptoms that are presented by a client may have medical causes can lead to lack of problem resolution at best and to life-threatening situations at worst. For example, hyperthyroidism, a physical disorder with symptoms similar to bipolar disorder, agitated

depression, or anxiety, is exacerbated by tricyclic antidepressants and by lithium. Making the incorrect diagnosis of bipolar disorder, for example, can aggravate the physical disorder and result in worsening of symptoms (Morley & Krahn, 1987). It is essential for the responsible mental health care provider to have some awareness of possible differential medical diagnoses that can be the cause or superior explanation for psychological symptoms related by clients. Information to that effect is summarized in Table 3-7. At a minimum, clinicians need to memorize the psychological disorders and symptoms that are listed in this table; preferably, they also should recall some of the differential diagnostic possibilities for each of these psychiatric diagnoses to be able to conduct an intelligent and informed intake interview. It must be noted that the information in Table 3-7 is neither comprehensive, nor complex from a medical perspective. It reflects the minimum of knowledge a nonmedical mental health care provider must have to be able to initiate intelligent medical referrals when they are necessary. The medical aspects of Table 3-7 are not presented to be all-inclusive, decisive, or diagnostic. Only a medical provider is able to appreciate the potential complexity of the physical aspects of certain presenting problems. It is important to note that many differential diagnoses rely not only on a simple medical exam, but actually require in-depth and protracted medical testing and evaluation. It is never possible for a nonmedical provider to determine medical illness decisively. For example, in the case of delirium or dementia, EEGs are necessary for proper diagnosis (Primavera, Giberti, Scotto, & Cocito, 1994); to rule out certain drug-induced (e.g., steroid-induced) depressions, adrenocorticotropic hormone response to corticotropin-releasing factor may need to be measured (Pies, 1995); to assess whether hyperthyroidism may account for the psychological symptoms presented, a thyroid panel may need to be prepared (Klonoff & Landrine, 1997). Table 3-8 lists a large number of the most common physical disorders that can have psychological manifestations and the types of action medical providers need to take to rule them out. For each disorder, the most common psychological symptoms are listed as well. Table 3-8 is based largely on Morrison's (1997) excellent text called *When Psychological Problems Mask Medical Disorders*, supplemented by Klonoff and Landrine's (1997) text entitled *Preventing Misdiagnosis of Women*, two books that need to be part of every nonmedical clinician's library. Clearly, there is overlap between Tables 3-7 and 3-8; however, the inclusion of both tables makes for easier reference.

How does a nonmedical mental health care provider make the decision to refer for medical differential diagnosis, beyond using Tables 3-7 and 3-8? A number of suggestions can be made in this regard. First, whenever a client presents with inconsistent symptoms, the clinician should suspect a physical disorder. For example, if a client complains of fatigue, lack of appetite, and sexual disinterest, but claims no other symptoms consistent with depressive disorder, a medical evaluation is indicated given the lack of consistency that would support a pure psychiatric diagnosis of depression (i.e., beware of diagnosing an atypical depression without medical corroboration). Whenever a client who has been seen for a while suddenly develops new symptoms, more symptoms, or more severe symptoms, a medical referral is warranted. Morrison (1997) warns that mental health care professionals must "think outside the mental health box" (p. 2), especially with clients they have seen for some time. When sudden symptomatic changes occur, it is the responsibility of the counselor to begin to question whether the pure mental health diagnosis truly accounts for the entire clinical picture. Another cue to the need for medical referral can be unusual or changing appearance or mannerisms (Morrison, 1997). Examples may include features such as premature or nonmale pattern thinning of hair (e.g., hypothyroidism, malnutrition, or liver failure), darkening of skin (e.g., adrenal insufficiency or hypothyroidism), stiff or halting movements (e.g., fibromyalgia or Creutzfeldt-Jacob disease), shortness of breath (e.g., B1 deficiency or congestive heart failure), or tremors (e.g., Parkinson's disease, multiple sclerosis, or hy-

poglycemia). Any alarming symptoms, such as blood in sputum or stool, persistent headaches, or similar severe or sudden physical manifestations also always warrant a medical referral, even if they are not connected by clients to the psychological symptoms for which they presented to the counselor. Finally, Klonoff and Landrine (1997) suggest that "visual illusions or hallucinations always have an organic, rather than functional or psychiatric, etiology" (p. 59), and hence they always require a medical referral. In fact, these authors indicate (and I thoroughly agree) that basic physical exams should be required of all psychotherapy clients. The listings provided in Tables 3-7 and 3-8 suffice to keep mental health care providers alert and to flag those clients who are in greatest need for medical evaluation. Whenever a mental health care provider is uncertain about whether a medical referral for differential diagnosis may be needed, it is best to be conservative and to send the client for a medical exam.

The astute mental health care provider knows how best to facilitate a medical referral for differential diagnosis purposes. This process merely starts with having knowledge of the information contained in Tables 3-7 and 3-8. For all of the listed diagnostic categories that require medical differential diagnosis, psychotherapists or counselors need to ascertain during the intake interview whether the client has received medical evaluation recently (within the past one to three months) that has assessed the client for the medical disorders that are potentially involved in the current emotional or psychological symptoms. If clients indicate that they have indeed been examined by a medical provider who ruled out medical or physical causes for the symptoms currently presented, the clinician has the responsibility to request an ROI from the client to access this medical information (Klonoff & Landrine, 1997). Knowing which medical tests are typically used to rule out which medical disorders greatly facilitates the counselor's judgment as to whether the medical exam of the client was sufficient and targeted the same symptoms presented currently (see Table 3-8). The ROI will permit the counselor both to

TABLE 3-7   Psychiatric Diagnoses Requiring Medical Differential Diagnosis*

DIAGNOSTIC CATEGORY	POTENTIAL DIFFERENTIAL DIAGNOSES	SAMPLE REFERENCE(S)
Anorexia Nervosa	Crohn's Disease hypopituitarism systemic lupus erythematosus weight loss due to other mental disorder weight loss due to other medical disorder	Hoffman & Halmi (1993) Klonoff & Landrine (1997)
Delirium	drug toxicity or withdrawal metabolic disease psychosocial trauma or stress postoperative and postictal states CNS trauma infection	Katz (1993)
Delusional Disorder	metabolic/endocrine disorder (e.g., thyroid disturbance, CNS lupus, hypopituitarism, Cushing's syndrome) neurological disorders (esp. temporal lobe epilepsy) substance use Wilson's disease	Breier (1993) Klonoff & Landrine (1997)
Dementia	primary dementia (e.g., Alzheimer's disease, Pick's disease, Creutzfeld-Jacob disease) secondary dementia 　　psychiatric (e.g., depression, schizophrenia) 　　endocrine disorder 　　infections (including HIV infection) 　　tumors (mainly in central nervous system) 　　neurologic disorder (e.g., Huntington's Chorea, Parkinson's disease, hydrocephalus, nutritional deficiencies, palsy, subdural hematoma) 　　vascular disorders 　　toxicity 　　head trauma	Loebel, Dager, & Kitchell (1993)
Generalized Anxiety Disorder	drug-induced (e.g., caffeine, steroids, lidocaine, thyroid replacement, dopamine) cardiovascular disease (e.g., arrhythmias, coronary artery disease, hypertension, mitral valve prolapse) respiratory disease (e.g., asthma, hyperventilation, chronic obstructive lung disease, pulmonary embolus) endocrine/metabolic disorders (e.g., hypoglycemia, hyper- or hypothyroidism, hyponatremia) neurological disorders (e.g., tumors, infection, complex partial seizures, migraines) peptic ulcers and ulcerative colitis	Cowley (1993) Wise & Griffies (1995)
Major Depression	malignancies (e.g., lymphoma, hematoma) CNS impairment (e.g., tumor, uremia, demyelination, hypoxia, hepatic encephalopathy) drug-induced (e.g., steroids, narcotics, hormonal agents, anticonvulsants, histamine blockers) infections (e.g., hepatitis, mononucleosis, syphilis) deficiencies (e.g., niacin, electrolytes, B12, folic acid, pyridoxine) and malnutrition	Tollefson (1993) Ammon-Cavanaugh (1995)

Continued

TABLE 3-7   Psychiatric Diagnoses Requiring Medical Differential Diagnosis* (continued)

DIAGNOSTIC CATEGORY	POTENTIAL DIFFERENTIAL DIAGNOSES	SAMPLE REFERENCE(S)
Major Depression (continued)	endocrine (e.g., hypo- and hyperthyroidism, diabetes, pituitary insufficiency, Cushing's syndrome, Addison's disease)	
Mania	hypo- or hyperthyroidism	Post (1993)
	diencephalic or frontal stroke	Fava, Morphy, & Sonino (1994)
	multiple sclerosis	
	complex partial seizures	
	drug-induced (e.g., stimulants, steroids)	
	brain tumors	
Organic Brain Syndrome	parasitic disease	Weiss (1994)
	infection	
Panic Disorder	cardiovascular disease (esp. mitral valve prolapse)	Sheehan & Ashok (1993)
	respiratory disease	
	neurological disease	Klonoff & Landrine (1997)
	endocrine disorder, pheochromocytoma	
	substance-induced	
Psychogenic Amnesia	organic amnestic disorder	Davis (1993)
	epilepsy	
	post-concussion amnesia	
	substance-induced amnesia (e.g., alcoholic blackout)	
Psychogenic Fugue	organic mental disorder	Davis (1993)
	complex partial seizures	Klonoff & Landrine (1997)
	malingering	
Schizophrenia	epilepsy, partial complex seizures	Potkin, Albers, & Richmond (1993)
	CNS tumor or infection	
	CNS degenerative disease (e.g., Huntington's)	Klonoff & Landrine (1997)
	B12 and/or folic acid deficiency (including pellagra)	
	endocrine/metabolic disease (e.g., Cushing's disease, Addison's disease, hypothyroidism)	
	toxicity (e.g., heavy metal poisoning)	
	multiple sclerosis	
Sexual Dysfunction	neurophysiological factors	Heiman (1993)
	side effect of drug- or medication-use	
	general medical illness	
Sleep Disorders		
Insomnia	organic factors (e.g., Parkinson's Disease)	Gillin (1993)
	cardiovascular insufficiency	
	respiratory disease	
Parasomnia	endocrine disorders	Gillin (1993)
	diabetes mellitus	
	vascular disorder	
	neural disorders	
	epileptic seizures	
Substance Use		
Intoxication	organic hallucinosis	Schuckit (1993)
	psychosis	
Delirium	(see above)	

*My sincerest appreciation goes to Dr. Stan Mlynczak for his invaluable contributions to this table.

TABLE 3-8   Physical Disorders with Psychological Symptoms

DISORDER	PSYCHOLOGICAL SYMPTOMS	PHYSICAL SYMPTOMS	MEDICAL TESTS
Adrenal Insufficiency (Addison's Disease)	Fatigue, apathy, depression, social withdrawal, anxiety, suicidality, psychosis, poverty of thought, recent memory impairment	Weakness, darkening skin, nausea, abdominal pain, fainting, vomiting, weight loss, anorexia (loss of appetite)	History of salt cravings; urine or sputum test measuring cortisol levels
Amyotropic Lateral Sclerosis (Lou Gehrig's Disease)	Depression, dementia	Muscle weakness, weight loss, ataxia (inability to coordinate voluntary muscle movement), dysarthria (inability to articulate words), cramping	Electromyography (to show muscle twitching)
Brain Abscess	Lethargy, cognitive changes and symptoms	Headache, fever, stiff neck, seizures, nausea, vomiting, focal neurological symptoms	CT scan; MRI
Brain Tumor	Loss of memory, cognitive changes, dementia, depression, psychosis, dissociation, personality changes	Headaches, vomiting, dizziness, seizures, focal neurological symptoms	CT scan; MRI; Brain biopsy
Carcinoid Syndrome	Flushing of the face and body (blushing)	Diarrhea, abdominal pain, blood-containing stool	Urine sample to assess for high levels of breakdown products of serotonin
Cardiac Arrhythmia	Anxiety, delirium	Fatigue, dizziness, delirium, palpitations	Electrocardiogram
Chronic Obstructive Lung Disease (e.g., emphysema)	Anxiety, panic, depression, insomnia, delirium	Cough, shortness of breath, tremor, headache, dark skin hue	Pulmonary function studies; blood-gas determination
Congestive Heart Failure	Anxiety, panic, insomnia, delirium, depression	Shortness of breath, fatigue, edema, cold, weakness, cyanosis	Chest X-ray; echocardiogram
Cryptococcus	Irritability, disorientation, mania, dementia, psychosis	Headache, fever, stiff neck, blurred vision, nausea, staggering gait	Search for the causative yeast organism in cerebrospinal fluid bathed in India ink
Cushing's syndrome	Emotional lability, depression, anxiety, loss of libido, delirium, irritability, paranoid delusion, suicidality (high risk)	Hypertension, amenorrhea (cessation of menstrual period), oily skin, increased body hair, weakness, facial and truncal obesity, buffalo hump	Physical exam; corticosteroid level in 24-hour urine specimen; history of steroid-containing substances
Diabetes Mellitus	Fatigue, lethargy, panic, depression, poor concentration, delirium	Increased hunger, thirst, and urine output; rapid weight loss, blurred vision	At least two abnormal glucose tolerance tests
Fibromyalgia	Chronic fatigue, depression, anxiety	Muscle pain, stiffness, and tenderness	Diagnosis based on symptoms

Continued

Table 3-8 Physical Disorders with Psychological Symptoms (continued)

DISORDER	PSYCHOLOGICAL SYMPTOMS	PHYSICAL SYMPTOMS	MEDICAL TESTS
Head Trauma	Personality change, delirium, dementia, amnesia, mood swings, psychosis, anxiety	Headache, dizziness, fatigue, paralysis, seizures, anosmia (loss of sense of smell)	Skull x-ray; MRI; CT scan
Herpes Encephalitis	Forgetfulness, anxiety, psychosis	Fever, headache, stiff neck, vomiting, focal neurological symptoms	Electroencephalogram; brain biopsy; CT scan
Homocystinuria	Mental retardation, dementia, behavioral problems	Impaired vision, shuffling gait, blotchy skin	Blood or urine test to check for elevated levels of the amino acids homocysteine and methionine
Huntington's Disease	Apathy, depression, irritability, impulsive behavior, personality changes, cognitive changes, suicidality, dementia	Insomnia, restlessness, ataxia, inarticulate speech, good appetite with weight loss, clumsiness, writhing motions of the limbs	Family history of this fatal neurological disease; genetic testing
Hyperparathyroidism (Hypercalcemia)	Personality change, depression, anxiety, suicidality, delirium, psychosis; often mistaken for hypochondriasis	Urinary tract infections, weakness, tiredness, anorexia, nausea, vomiting, thirst, constipation, muscle and abdominal pain	Blood test to establish high serum calcium and parathyroid hormone levels
Hypertensive Encephalopathy	Paranoia, delirium	Headache, nausea, paralysis, vomiting, visual impairment, seizures	Measurement of blood pressure (presence of hypertension)
Hyperthyroidism	Agitated depression, depression, anxiety, panic, delirium, psychosis; often mistaken for bipolar disorder	Goiter, red and puffy eye lids, bulging eyes, weakness, palpitations, hunger, tremor, warm, increased appetite with weight loss, diarrhea	Thyroid panel (blood test) to check for elevation of serum thyroxine levels and drop in thyroid stimulating hormone
Hypoglycemia	Anxiety, depersonalization, lethargy, fatigue	Sweating, palpitations, tremulousness, headache, confusion	Food diary; 5-hour fasting glucose tolerance test
Hypoparathyroidism (Hypocalcemia)	Irritability, mental retardation, depression, anxiety, paranoia, delirium, dementia	Numbness, tingling, and spasms in hands, feet, and throat; headaches; thin, patchy hair; poor tooth development	Blood test to establish low serum calcium and parathyroid hormone levels
Hypopituitarism	Apathy, indifference, fatigue, depression, decreased libido, drowsiness; often mistaken for dependent personality disorder or psychotic depression	Waxy skin, loss of body hair, inability to tan, loss of appetite and weight, loss of nipple pigmentation, premature wrinkles around eyes and mouth	X-ray, CT scan, or MRI to establish structural pituitary abnormality; and blood test to establish hormonal deficiencies
Hypothyroidism	Apathy, depression, suicidality, slowed cognitive function, dementia; mistaken for rapid-cycling bipolar disorder	Dry and brittle hair, dry skin, hair loss, edema, cold intolerance, appetite loss with weight gain, goiter, constipation, hoarseness, hearing loss, slow heartbeat	Blood test to establish drop in serum thyroxine and elevation in thyroid stimulating hormone levels; measurement of basal body temperature on five consecutive mornings

TABLE 3-8 Physical Disorders with Psychological Symptoms (continued)

DISORDER	PSYCHOLOGICAL SYMPTOMS	PHYSICAL SYMPTOMS	MEDICAL TESTS
Lyme Disease	Depression, psychosis, anxiety, mild cognitive symptoms	Headache, fever, chills, fatigue, stiff neck, malaise, achiness	History of tick bite; serum antibody response to B. Burgdorferi
Meniere's Syndrome	Anxiety, panic, depression, poor concentration	Dizziness, nausea, vomiting, tinnitus (ringing in the ears), nystagmus (rapid, involuntary eyeball oscillation), deafness	Diagnosis based on symptoms
Mitral Valve Prolapse	Panic (do not use anxiolytics)	Chest pain, fainting, palpitations, breathlessness	Echocardiogram
Multiple Sclerosis	Depression, mania, sudden emotionality, cognitive impairment, dementia; misdiagnosed as somatization or histrionic personality disorder	Ataxia, numbness, weakness, fatigue, visual problems, incontinence, trouble walking, paresthesias (tingling or prickling of the skin)	MRI to show areas of plaque; birth place 40° North and 40° South; hot bath test (weakness and faintness after a hot bath)
Myasthenia Gravis	Anxiety, memory loss, minor cognitive symptoms	Muscle weakness	Tensilon Test (injection of edrophonium to check for briefly improved muscle strength)
Niacin Deficiency (Pellagra)	Depression, anxiety, delirium, dementia	Weakness, anorexia, headache, diarrhea, red and rough skin	Food diary; based on symptoms; urine test
Pancreatic Cancer	Depression, initial, crying spells, suicidality, anxiety	Weight loss, weakness, abdominal pain, insomnia, hypersomnia	Ultrasound, CT scan, or endoscopic retrograde pancreatography; needle biopsy
Parkinson's Disease	Depression, anxiety, impaired attention, cognitive deficits, paranoia; visual hallucinations as side effect of medications	Tremor, muscle rigidity, decreased mobility, masked facies, trouble walking, poor fine motor coordination	Based on symptoms and physical exam
Pernicious Anemia	Forgetfulness, depression, dementia, psychosis	Anemia, dizziness, tinnitus, glossy tongue, palpitations	Blood test
Pheochromocytoma	Anxiety, panic	Headache, sweating, palpitations, nausea, high blood pressure	24-hour urine test for high catecholamine levels
Porphyria	Depression, mania, euphoria, anxiety, delirium, psychosis	Abdominal pain, muscle weakness, tremors, dark urine, vomiting, seizures, sweating	Blood or urine test to check for high levels of porphobilinogen
Posterolateral Sclerosis	Anxiety, weakness, memory impairment, psychosis; mistaken for conversion disorder	Heavy limbs, stocking and/or glove sensory loss alteration in reflexes	Electromyography
Prion Disease	Anxiety, fatigue, poor concentration, slowed mental function	Difficulty walking, tremors, muscle rigidity, hypokinesia (decreased muscle movement)	Electroencephalogram, history of ingestion of infected meat

Continued

TABLE 3-8 Physical Disorders with Psychological Symptoms (continued)

DISORDER	PSYCHOLOGICAL SYMPTOMS	PHYSICAL SYMPTOMS	MEDICAL TESTS
Progressive Supranuclear Palsy	Slowed mental function, forgetfulness, apathy, labile mood	Double vision, unsteady gait, stiffening of muscles	CT scan showing atrophy of pons and midbrain
Protein Energy Malnutrition	Apathy, lethargy, cognitive changes	Weight loss; loss of skin elasticity; dry, thin hair; low body temperature, heart rate, and blood pressure	Food diary; physical exam; and blood test for low serum protein levels
Sleep Apnea	Insomnia, depression, drowsiness, irritability, poor concentration	Snoring, morning headache, nocturia (nighttime urination)	Sleep polysomnography
Syphilis	Personality changes, fatigue, irritability, grandiosity, cognitive symptoms, psychosis	Ulcerous chancre, fever, headache, sore throat, skin rash, swollen lymph nodes	Serum screening test and serum fluorescent treponeme antibody absorption test
Systemic Lupus Erythematosus	Severe depression, cognitive symptoms, anorexia, psychosis (thorazine exacerbates symptoms)	Muscle and joint pain, butterfly rash, fatigue, fever, loss of appetite, nausea, vomiting, weight loss	Blood test to establish elevation of antinuclear antibodies in blood
Thiamine Deficiency (Beriberi)	Fatigue, irritability, anxiety, delirium, amnesia	Shortness of breath, edema, rapid heartbeat, nystagmus, trouble walking, fever, vomiting	History of alcoholism; food diary; MRI; CT scan; blood and urine tests
Wilson's Disease (Inherited Copper Toxicosis)	Anxiety, personality change, irritability, anger, loss of inhibition, psychosis, depression, cognitive symptoms	Dysarthria, tremor, spasticity, rigidity, trouble swallowing, dystonia (poor tonicity of tissue), drooling	Liver function test (excess copper); blood tests (deficient copper-protein ceruloplasmin); MRI; CT scan

verify that, at the time of the physical exam, the medical professional was indeed aware of the psychological symptoms with which the client is currently presenting, and also to ascertain that the medical provider indeed ruled out medical or physical concerns. This thoroughness on the part of the therapist is not only suggested because the client may have misunderstood the medical provider, or because the medical provider may have misunderstood the client, but also because unfortunately not all clients are always entirely truthful and not all medical providers are equally astute and qualified. In the case of psychological symptoms that are strongly associated with medical differential diagnoses, it is only after the clinician has confirmed the client's information directly with the medical provider that the clinician can rest assured that medical issues have been attended to. Making the correct medical diagnosis is just as important as making the correct psychiatric diagnosis.

If the client presenting with potential differential diagnosis needs has not had recent contact with a medical provider, a referral is absolutely necessary. The referral process is much like the process outlined above for psychotropic medication referral. The client is made aware of the need for the medical referral, prepared for the medical contact, given a referral name if she or he does not have a regular medical provider, and asked to sign an ROI to allow communication between the medical and mental health care provider. If the medical provider rules out physical or medical causes for the client's presenting concerns, the clinician can then move on to making a proper mental health care treatment plan. If, however, the medical provider indicates that there is a medical diagnosis, the two professionals should collaborate to determine the best course of action. The symptoms are likely to be most successfully treated by the medical provider; however, the psychological care provider may best treat the sequelae of having a medical disorder that has emotional consequences. This psychological treatment will of course be different, in that it clearly has different goals in mind (see Brems, 1999a, for goal setting and related treatment planning). A medical referral never assumes that the clinician may lose the client to the medical provider; in most cases that require a medical referral, the differentiation is between whether the client can be treated solely by a mental health care provider or treated jointly by a medical doctor and a counselor. Not making a referral for medical evaluation for fear of losing the client to the medical provider is unethical, unprofessional, and unconscionable.

Psychological Disorders and Symptoms That May Result in Medical Complications

A related, but separate, issue is posed by clients who present with psychological symptoms or disorders that may not require a medical differential diagnosis per se, but who are referred because of possible physical and medical implications of the psychological symptoms in and of themselves. In other words, some clients present with psychiatric disorders that, although properly diagnosed and treated as psychiatric in nature, still have medical implications and need medical intervention. A client who presents with such disorders is referred to a medical provider after the clinician has discussed the potential physical implications of the symptoms with the client. Then the client is prepared for the referral and given the names of a few referral sources, if there is no regular physician. If the client is already

under the care of a physician or other medical provider, appropriate ROIs are collected to communicate with this medical provider and to coordinate treatment. The model of collaboration between mental health and medical health care providers is very similar to that presented for psychotropic medication referral. The two providers will work separately with the client, but they will communicate from time to time to ascertain that treatment is compatible and that no splitting or devaluing of the other provider occurs on the part of the client. Each provider must be respectful of the contribution of the other professional, and the well-being of the client must be the guiding principle for interaction and collaboration. The psychotherapist is more likely than the medical provider to have regular contacts with the client, and hence receives some education from the medical provider about issues that could be easily monitored by the psychotherapist. For example, if the client is placed on certain medications, the therapist can assist with monitoring for effectiveness as well as side effects; if the client needs to follow certain behavioral regimens, the mental health care provider can assist with monitoring and possibly even with compliance issues. As such, the two providers together may devise a better treatment strategy for the client than either provider would have been capable of individually.

One of the most important examples of a disorder that requires much medical monitoring and collaboration is anorexia nervosa. A psychiatric disorder that requires mental health care, this disorder can also be a life-threatening medical illness because of the physical dimensions and side effects of the psychological symptoms involved. Clients with anorexia nervosa are vulnerable to the development of a variety of medical complications, ranging from extreme weight loss to endocrine abnormalities and amenorrhea due to hypothalamic disruption or abnormal thyroid functioning. If these medical concerns are severe, they can be life threatening and may require inpatient treatment before the client is ready to be treated by an outpatient mental health care provider. Further, some medications have been found useful in promoting weight gain among anorexic clients by improving appetite and caloric consumption (Hoffman & Halmi, 1993). Similarly, clients presenting with bulimia nervosa are always in need of medical and dental referrals due to physical complications secondary to the binge-purge cycle in which they engage. For example, they may experience dehydration, electrolyte imbalances, gastrointestinal problems (e.g., bleeding ulcers, pancreatic dysfunction), cardiovascular disease (including fatal cardiomyopathy), renal function impairment, and endocrine disorders due to interference with hormone production (Zwaan & Mitchell, 1993). Medical attention is also indicated, because bulimic clients occasionally respond to antidepressant medication (see Table 3-1) and may have other overtly unrelated medical illnesses (Ghadirian, Englesman, Leichner, & Marshall, 1993).

A range of sleeping disorders exists, all of which generally require medical attention to rule out medical or physical reasons for, and consequences of, their existence. Clients presenting with narcolepsy, hypersomnia, sleep apnea, or parasomnias generally require medical attention in addition to psychotherapy (Gillin, 1993). Clients presenting with insomnia or dyssomnia that are not due to a depressive disorder (e.g., major depression) also need to receive medical attention, often because they have a tendency to self-medicate with over-the-counter sleeping pills that wreak havoc with their sleep cycles. Such clients can on occasion benefit from benzodiazepines and sedative antidepressants (Gillin, 1993). Medical referral can thus be combined with psychotropic medication referral.

Various dissociative disorders require medical attention, not only for possible differential diagnosis or psychotropic medication referral, but also because sodium amytal interviews may help distinguish psychiatric from medical diagnosis and intervention in cases of psychogenic fugue and psychogenic amnesia. Further, in cases of dissociative identity disorder (DID), medication may be helpful not for the DID diagnosis, but for ancillary symptoms (possibly related to depression, anxiety, or medical disorder).

Disorders such as conversion, hypochondriasis, somatization, and somatoform pain disorder cannot truly be diagnosed by a nonmedical provider. If a mental health care provider suspects such a diagnosis, a medical referral is essential for confirmation (Katon, 1993). Medical pathology in these cases needs to be ruled out by a medical provider. Fortunately, these disorders are more likely to come to the attention of medical providers to begin with as they involve somatic complaints (by definition, somatic complaints for which there is no physical or medical basis or pathology). The great concern about these disorders are migratory symptoms (if one somatic pathology is ruled out, the client often develops new somatic symptoms in another part of the body) and iatrogenic problems that result from multiple unnecessary surgeries and other medical interventions. The collaboration between medical and psychological care providers is essential for proper treatment of such clients who otherwise may be treated by one medical care provider after another, without proper psychiatric diagnosis.

Substance use disorders often require referral for medical care because of management of complications in the medical realm (Schuckit, 1993; also see Chapter 2). Clients may need medical attention during withdrawal and detoxification, especially if the possibility of delirium tremens exists. Such withdrawal can only be managed safely in a medical setting where a quick medical response is possible. In fact, any presentation of delirium requires urgent medical intervention "to prevent irreversible deterioration or death" (Katz, 1993, p. 68). Substance use disorders may also require medical attention if the client is still actively using (i.e., for reasons other than withdrawal and detoxification) because issues of acute toxicity may require medical management (Schuckit, 1993). Further, evidence is accumulating that certain medications can assist with the reduction of drug or alcohol cravings.

Various psychological symptoms (not necessarily DSM-IV diagnosable disorders) may also suggest that a medical referral could be valuable for the client. Prolonged and intense experience of stress can result in a variety of medical consequences, including adrenal insufficiency or immune dysfunction (Leonard, 1990; Vollhardt, 1991). Whenever psychological symptoms are accompanied by somatic complaints, a medical referral is indicated, if for no other reason than to rule out long-lasting medical effects and complications (e.g., irritable bowel syndrome) (Walker, Roy-Byrne & Katon, 1990). On the other hand, it is also possible that somatic complaints cover up underlying psychological symptoms (Terre & Ghiselli, 1995) or precipitate psychiatric symptoms (Foster & Oxman, 1994). Proper treatment planning depends upon understanding that clients who indicate that they have a diagnosed physical illness may actually suffer from psychological side effects of this illness (as opposed to a discrete DSM-IV disorder) (Dagadakis, 1993; also see differential diagnosis earlier in this chapter). This issue may be more important to the medical practitioner who must ascertain that psychological symptoms secondary to medical illness receive treatment and consideration, though they may affect the nonmedical service provider as well.

For example, a client who comes to treatment complaining of agitation, anxiety, and confusion, and who reveals during the medical portion of the intake interviews having been diagnosed with hypoglycemia (a metabolic disorder resulting in an abnormal decrease of blood sugar level), may not actually suffer from a discrete anxiety disorder. Instead, the clinician may need to recognize that agitation, confusion, and anxiety are frequent psychological symptoms of the client's diagnosed medical illness (Dagadakis, 1993). Similarly, a client presenting with symptoms of mood swings and depression who has been diagnosed with porphyria (a hereditary disorder that results in the abnormal metabolism of porphyrin) may not have a discrete depressive disorder, given that these affective symptoms are often a distinct aspect of this medical disorder

(Dagadakis, 1993). This does not mean that the nonmedical provider would not have a role in the treatment of these clients' psychological symptoms; however, it would imply that the clinician would make an error in treatment planning should she or he approach the clients' treatment from the perspective of treating a discrete psychiatric condition, drawing conclusions about their overall state of mind or personal and interpersonal dynamics, and searching for psychological explanations of symptoms.

Psychological Disorders and Symptoms with Frequent Medical Comorbidity

Finally, various psychological symptoms and disorders have a strong correlation with medical illness. The high levels of medical and psychiatric comorbidity present for disorders such as major depression, for example, are often not explainable in terms of cause and effect sequencing. In other words, it is not clear whether the psychiatric disorder made the client vulnerable to the development of an additional (apparently unrelated) medical disorder or whether the medical disorder resulted in the psychological symptoms and disorders. However, when such clients present for treatment, it is important to recognize the coexistence of illnesses and to attend to both. Clearly, a nonmedical mental health care provider is not able to manage such a client independently. Similarly, a conscientious medical provider recognizes the necessity of involving a mental health care provider to assist in the management of the psychiatric illness. The collaborative relationship that could and should be developed is identical to that described previously in this chapter. Only a collaborative team of health care and mental health care providers will be able to manage clients with comorbid medical and psychiatric illnesses optimally, efficiently, and cost-effectively (Simon, VonKorff, & Barlow, 1995).

For example, depression is highly correlated with medical illnesses such as myocardial infarction, stroke, epilepsy, cancer, and migraines (Cassem, 1995; Stevens, Merikangas, & Merikangas,

1995). The presence of depression complicates medical management of the physical illness and the presence of the physical illness often complicates the treatment of depression (Katon, 1996). Similarly, clients with eating disorders and clients with posttraumatic stress disorder are vulnerable to related, but also unrelated (i.e., comorbid), medical illnesses; failure to treat such medical conditions increases the client's psychological vulnerability (Ghadirian, Englesman, Leichner & Marshall, 1993; Hamner, 1994). Clients with major depression, generalized anxiety disorder, and panic disorder have significantly increased rates of medical illness and often place a greater strain on the medical delivery systems than medical patients without such comorbid psychiatric illness (Katon, 1996; Rogers, White, Warshaw, & Yonkers, 1994; Roy-Byrne, 1996; Zaubler & Katon, 1996), a reality that may be true even more so among elderly populations (Hocking & Koening, 1995; Katz, Streim, & Parmelee, 1994).

Another important example of psychopathology with high medical comorbidity is substance use. It is possible that higher rates of other psychopathology among drug users (e.g., depression, anxiety; see Brems & Johnson, 1997) are at least partially related to the fact that drug users are often afflicted with a variety of cooccurring physical diseases, illnesses that often have symptoms of psychopathology or emotional consequences in and of themselves. Specifically, substance use is highly correlated with the development of infectious diseases such as AIDS (e.g., 33% of adult AIDS cases can be traced to injection drugs use; 56% of children with AIDS have mothers who are or were injection drug users [Centers for Disease Control and Prevention, 1993]); tuberculosis (especially among drug users who are HIV positive, are homeless, and or are malnourished [Bartwell & Gilbert, 1993; Selwyn & Merino, 1997]); hepatitis A and B (Novick, 1992), and sexually transmitted diseases (Crowe & Reeves, 1994; Novick, 1992). Further, substance-using individuals tend to be at higher risk for poor malnutrition, poor general health and health care, brain impairment (due to substance

use per se as well as due to higher rates of head injuries among substance users), infections in general, and higher levels of stress (Crane, 1991; Crowe, 1990; Crowe & Reeves, 1994).

PSYCHOLOGICAL SYMPTOMS AND DISORDERS AS SIDE EFFECTS OF PRESCRIPTION AND OTC MEDICATIONS

A final issue that needs to be kept in mind by the responsible nonmedical practitioner is possible psychological or psychiatric side effects of medications or drugs used by the client. The clinician must consider the possibility of emotional or behavioral side effects of prescription and over-the-counter (OTC) medications (e.g., anithistamines leading to agitation or sleep disturbances), side effects of prescription or OTC medications for physical symptoms or medical disorders that may mimic psychological disorders (e.g., caffeine in drugs like fiorinol leading to anxiety or even panic attacks; interferon for hepatitis C leading to depression), and consequences secondary to legal (e.g., alcohol and in some states marijuana) and illicit drug use (e.g., cocaine, opiates, and so forth; see Chapter 2). The latter issues were covered in some detail in Chapter 2 and will not be revisited here, except to point out once again the importance of considering substance use among counseling and psychotherapy clients. The former two issues, however, psychological or behavioral symptoms and disorders as secondary to prescribed and OTC medication use, deserve some attention.

The fact that prescription and OTC medications can have psychiatric side effects serves again to highlight the great importance of a thorough assessment of medical aspects of the client during the intake exam. It is critical to inquire not only about diagnosed illness and undiagnosed physical symptoms, but also about use of medications, both prescribed and over-the-counter. This is so because even nonpsychiatric medications have effects on the central nervous system (CNS) and hence may influence the client's mental and affective state. Nonpsychiatric medications "may cause CNS effects because they: (1) cross the blood-brain barrier routinely, (2) affect susceptible individuals who have an idiosyncratic sensitivity to them, (3) produce secondary effects that then have CNS consequences, and (4) have effects that are dose related and occur in many individuals" (Gerner, 1993, p. 464). Not surprisingly, psychological and behavioral side effects from nonpsychiatric medications are difficult to predict because clients tend to respond to medications highly idiosyncratically (Brown & Stoudemire, 1998). However, there are a few groups of people who are more vulnerable to such reactions. They include the elderly, clients with organic impairment secondary to disease or trauma, clients taking multiple drugs, clients who are physiologically stressed, and clients who were already diagnosed with a DSM-IV disorder prior to taking the medications (Gerner, 1993). Clients themselves cannot generally be relied upon to make a connection between medications taken and changes in mental functioning, behavior, or mood; hence, it is up to the clinician to recognize this link and draw the proper conclusions. Because nonmedical providers cannot do so definitively, they learn when to develop the suspicion that medications that warrant a medical referral may be involved.

Medical referrals in this instance are relatively straightforward and are directed to the prescribing physician (except in cases of OTC medication use when the clinician would refer as described above). Clients are generally amenable to such referrals because they can usually understand that there may be a connection between their change in mood, mental functioning, or behavior, once the possible link is pointed out to them. The therapist or counselor would not make a definitive treatment plan until communication with the prescribing physician has taken place and a clearer picture has been developed about the connection between the client's current psychological presentation and the medications taken. Table 3-9 lists common nonpsychiatric medications that

can have psychological side effects. This listing is by no means all-inclusive or definitive. However, it suffices for most nonmedical providers and is presented mainly to raise awareness among clinicians that medication regimens need to be considered when conceptualizing a client and developing a treatment plan.

DOCUMENTATION AND RECORD KEEPING

All medical data the client provides are recorded as part of the regular intake report, the intake report addendum, or a progress note, as most appropriate. Sufficient detail is written down to clarify that the mental health care provider engaged in a thorough inquiry and also to lay the foundation for any referrals that might be made. The clinician's decision-making process about the need or lack thereof for a medical evaluation is best carefully documented in case of future inquiries. Two other important aspects of documentation in the context of medical referrals are proper releases of information to allow for free communication among providers and the proper recording of discussions about referral needs and processes with clients in progress notes. Not all clients are willing to follow through on referrals; if providers continue to work with such clients, they retain responsibility for their improvement. Should an adverse condition arise due to the fact that the client did not seek a medical consultation against the mental health care provider's advice, it will be helpful for the clinician to be able to provide evidence through documentation that such a referral was indeed suggested, but was rejected by the client.

Whenever a referral is successfully initiated, it will be important to document this fact in the client's chart. Such progress note entries are then supported through appropriate ROIs, the originals of which are kept in the client's chart, and copies of which are mailed to the other provider(s). Any communications with other providers also need to be documented in the

TABLE 3-9 Medications with Possible Psychological Side Effects*

MEDICATION	SYMPTOMS
CARDIOVASCULAR AGENTS	
calcium channel blockers (e.g., Norvase, Procardia, Isoptin, Cardizem, Vascor)	mood changes (depression or mania), dizziness, lethargy, sedation
adrenergic agents (antihypertensive agents; e.g., b-blockers, a2-agonists)	depression, decreased sex drive
diuretics (e.g., lithium, Midamor, Diamox, thiazines)	dizziness, sedation, weakness, fatigue, lethargy
cardiac glycocides (e.g., digitalis)	sedation, apathy, depression, visual illusions and hallucinations
vasodilators (e.g., minoxidil, diazoxide, tolazoline)	dizziness, fatigue
vascular relaxation agents (e.g., nitrates, nitrites)	agitation, anxiety, delirium, psychosis, restlessness, hypomania
antiarrhythmia agents (e.g., lidocaine)	lethargy, confusion, sedation, decreased sex drive, insomnia; also see local anesthetics
GASTROINTESTINAL AGENTS	
diarrhea treatments	
adsorbents (e.g., Kaopectate, Pepto Bismol)	behavior changes, sleep disturbance, short-term memory problems, anxiety, depression, hallucinations
anticholinergic agents	see CNS agents below
opiates	see CNS agents (analgesics) below; also see Chapter 2
vomiting treatments	
anticholinergic agents	see CNS agents below
prochlorperazine	akinesia, akathisia; also see Antipsychotics in Table 3-5
cannabinoids (e.g., Cesamet, Marinol)	anxiety, panic, confusion, perceptual disturbances; also see Cannabis in Chapter 2
antihistamines	see other agents below
bowel movement agents	
prokinetic agents (e.g., Metoclopramide)	akinesia mimicking depression due to flat affect, loss of interest and psychomotor retardation; akathisia resulting in restlessness, agitation, anxiety, worry
laxative agents (e.g., mineral oils, bulking agents)	no direct effects; indirect effects via vitamin or mineral deficiencies due to malabsorption
AGENTS AFFECTING THE ENDOCRINE SYSTEM	
glucose-regulating agents (e.g., insulin, sulfonylureas)	indirect effects via induction of hypoglycemia: confusion, dizziness, fatigue, anxiety
thyroid hormones (e.g., levothyroxine)	anxiety, restlessness, agitation, poor sleep (also see Tables 3-7 and 3-8 for hyperthyroidism)
antithyroid agents (e.g., iodides, thioamides)	confusion, disorientation, amnesia, hallucinations, reduced sex drive, sleep disturbance (also Table 3-7 and 3-8 for hypothyroidism)
anabolic steroids	depression, mania, psychosis
immune suppressants	
corticosteroids (e.g., prednisone, dexamethasone)	depression, mania, psychosis, exacerbation of pre-existing psychiatric conditions, delirium, lethargy, insomnia, euphoria
cyclosporine	paranoia, hallucinations, anxiety, depression, delirium

TABLE 3-9 Medications with Possible Psychological Side Effects* (continued)

MEDICATION	SYMPTOMS
AGENTS AFFECTING THE ENDOCRINE SYSTEM	
reproductive hormonal agents	
antiestrogens (e.g., danazol, tamoxifen, clomiphene)	depression, anxiety, fatigue, dizziness, sleep disturbance
estrogens (e.g., Premarin, Feminone, mestranol)	decreased sex drive, depression, anxiety, seizures, psychosis
progestins (e.g., Provera, Norplant)	dizziness, depression, anxiety, irritability, change in sex drive
oral contraceptives (i.e., combinations of estrogens and progestins)	as in estrogens and progestins
androgens (i.e., testosterone analogs)	hypomania, mania, depression, psychosis
gonadotropin-releasing agents (e.g., buserelin, gonadorelin)	dysphoria, emotional lability, decreased sex drive, fatigue, anxiety, weakness
AGENTS DIRECTLY AFFECTING THE CENTRAL NERVOUS SYSTEM (CNS)	
analgesic pain medications	also opiates in Chapter 2
naturally occurring opiates (e.g., morphine, codeine)	sedation, dizziness, mental slowing, irritability, dysphoria, euphoria, panic
phenanthrene alkaloids (e.g., Dilaudid, Percocet, Percodan)	sedation, dizziness, disturbed dreams, hallucinations, other psychotic symptoms
phenylpiperidine opiates (e.g., Lomotil, Innovar, Imodium)	dizziness, sedation, fatigue, confusion, anxiety, somnolence, seizures, hallucinations
opioid antagonists (e.g., Talwin, naltrexone)	symptoms of acute opioid withdrawal, anxiety, dysphoria, dizziness, fatigue, insomnia, somnolence, suicidality
mixed opioid agonist-antagonist (e.g., Stadol, Talacen, Nubain)	sedation, dizziness, confusion, somnolence, insomnia, asthenia, hallucinations, seizures
anesthetic agents	
inhalational agents (e.g., ether, nitrous oxide, chloroform)	anxiety, euphoria, nightmares, flashbacks, temporary impairment in attention and short-term memory
intravenous agents (e.g., Ketalar, Pentothal, Amidate)	visual hallucinations, dissociation, vivid dreams, decreased inhibitions
local anesthetics (e.g., lidocaine, Tonocard, Mexitil)	lethargy, mild confusion, sleep disturbance, sedation
anticonvulsants	also see antimania medications in Table 3-6
barbiturates (e.g., primidone, phenobarbitol, mephobarbital)	mood disturbances (depression and mania), insomnia, nightmares, sedation
bromides	depression, seizures, delirium, psychosis
oxazolidindiones (e.g., Tridione, Paradione)	sedation, extreme light sensitivity (hemeralopia), anxiety, irritability
succinimides (e.g., Celontin, Milontin, Zarontin)	depression, suicidality, hyperactivity, sleep disturbance, psychosis
valproic acid (e.g., Depakote)	sedation, depression, confusion, hallucinations
carbamazepine (e.g., Epitol, Tegretol)	sedation, depression, irritability, agitation, ataxia, psychosis

Continued

TABLE 3-9   Medications with Possible Psychological Side Effects* (continued)

MEDICATION	SYMPTOMS
AGENTS DIRECTLY AFFECTING THE CENTRAL NERVOUS SYSTEM (CNS)	
skeletal muscle relaxants (e.g., Baclofen, Dantrium)	depression, delirium, schizophrenia is exacerbated; withdrawal leads to anxiety and insomnia
antiparkinsonian agents (e.g., Cogentin, Parsidol, Norflex)	sedation, depression, anxiety, insomnia, confusion, hallucinations; worsens preexisting psychotic disorders
anticholinergic agents (e.g., atropine, scopolamine, trihexyphenidyl)	impairment in memory, euphoria, confusion, depression, visual hallucinations, delirium, disorientation
dopaminergic agents (e.g., L-Dopa, levopoda)	dizziness, insomnia, poor concentration, hallucinations, abnormal dreaming, anxiety, depression, delirium
ergot derivatives (e.g., Permax, Parlodel, Ergostat)	dizziness, sedation, fainting, lightheadedness
catecholamine-depleting agents (e.g., reserpine, tetrabenazine)	sedation, dysphoria, anxiety, insomnia, parkinsonism
botulinum toxin (i.e., Botox)	weakness, lightheadedness, dyspnea (labored breathing)
VARIOUS OTHER AGENTS	
antihistamines	
histamine-2 receptor blockers	depression, confusion
histamine-1 receptor blockers (e.g., phenothiazine, diphenhydramine, alkylamines)	sedation, dysphoria, agitation, confusion, hallucinosis
disulfiram (i.e., Antabuse)	psychosis, delirium, depression
antinociceptive medications	
pentazocine	hallucinations, illusions, nightmares, confusion, depression
acetylsalicylic acid (i.e., aspirin)	chronic, high dose use only: confusion, agitation, bizarre behavior, coma
nonsteroidal antiinflammatories	anhedonia, apathy, poor concentration
bariatric medications (i.e., weight-loss agents)	anxiety, paranoia, agitation, hallucinations
bronchodilator sympathomimetics	anxiety, tremors, tachycardia, psychosis
antibiotics	fatigue, malaise, poor concentration, depression, panic, hallucinations, paranoia

*This table is based largely on Gerner (1993) and Brown and Stoudemire (1998). Capitalized medications indicate brand names; noncapitalized medications indicate generic names.

client's chart through separate entries that can be filed with progress notes.

Occasionally, when referrals are initiated, the referral target requests information above and beyond that contained in a standard screening summary, intake report, or progress note. In such instances, the referral source may need to prepare a special referral report or referral note for the referral target. Whenever this is the case, the original of this document is retained in the client's chart; the copy is mailed to the requesting provider. If any records are received from the referral target (e.g., medical reports), these documents become a permanent part of the client's chart. However, these documents generally cannot be released by the receiving clinician. Instead, if another provider needs them, this individual will have to request them from the practitioner who originally prepared them. This can be accomplished only with a properly signed ROI.

Any requests for records, whether by the referral source or target, are made in writing and a copy of the request is filed in the client's chart. Requests need to be mailed or faxed (confidentially) to the provider from whom the record is requested and accompanied by a signed and current ROI. It is important to remember that ROIs do have expiration dates; this detail must be attended to since records cannot be released if the ROI has expired.

SUMMARY AND CONCLUDING THOUGHTS

Providers in both areas often ignore the mind-body interface. This is unfortunate because the close collaboration between medical and nonmedical providers can greatly enhance the quality of treatment (and as such the quality of life) of clients who are truly in need of coordinated medical and psychological intervention. Learning to recognize when a medical referral may benefit a client is an important aspect of becoming a better counselor or psychotherapist. All nonmedical mental health care providers need to have rudimentary awareness of psychotropic medication, differential medical diagnosis, medical comorbidity, medical consequences of psychological disorder or symptoms, and psycholog-

ical consequences of medication and drug use. Psychotherapists do not need to remember volumes of medical information to become efficient referral sources; they merely need to raise their awareness of physical issues and of the fact that mind and body do indeed interact.

In addition to collaborating with medical providers and being sensitive to the holistic quality of humans, whose mind and body clearly interact, clinicians also need to be aware that another professional may be able to provide valuable input regarding a given client. Nothing dooms clients and clinicians for treatment failure more so than the unwillingness or inability of the mental health care provider to make proper referrals for additional assessment when the limits of personal expertise are exceeded. Beyond the medical area, if concerns arise in the area of speech or language, referrals can be initiated to speech therapists. If concerns arise about a client's eligibility for social services (e.g., AFDC, SSI), and the current provider is unfamiliar with eligibility determination and application procedures, a referral to a social worker is in order. Concerns about nutrition and eating or dieting habits can certainly be made to traditional medical providers; however, it is often also helpful to draw upon the specialized expertise of nutritionists or nutrition counselors. There is an almost limitless source of information, support, input, and assistance in most communities; therapists and counselors must not hesitate to assist their clients in making use of such resources.

4

The Challenge of
Psychological Complexity

One's own self is well hidden from one's own self; of all mines of treasure,
one's own is the last to be dug up.

NIETZSCHE

At times, upon completion of an intake inter-view, questions remain about a client's level or pattern of functioning in a certain area or about a client's overall presentation of self. Some clients present with such complexity that it is difficult to organize all of the collected information into a cohesive and meaningful picture of the person. If information in certain areas is lacking or confusing, psychological assessment may be the best avenue to collect additional data that otherwise would remain elusive. This chapter details the referral process for psychological testing. This process may need to be initiated for the following several reasons:

- The need for clarification about intellectual functioning and cognitive processing, not related to psychosis or brain disease, but related to day-to-day performance and academic achievement (referral for psychological assessment with focus on cognitive testing [e.g., intelligence, achievement, and academic testing])

- The need for clarification about cognitive and perceptual processes, such as evidenced

in psychosis or organic brain disease (referral for neurology or neuropsychological assessment)

- The need for clarification about coping ability and problem-solving style in specific circumstances (psychological assessment with focus on cognitive and personality variables [e.g., intelligence testing, objective personality assessment, and projective assessment])

- The need for clarification about personality functioning and related intrapsychic and interpersonal dynamics within a context of community and culture (psychological assessment with focus on objective and projective assessment)

On occasion, a clinician, skilled in psychological testing and working with a client who might benefit from psychological assessment, may have to decide who should conduct this assessment. Most generally, it is best for the clinician who provides the client's therapy or counseling *not* to test the client. The type of relationship involved in therapy or counseling and the type of relationship that is developed in a testing situation are sufficiently distinct from one another that the two relationships are best kept separate and unconfounded. Thus, even a skilled examiner may be faced with the need to make a psychological testing referral and to follow the procedures outlined below.

Human behavior is complex and specific, comprehensive and detailed. Psychological testing can provide information about a variety of human contexts and can answer questions in a wide range of areas, integrating the details into a meaningful whole. Table 4-1 outlines the areas of functioning that can be addressed by psychological testing, that is, the types of questions that can be answered through a psychological testing referral. It is for the clarification of these aspects of human functioning that psychological testing can be of greatest assistance.

Psychological assessment can facilitate the conceptualization process tremendously, providing a thorough description of the client that is both detailed (i.e., addresses specific aspects of functioning) and integrated (i.e., combines all details into a meaningful whole that serves to explain the client's current state of being). Psychological assessment is more than the mere administration of a few psychological tests; much thought goes into test selection, and test administration itself is always embedded in a larger context. Responsible psychological assessment generally includes (at a minimum) a records review of referral information (with appropriate releases of information, of course), an interview with the client that is similar to (though shorter and more focused than) the intake interview, observation of the client (in session and/or out of session), test administration, interpretation of test and observational data, feedback to the client and the referral source, and a written report (Krull, 1997). Interpretation considers the larger context of the client and therefore requires the examiner to carefully collect and consider information about the client's overall personal history. Once clinicians appreciate that responsible testing is such a complex process, they will be in a better

position to choose referral targets (i.e., examiners to whom they refer their clients). Needless to say, the examiner to whom a therapist refers must be well versed in psychological testing and must not practice beyond the bounds of professional expertise and training.

THE REFERRAL PROCESS

The referral process for psychological assessment most concretely starts with a question in a counselor's or therapist's mind about a client that cannot be answered through data collected thus far. Whenever such a situation arises, and the question is psychological (as opposed to medical) in nature, a referral for psychological testing can be useful. To ascertain that psychological testing will be maximally successful, the referring clinician (referral source) makes a careful choice about the examiner to whom the client will be referred (referral target), prepares the client for the testing and feedback process, formulates a referral question, and participates in the feedback process.

Choosing a Referral Target

Choices about referral targets are driven by familiarity with the credentials, experiences, and interpersonal skills of potential assessment providers in a community. Credentials always include a doctorate in clinical or counseling psychology from an institution that specifically trains its graduates in the practice of psychological testing. Not all doctoral training programs emphasize assessment; hence, not all doctoral level providers are equally skilled in test administration and interpretation. Because psychological assessment is a complex process that lasts from the moment of referral to the moment of feedback, a referral cannot be taken lightly. The impact of a psychological assessment referral on a client must not be underestimated. A referral suggests that the referral source has a significant concern about the client that she or he cannot answer, signaling to the client the recognition of

TABLE 4-1   Areas of Human Functioning Tapped by Psychological Tests

CATEGORY OF FUNCTIONING	DEFINITION OF INCLUDED CONCEPTS
Cognitive Functioning	
Intellectual Functioning	ability to use prior learning and/or innate intellectual problem-solving skills adaptively and creatively; intelligence quotient (IQ), planning ability, flexibility of thought, abstract reasoning, capacity for synthesis, adaptation to new learning situations
Processing and Reality Testing	functions that affect how a person perceives, processes, and reacts to reality; memory, planning, thought process, concept formation, learning, judgment
Achievement	previous learning and retention of presented material; abilities and knowledge in scholastic, vocational, and other areas such as reading, writing, mathematics, science, humanities, etc.
Aptitude	potential for new learning and performance in a variety of areas, often related to job or leisure skills; skill in areas such as science, math, humanities, physical ability, etc.
Neuropsychology	strengths and weaknesses in specific areas of cognitive functioning in a developmental and skills context; visuomotor skills (coordination and speed), memory, abstract thinking, flexibility of thought, motor speed and manipulation (coordination), perceptual disturbances and deficits
Manifestations of Personality	
Emotional Functioning	identification of affect and emotions along with awareness and the ability to label, accept, and express these emotions (e.g., anxiety, depression, anger); range of expressiveness (e.g., flat, blunted, restricted); management of affect and emotions; sources of affect and emotion
Personal Functioning	level of self-integration and its translation into living skills and self-perception (e.g., self-esteem, self-efficacy, coping and problem-solving skills and confidence, psychological defenses, goal-directedness)
Interpersonal Functioning	behaviors, conflicts, attitudes in a variety of interpersonal contexts (e.g., family, work settings, friends, unfamiliar situations); interpersonal style, patterns, and dynamics; behavioral strength and weaknesses in social settings; social skills and deficits
Behaviors	behavioral self-expression, such as eating, sleeping, substance use, assertiveness skills; impulse control issues, such as planning, judgment, delay of gratification, cognitive control (strengths, resourcefulness, flexibility), fear and/or awareness of consequences
Risk Situations	affects, attitudes, behaviors and other self-expressions that result in high-risk situations (e.g., suicidality, hopelessness, violence, child abuse potential, substance abuse and dependence)
Interests	vocational and leisure interests (e.g., interests in a variety of job-related activities, interests in a variety of leisure activities)
Issues Related to Pathology	
Diagnosis	current and past symptom manifestation for use in DSM-IV diagnosis on axes 1 and 2; differentiation of acute versus chronic conditions; contextual factors useful for DSM-IV axes 3 and 4; assessment of global functioning as used for DSM-IV axis 5
Etiology	exploration of personal and interpersonal dynamics in past and present that can assist in the explanation of current and past symptom manifestation; social, familial, economic, cultural, educational, professional, and other circumstances contributing to the development or maintenance of the presenting concern
Prognosis	prediction of behavior and exploration of likelihood of change; assessment of past adaptations and changes; prediction of outcome of treatment; estimation of likelihood of success of therapeutic intervention
Treatment Planning	integration of information for purposes of implications for therapeutic relationship building, strategy selection, intervention speed and length, and special needs or circumstances to be attended to and addressed

great distress (Kellerman & Burry, 1991). The consequences that arise from psychological assessment can be substantial, depending on the specific purpose of the testing. Most generally speaking, there are five primary uses of psychological tests: (1) classification, which refers to the use of testing for certifications of various forms (such as certifying someone as eligible to receive disability insurance), screening, placement, and selection; (2) diagnosis and treatment planning; (3) self-knowledge; (4) program evaluation; and (5) research (Gregory, 1996). In the context of therapy, client and clinician will be most concerned with the issues of classification, diagnosis and treatment planning, and self-knowledge.

It is important to be sufficiently familiar with referral targets to know whether they are capable of psychological testing that is commensurate with the purpose of testing for which the client is referred (i.e., classification, self-knowledge, or diagnosis and treatment planning). If a given psychologist is primarily a researcher who has been providing psychological testing in the context of research and program evaluation, she or he may be an excellent tester, but not a skilled clinician who can phrase findings in the context of a therapy client's needs. It would be better to find a practitioner who has been conducting psychological assessments for the purposes specified in the context of psychotherapy or counseling. In addition to having proper credentials and experiences, it is also helpful if the referral target has interpersonal skills that are consistent with the process of psychological testing. Many of these skills overlap with the therapy skills that were outlined in Chapter 1 as useful to the counselor or therapist. However, a psychological examiner also must be organized, structured, able to motivate a client, capable of extraordinary sensitivity to optimize the client's test-taking behaviors and attitudes, and responsible with regard to completing and submitting much-needed reports. It is best to develop a relationship with a psychological examiner in the community and to be loyal to a clinician who has proven to have the

necessary skills and credentials. The examiner must also be able and willing to communicate openly with the referring clinician, both taking and making phone calls. The client should sign releases of information for the examiner who will conduct the testing so that the two providers may communicate freely. The referral process is greatly facilitated by such communication because it reduces unnecessary duplication of services. Finally, the examiner must evidence skill in providing feedback, both verbally and nonverbally. Testing that is most useful has direct implications for understanding the client, planning treatment, and appreciating prognosis. Both client and referring clinician need to feel satisfied that the amount of time, energy, and money invested in the assessment process was well worth it (more about this in the feedback section of this chapter).

Formulating the Referral Question

Although it is the examiner's responsibility to make sure that a referral question is appropriate and has sufficient clarity to guide an assessment, a clinician who already knows how to formulate a referral question greatly facilitates the referral process (Krull, 1997). Psychologists who receive a referral for testing have the right to reject it if the referral question appears inappropriate to what they can provide. Rejections can result for three primary reasons: (1) examiners may not be qualified to provide the type of testing required by the referral question; (2) the client may not be appropriate for testing (e.g., a psychotic client cannot be tested with tests that require some amount of reality contact, judgment, attention, and concentration); or (3) the examiner cannot guarantee test completion in a reasonable time frame (Krull, 1997).

When referring a client, the clinician cannot simply ask for psychological testing. Such a request is too broad and lacks the kind of specificity needed for the examiner to make choices about test selection, test scheduling, and other client- and referral question-related variables. In formulating the referral question, clinicians need to have at least two issues in mind. First, they must ponder what type of information might be helpful to the diagnostic and treatment planning process and, second, what questions need to be answered about the client to plan therapy and interact with the client (Krull, 1997). In other words, the clinician needs to attempt to pose referral questions in terms of specific questions or decisions that have to be made about the client (Groth-Marnat, 1997). Referral questions need to be specific, but not so minute as to decrease the degrees of freedom the examiner has to make proper test selection decisions (e.g., they do not include requests for specific tests). The clinician needs to understand that the examiner needs some latitude in test selection to gather enough context for test interpretation and formulation of recommendations. "Clinically relevant referral questions are of five general types: (1) They ask for a description or formulation of the pattern of current behavior; (2) they ask about the causes of behaviors observed; (3) they ask about the changes that can be anticipated in these behaviors over time; (4) they ask for ways in which these patterns may be modified; or (5) they ask about patterns and areas of deficit" (Beutler & Berren, 1995, p. 7). In other words, most referral questions concern themselves with diagnosis, etiology, prognosis, treatment, and/or functional impairment, at times also requesting information about severity in each of these areas and about interrelationships among them (e.g., how etiology affects prognosis or how functional impairment might influence treatment).

In formulating any one or more of these five types of referral questions, the clinician can also consider what the current goals of therapy are, what information he or she needs from the examiner to assist with the accomplishment of these goals, whether testing might answer these questions and gather this information, and what questions testing can feasibly answer (Seligman, 1996). Regardless of the thoroughness and thoughtfulness reflected by the clinician's referral question, many times the examiner will still choose to call the referral source to discuss the

referral (Beutler & Berren, 1995), especially if this is the first time the two providers have worked together. This is done because the examiner wants to double-check the context of the referral and clarify that the clinician has communicated the testing needs accurately and reliably. Once two providers have established a close and consistent working relationship such checking becomes unnecessary; however, a clinician who works with an examiner for the first time should be pleased, not offended, if the examiner calls to double-check the referral rationale. In fact, such thoroughness on the examiner's part suggests awareness that a clear and concise context for testing is crucial and also suggests recognition of the importance of having as thorough an understanding of the clinician's needs and client's circumstances as possible. Having a context for testing assists the examiner with test selection by revealing something about the client's likely level of compliance, ability to perform, resistance, propensity for decompensation, and ability to handle the stress of the testing situation.

A variety of referral questions can be formulated. To give the reader an idea of how to phrase referral questions appropriately, examples of comprehensive, yet still specific, referral questions from within each category of assessment (see Table 4-1) are provided in Table 4-2. This table demonstrates that referral questions are rarely phrased as single questions, but rather as questions that build upon one another. The samples also are meant to show that, although the questions are specific with regard to requesting detailed information necessary for the referring clinician's diagnosis and treatment planning, they do not prescribe the use of certain tests. The examples in Table 4-2 are not the only possible referral questions that could be asked within each assessment category. They are merely presented to give the reader an idea of how referral questions are phrased. The actual variety of referral requests is infinite and highly idiosyncratic. In addition, each examiner may have slightly different ideas about what constitutes a proper referral question. Thus, once a clinician has established a

TABLE 4-2   Samples of Appropriate Referral Questions

CATEGORY OF FUNCTIONING	SAMPLES OF POSSIBLE REFERRAL QUESTIONS
Cognitive Functioning	
Intellectual Functioning	What is the client's cognitive potential for adaptive problem solving, planning, and adaptation to new learning situations? What are the client's cognitive strengths and weaknesses and what is the client's general cognitive style and capacity? Are there areas that need remediation or special accommodation? Are the client's cognitive skills interfered with by noncognitive factors (e.g., by anxiety, drug use)?
Processing and Reality Testing	How does the client process, perceive, and react to the world? Is perception interfered with by cognitive and/or personality factors? What is the client's capacity for planning, judgment, and other higher cognitive functions? What is the impact of the client's cognitive and/or personality characteristics and skills or deficits on reality perception and testing?
Achievement	Has the client benefited from academic learning? Is the client's academic achievement commensurate with cognitive capacity and potential? Are there unexpected strengths or weaknesses and how can they be explained?
Aptitude	What is the client's potential for new learning and performance in a particular job- or leisure-related area? Is the client capable of adequate performance in a given area?
Neuropsychology	Does the client have particular strengths or deficits in specific areas of cognitive functioning? Has the client's recent head injury affected cognitive and/or behavioral functioning? What are the client's strengths and weaknesses and how do these translate into rehabilitation needs?
Overall	What is the client's cognitive potential at the current time and how does this compare to performance prior to the head injury? What are the implications for the client's ability to process reality, to benefit from past learning, and to prepare for future job performance?
Manifestations of Personality	
Emotional Functioning	What are the client's primary affects and how does the client manage and express them? What is the severity of the client's affect and how appropriate is its expression and management? Is the client's affective awareness appropriate and is affective expression full range and appropriate to context?
Personal Functioning	What is the client's level of self-integration and how realistic is the client's self-perception? What are the client's strengths and weakness with regard to coping? What are the client's coping styles and patterns as well as preferred defenses and problem-solving situations? Is the client goal directed and clear about self-identity and self-perception?
Interpersonal Functioning	How does the client behave, feel, and react in interpersonal situations? Do these reactions vary depending on context? What is the client's general interpersonal style and what are the client's strength and weaknesses with regard to social skills?
Behaviors	Is the client capable of proper impulse control (i.e., neither too overcontrolled nor too impulsive)? What is the client's capacity for appropriate self-assertion? What are the client's general behavioral patterns? Are they rigid or functional and can they vary depending on context?
Risk Situations	Does the client present a significant suicide risk? How and in what situations? Does the client manifest violent tendencies? How best is the client's risk behavior managed? What is the likelihood of substance use and dependence in this client?

Continued

TABLE 4-2   Samples of Appropriate Referral Questions (continued)

CATEGORY OF FUNCTIONING	SAMPLES OF POSSIBLE REFERRAL QUESTIONS
Manifestations of Personality	
Interests	What are the client's preferred vocational and leisure interests? Are these interests compatible with aptitude, cognitive potential, and personality style?
Overall	Does the client's interpersonal style result in social isolation and does personal and emotional functioning interfere with more successful interpersonal relationships? Is the client's current expression of hopelessness an acute or chronic concern and does it reflect a general emotional adjustment or situational reaction?
Issues Related to Pathology	
Diagnosis	Is the client's depression sufficiently severe to warrant a diagnosis of major depression and should a referral for psychotropic medications be considered? Is the client's anxiety a situational reaction or a chronic adjustment? Does the client manifest traits, characteristics and symptoms that warrant a personality disorder diagnosis? Is the client more likely to be severely (psychotically) depressed or schizophrenic?
Etiology	What are the client's family dynamics and how have they contributed to the current symptomatic picture? How can the client's symptoms best be understood in terms of predisposing, precipitating, and reinforcing factors? What has contributed to the client's current symptomatic presentation?
Prognosis	Will the client show potential for improvement once the most severe symptoms have resolved? Is this client likely to become a danger to others?
Treatment Planning	Is the client's cognitive potential a factor in treatment planning? Does the client's current affective adjustment preclude the use of any particular treatment strategies? What is the best course of action for this client in terms of brief therapy intervention? Is this client capable of forming a therapeutic relationship of sufficient strength to make psychodynamically oriented psychotherapy possible?

working relationship with a particular examiner, it behooves both professionals to talk about how the examiner prefers referral questions to be phrased so that the clinician may adapt the referral procedure to the needs of this particular examiner.

Preparing the Client

Once counselors or therapists have decided that a client needs to be referred for psychological testing, and after they have chosen a referral target and formulated referral questions, they broach the idea of psychological testing with the client. Clients require thorough preparation for and motivation to comply with psychological testing lest it be a useless and less-than-reliable endeavor. In preparing a client for assessment, clinicians

- carefully point out the purpose of the testing (sharing the referral question in layperson's terms)

- expound upon the potential benefits of testing that will arise for the client

- provide details about the testing process per se (i.e., where, with whom, how, cost, time commitment)

- give choices about possible examiners (if applicable)

- delineate what type of feedback can be expected as well as how and where it will be delivered

- give the client an opportunity for questions about the testing process

Clinicians need to be cautious about two issues as they prepare the client for testing. First, a clinician must not overrely on psychological testing (Groth-Marnat, 1997). Psychological assessment is not the answer to everything and cannot be 100% conclusive and informative. Having too much faith in the testing process can lead to misconceptions, misunderstanding, and disappointments. Thus, in preparing the client, the therapist takes care not to mislead the client into thinking that psychological testing will have all the answers; in fact, psychological testing often serves to raise even more questions. Second, psychological testing will only be as good as the examiner (Groth Marnat, 1997). If the clinician chooses an unqualified examiner, test results will be less than valuable. If a client is being prepared for psychological testing with an examiner to whom the clinician has never referred before, because the clinician cannot reassure the client about the actual procedures and qualifications of this examiner, some special cautions may be necessary. Most importantly, the client needs to be told that the counselor is not familiar with the referral target. Thus, the clinician can neither guarantee the compatibility of examiner and client nor the nature and quality of the process and outcome. Preparation of a client is much easier if clinician and examiner have a preexisting relationship, have met one another, and have worked together on prior occasions. Only then can the referring counselor describe the examiner and testing process with authority and confidence.

Participating in the Feedback Process

The referral process is not complete until feedback has been received about the client from the referral target. Such feedback can be structured in a variety of ways. The examiner can decide to meet with the client alone to provide test results, can meet jointly with client and therapist, or can meet solely with the therapist. I endorse the method in which the examiner meets with client and therapist together to provide feedback about test results. The advantages of this approach are that both client and therapist receive the same verbal information and, because the client is present, he or she will not perceive that either

therapist or examiner is hiding anything. The disadvantage of this model is the time involved for the two providers and the potential expense this may incur for the client. This approach involves difficult decisions about where to meet and also may require travel time for at least one of the providers. It is preferable for the examiner to travel to the therapist's office. That way, after the examiner provides test feedback (which takes only about 20–30 minutes), the examiner can leave, allowing the client and counselor to begin to process the information. Receiving testing feedback can have a profound impact upon a client, and this time spent together processing what was heard can be extremely helpful for the client.

The disadvantages of the other two models are self-evident. If feedback is provided only to the client, the therapist relies on the client's report about what was said by the examiner. This is problematic even, or especially, if the therapist receives a testing report. The counselor will not know how certain pieces of data were communicated to the client, and misunderstanding and misperceptions may arise between client and therapist. Providing feedback only to the therapist is problematic because then the client never hears the information from the examiner, who is likely to be more objective. Further, the client may never receive adequate feedback because the therapist is not specifically trained in how to provide testing feedback, whereas the examiner presumably is. The advantages of these two models rest strictly in their greater convenience for the providers and their lesser expense for the client.

Regardless of how the feedback session is structured, the nature of feedback has certain universal features that make it optimally useful for client and therapist. First, it is delivered respectfully and without jargon (Hood & Johnson, 1997). Both client and therapist need to understand what the examiner is attempting to communicate. To facilitate this understanding further, the examiner accepts questions from the client and therapist and watches for the client's reactions to the information. All data and interpretations are placed in a greater social or sociocultural context and reflect the examiner's consideration of all aspects of the client's life and history. Feedback is descriptive, but neither offensive nor overwhelming; it is honest, yet sensitive to the client's feelings and reactions; it is truthful, comprehensive, and includes a thorough discussion of the client's strengths (Krull, 1997). Feedback must also be useful with regard to treatment implications and recommendations. It is not sufficient for the examiner to provide information about the client without indicating how this information will be helpful to treatment planning and the therapeutic process. In other words, feedback states probabilities, not certainties, and presents data contextually, not in terms of absolutes that could apply to anyone. The feedback must demonstrate that the examiner has considered all variables of an individual client's life. In sum, feedback must be helpful to the client by enhancing self-understanding and self-knowledge that will motivate the client for counseling; it must be helpful for the therapist by enhancing understanding of the client's dynamics and personality in a manner that will have clear and obvious implications for treatment planning and rapport building (Hood & Johnson, 1997). Such feedback has been shown to result in greater treatment gains among clients who received psychological testing (Hood & Johnson, 1997) and meets the true spirit of a psychological assessment.

Although a verbal feedback session is certainly the more important way of communicating test results, the therapist or counselor can also expect to receive a testing report from the examiner. This testing report is generally not written for the client's perusal, but rather it strictly reflects a professional communication between clinician and examiner. This report and the recommendations within it often become a permanent cornerstone in the client's treatment process. A good comprehensive psychological assessment report covers a multitude of issues about the client. Although it is impossible to describe one generic format of such a report, given the wide variation in referral questions and potential tests administered, a few consistent elements can be expected. A good report is individualized to the client and is not a simple interpretive computer printout about one test that was administered. Any clinician who receives such a report from a referral target needs to look for a different examiner for future referrals. A good report also provides background about the client, showing which aspects of the client were considered in making interpretations and recommendations. It also gives a thorough description of the client's test-taking behavior, along with a clinical judgment about the validity and reliability of the data that were obtained and the interpretations that are rendered. Only after dealing with these preliminary issues does the report detail test findings, providing clear and concrete answers about the referral question. Finally, and most importantly, test findings are used to arrive at treatment and related recommendations for the client and referring clinician. The usefulness of the report will give the clinician a good idea about the skill of the examiner. A report that has no impact on the clinician's understanding of the client, treatment planning, or prognosis was not worth the referral. Clinicians need to learn to become good consumers of psychological assessment. Referring a client for psychological testing is a large time, energy, and money investment, and few things are more destructive to therapeutic rapport than sending a client to a professional who wastes all of these resources.

ESSENTIAL KNOWLEDGE ABOUT PSYCHOLOGICAL TESTING

For clinicians to formulate a referral question, prepare a client properly for testing, and have realistic expectations about possible feedback, they must have a certain amount of knowledge about testing. This knowledge can be used to explain to the client what will happen and why it will happen and then to accurately answer questions the client might have. At a minimum, a counselor needs to be aware of the philosophical underpinnings of psychological testing and their cultural implications and must also have some rudimentary knowledge about the process of and instruments used for psychological assessment.

Philosophical Underpinnings

As originally conceived, testing was an enterprise based on positivism, the belief that it is possible to find incontrovertible truths about people that apply to everyone equally, and on the idea that science is the only and best way to accumulate such knowledge (e.g., Kaplan & Saccuzzo, 1997; Rogers, 1995). Testing, in its rudimentary beginnings was, by definition, of benefit to the institutions that used its results, not to the individuals who were subjected to it. Early testing benefited the military (see Brems, Thevinin, & Routh, 1991), hospitals, and schools and universities that made use of the results to make life decisions for examinees (Rogers, 1995). Originally, testing as a philosophy was firmly rooted in the hereditarian camp of human differences. It followed in the tradition of Sir Francis Galton, who argued that individual differences were innate and therefore not amenable to remediation or change. This approach stands in direct contrast to current multicultural thinking and sensitivity, which is more in line with the philosophical school of thought of John Stuart Mills, who believed that individual differences were the consequence of exposure to learning opportunities and, hence, alterable if necessary or desirable. The two approaches continue to split scientists even today in the nature versus nurture debate about human behavior, cognition, affect, and illness.

The traditional, scientific, hereditary-based approach to testing can easily force a value of assimilation because it makes assumptions about group means (and standard deviations) as a criterion for health or normality. Deviation from the mean is often interpreted as "abnormal" and as indicative of a person's need to change (Dana, 1993). This pure science approach has patently ignored social, cultural, and sociocultural contexts of both the examinee and the examiner (Gould, 1981). The traditional focus on science has perpetuated Eurocentric biases inherent in all

early psychological tests that were developed, standardized, and used for White middle-class Americans. In other words, testing as originally conceived was Eurocentric, ethnocentric, and anything but pluralistic (Quinn, 1993). The dire social consequences of this ethnocentric approach for individuals who are neither White, nor middle class, have of course been documented widely in the popular media as well as in the professional literature (e.g., Darou, 1992; Padilla, 1988; Solomon, 1992; Valencia & Guadarrama, 1996) and have been subject to lawsuits and legislative action (Gregory, 1996; Kaplan & Saccuzzo, 1997). Unfortunately, despite their good intention, resultant laws have not always been entirely successful in fulfilling their purpose. For instance, Public Law 94-142, the Education for All Handicapped Children Act (later extended to preschoolers by PL 99-457), not only provided for federal funds for the assessment and treatment of physically or cognitively challenged children but also had provisions for test selection. It specifically demanded that tests be used that are appropriate to the social needs of the individual being assessed and legislated the use of racially and culturally nondiscriminatory testing procedures. It even stipulated that children should be assessed in their native or primary language (Salvia & Ysseldyke, 1988). Although this law spurred the development and use of a wide range of tests adjusted for special physical needs (e.g., tests accommodating visually-, auditorily-, or motor-impaired children), it did little for the development and use of culturally relevant tests. Instead, culture remains a component of humanity that has seen little accommodation by the tests that are currently used by psychologists around the nation.

Nevertheless, strides have been made, as mental health professionals have become more sensitive to cultural issues in general and to cultural issues in testing in particular, and as they have begun to subscribe to values of pluralism. Indeed, if the original historical realities of testing were practiced today, testing would be doomed due to society's increased awareness of the need for individual differences, pluralism, cultural sensitivity, and competence. Fortunately, the tide has begun to turn with many psychologists' efforts to make testing a more sensitive, positive business for the consumer (i.e., the client or test taker). It is indeed possible to use tests that were developed based on the ethnocentric and positivistic historical beliefs in a manner that is more culturally sensitive, appropriate, and competent. With appropriate modifications, new test development, and new approaches to testing procedures, it is possible to use testing without subscribing to biological determinism. For it would be a shame to "throw the baby out with the bath water" and lose the positive consequences that testing can create for a client (see Messick, 1980). Testing has to follow neither pure biological determinism nor endorse pure environmental determinism; it can be socially appropriate, be used in a culturally competent manner, and be supportive of individual differences and needs. This is most likely accomplished through assessment as opposed to simple psychological testing. Specifically, Rogers (1995) defined psychological testing to be "the technical aspects of tests— whether they measure what they are supposed to, whether they produce consistent results, and so on. The manner in which tests are used to make decisions in concert with other sources of information [on the other hand, is] considered psychological assessment" (p. 48). The model that is endorsed here, according to this definition, encompasses both testing (the technical aspects) and assessment (the human aspects). Testing alone, or use of one isolated instrument without consideration of personal contexts or ancillary sources of information, can rarely, if ever, be a culturally sensitive or humanly empathic process and is to be avoided by examiners.

In a philosophical model of testing that has been modified to be empathic and culturally sensitive, the examiner pays close attention to all aspects of assessment and testing through which bias can be introduced. In making referrals, the referral source is best served by seeking a referral target who has clearly demonstrated an empathic

and culturally sensitive approach. Most importantly, referring clinicians can interview potential examiners to assess whether they have carefully considered the impact that the following variables will have on the reliability and validity of their assessment and to determine whether they have developed procedures to address them.

Variables to Consider to Conduct Empathic and Culturally Sensitive Assessment

- optimal examiner-client matching on as many variables as is reasonably possible (e.g., gender, ethnicity, personality style, speed of work)

- reassurance of the client that the assessment is conducted for the purposes and benefits of the client, not the sole purposes and benefits of the referral source or target

- reassurance that testing is conducted for the purpose of documenting what the client is capable of doing, how the client prefers to function, and how the client perceives and interacts with the world

- reassurance that testing does not take a deficit approach (i.e., is not conducted to see what clients are incapable of, how they cannot function and what they are doing wrong)

- procedures to establish a maximally trusting and empathic relationship between examiner and client

- procedures to minimize misunderstanding and miscommunication between examiner and client

- adaptation of examination style and environment to the specific needs and background of each individual client

- clarification and consideration of the client's expectations about the assessment process, including its purpose, goals, involvement of others, outcomes to be expected, and similar issues

- procedures that work toward common goals, a sharing of the same task orientation

- careful multimethod, multisource assessment instrument selection and testing process

- selection of tests that have evidence of factor stability across cultures, predictive equivalence across cultural groups, proper translation with back translation if the instrument was originally in a language different than the client's, local norms, and representative items

- selection of tests with stimulus materials that are culturally appropriate and without stereotypes about cultural subgroups

- inclusion of a careful, culturally relevant and sensitive interview to create an interpretive context

- use of local norms whenever available

- inclusion of world view and acculturation assessment of minority clients (as defined in Chapter 1)

- clarification and consideration of the client's expectations about assessment feedback

- clarification and consideration of the client's expectations about recommendation based on the assessment

- contextual and culturally relevant and sensitive interpretation of data

- exclusion of use of computer-generated interpretive reports

- cultural knowledge regarding diagnosis and relevant recommendations

- openness to making nontraditional recommendations that reflect the client's specific social, cultural, familial, economic, and other circumstances

If a referral target can speak intelligently about *all* of these issues and can give examples of how he or she deals with them while with clients, especially clients who are ethnically or culturally different from the examiner, some faith can be placed in that professional. Nevertheless, a first time referral needs to be followed up with a critical review of the assessment process and the ensuing report. Only if the referral source is satisfied with the results of the first referral to a given examiner, should that professional be used again in the future.

To recapitulate, the assessment philosophy that is endorsed here as most likely to produce useful and helpful results asserts that testing is best understood, not as a science (although it has its roots in pure science), but rather as a unique integration of science and art. Despite being collected with tools that have to be scientifically evaluated and constructed, findings derived from assessments are not hard and fast, but rather they are educated guesses and attempts to come to terms with another human being's reality and way of being in the world. Tests are merely tools to facilitate understanding and to provide an additional means of communication. Testing never provides Truth. The problem of approaching testing as a truth-finding endeavor solves itself once one considers assessment in a cross-cultural context. In this context, testing becomes a joint endeavor in which examiner and examinee search for possible answers, presenting as many alternatives, perspectives, and opinions as possible without declaring any of these possibilities the Truth. If used fairly (and this is of course the difficulty), assessment in any context, but especially in a cross-cultural one, enhances communication, facilitates understanding, and reinforces the individuality of each and every human being, without imposing external standards and without implying deficits or pathologies. The essence of this philosophy can be summarized in four simple sentences:

1. Examiners view and understand all human beings in their own unique personal, social, familial, economic, cultural, and broader context.

2. Examiners let go of their own ethnocentrism (regardless of their cultural backgrounds) and become well versed in the cultures of all clients with whom they anticipate contact.

3. Examiners are flexible, willing to adapt, empathic, knowledgeable, nonjudgmental, nontraditional, and ever aware of the social consequences of their trade.

4. Examiners recognize that tests are only as good as the examiners who administer them.

Assessment Instruments and Process

A clinician needs to know the types of tests available to psychological examiners, the approximate administration time for each test, the reasons why a particular test may be chosen (i.e., the concepts measured), and the type of information that can be derived from tests. This information is summarized in Tables 4-3 to 4-5. The details contained in these tables are helpful in preparing the client, although they certainly do not qualify the untrained clinician to administer the tests directly or even to determine which tests to select. Knowing something about psychological tests merely facilitates the referral process by allowing the therapist to speak with the client about the testing process more knowledgeably and to formulate referral questions more intelligently.

Test selection for assessment purposes is a complex process that requires skills and experience not possessed by the nontesting trained clinician (Beutler & Berren, 1995). Instruments are chosen with a number of issues in mind. First, examiners are familiar with the psychometric properties of each test they are considering for use (Groth-Marnat, 1997). The two most important aspects of tests in this regard are their reliability and validity. Simply put, reliability refers to a test's likelihood to return the same score for the same client across time and situations upon retesting; it is essentially a measure of the amount of error that can be introduced when recording a particular client's characteristics, skills, or trait. Validity refers to the test's ability to measure what it purports to measure. In other words, validity coefficients express the amount of faith an examiner can have in the test's ability to truly reflect the trait that the examiner expects the test to measure (e.g., getting a measure of the client's level of depression when administering a test such as the Beck Depression Inventory). In addition to considering test reliability and validity, an examiner also recognizes any skills or characteristics of the client that may render a certain test

inappropriate. For example, an examiner working with Alaska Native clients may need to reconsider the use of a typical intelligence scale (such as the Wechsler Adult Intelligence Scale, Third Edition) if this test does not include individuals of a similar cultural and ethnic background as this client. Cultural sensitivity is a critical issue in test selection, and a referring clinician must point out to an examiner when special cultural considerations are present with a referred client (Brems, 1999b; Dana, 1993; Suzuki, Meller, & Ponterotto, 1996). Other related issues include clients who have physical limitations that may interfere with test taking demands. For example, a client with a visual disturbance cannot be tested with many standard achievement or aptitude tests, which require normal vision. The examiner also considers the referral question in making test selections and therefore chooses tests that are both relevant to the referral issue and able to provide answers to the question posed by the clinician. However, many examiners have a core of tests they use routinely with all of their testing clients, which they supplement with additional tests as needed based on the referral question or client characteristics (Beutler & Berren, 1995). This approach is a sound one because it provides the examiner with an internal set of norms and a frame of references that can be used to compare this client to others.

Beyond test selection, special training is also required in the administration of a multitude of psychological tests. Although some instruments look surprisingly easy to administer and are tempting for the untrained provider to use, most psychological instruments require the examiner to have explicit training in test administration and the special issues involved in rapport building for testing purposes. Even an instrument that looks easy to administer (e.g., an MMPI-2) may pose special challenges and may raise questions from a client that an inexperienced or nontrained examiner would not be able to address appropriately. In other words, nontrained providers must not be tempted to use psychological tests themselves, regardless of how easy it may seem to administer the tests. Most importantly, test interpretation is a special skill that is developed only through extensive education and supervised experience, and it cannot be undertaken by the nontrained clinician. The level of complexity in terms of test administration and interpretation varies widely across tests. Some of the most difficult tests to administer are the intelligence tests, which require very specific procedures lest the examiner invalidate the testing results. Some of the most complex tests with regard to interpretation are the projective tests, particularly the Rorschach Inkblot Test, because these tests require careful scoring and contextual interpretation.

Tables 4-1 and 4-2 suggest that psychological testing is conducted in a wide range of areas related to human functioning. There are thousands of psychological tests, tapping hundreds of human behaviors, affects, modes of functioning, strengths, weaknesses, deficits, and so forth. What all tests have in common is the fact that each is "a standardized procedure for sampling behavior and describing it with categories or scores" that can then be used to predict and compare the client's performance (Gregory, 1996, p. 33). Most simply put, there are two large groupings of tests, namely, abilities tests and personality tests (Kaplan & Saccuzzo, 1997). Each of these two categories is subdivided, resulting in increasingly complex and detailed assessments of human performance, functioning, and experience. Abilities tests include a minimum of four broad subcategories: achievement tests, aptitude tests, intelligence tests, and neuropsychological tests; personality tests include two broad subcategories of structured (also called objective) or unstructured (also called projective) tests. Structured tests are further broken down into multitrait versus single-trait instruments. Personality tests (both structured and unstructured) can measure emotional functioning (often with diagnostic implications), personal functioning, interpersonal functioning, interests, behaviors, and risk situations. Personality tests (often in conjunction with abilities tests) are used to draw inferences and make decisions about pathology, diagnosis, and treatment planning. This classification is visually displayed in Figure 4-1.

It must be noted that this discrete categorization is somewhat artificial and arbitrary. Especially in the personality area, test instruments more commonly measure overlapping content areas. For example, a Rorschach Inkblot Test can provide information about all aspects of personality, ranging from emotional functioning to personal functioning to interpersonal adjustment to behaviors and risk assessment. The overview of psychological tests in Tables 4-2 to 4-5 reflects only primary or broad subcategories. Additional breakdown with regard to personality assessment is achieved through test interpretation, not necessarily test administration. In other words, even though referral questions can be phrased in terms of the smaller personality subcategorizations (i.e., emotional functioning, personal functioning; see Table 4-2), personality testing per se will rarely address just one of these aspects of personality, usually inadvertently tapping more than one area (there are exceptions to this rule since there are a number of single-trait personality inventories that are designed to measure just one trait; however, their reliability and validity often leaves much to be desired and hence they are rarely used in isolation). Once testing is complete, the examiner will then phrase feedback and recommendation to address the referral question in the specific subcategory of interest.

Abilities Tests. Most abilities tests, whether they are achievement tests, aptitude tests, intelligence tests, or neuropsychological test batteries, are complex instruments that have to be administered according to a specific administration protocol, iolation of which results in the invalidation of test results. Abilities tests can be fairly brief, if a single area of cognitive functioning is measured; or quite extensive, if a more global assessment is desirable. A wide range of abilities tests are available and the purchase of most of them is restricted to examiners who can document education and experience with these instruments. The level of complexity of these tests varies widely, with some achievement tests being very easy to administer and with all individually administered intelligence and neuropsychological tests requiring extensive training

FIGURE 4-1 Classification and Overview of Psychological Tests

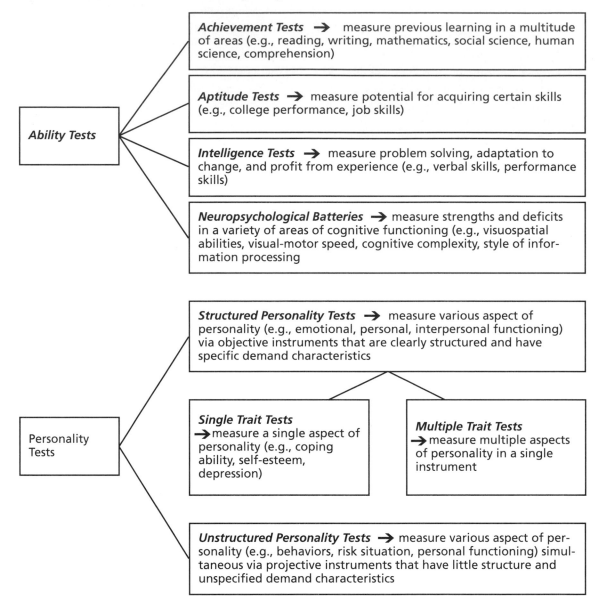

and background. Interpretation of all cognitive abilities tests requires experience and extensive education, especially in circumstances when information from various testing sources must be integrated and rendered meaningful. Information about the most commonly used cognitive abilities tests is contained in Table 4-3. This table does not cover aptitude tests, which are most commonly administered in group settings by nonpsychological examiners in the context of career planning and development (e.g., the Graduate Record Examination and the Scholastic Assessment Test)

TABLE 4-3   Overview of Samples of Abilities Tests

REFERENCE (PUBLISHER)	ADMINISTRATION AND TIME REQUIREMENTS	TYPES OF SCORES	DESCRIPTION OF ITEMS OR SUBSCALES
ACHIEVEMENT TESTS			
WOODCOCK JOHNSON TEST OF ACHIEVEMENT, REVISED (WJ-TA-R)			
Woodcock & Mather, 1989a (Riverside Publishing Company)	individually administered by a highly trained examiner; 30–60 minutes	standard scores, percentile ranks, grade equivalents, age equivalents, discrepancy scores, clinical cut-offs	24 subtests grouped into 5 standard clusters: **Broad Reading, Broad Mathematics, Broad Written Language, Broad Knowledge**, and **Skills**; and 6 supplemental clusters: **Basic Reading Skills, Reading Comprehension, Basic Mathematics Skills, Basic Writing Skills**, and **Written Expression**
WECHSLER INDIVIDUAL ACHIEVEMENT TEST (WIAT)			
Wechsler, 1992 (Psychological Corporation)	individually administered by a highly trained examiner; 50 minutes	standard scores, grade and age equivalents, percentile ranks, stanines	8 subtests measuring **Basic Reading, Mathematics Reasoning, Spelling, Reading Comprehension, Numerical, Operation, Listening Comprehension, Oral Expression**
WIDE RANGE ACHIEVEMENT TEST, THIRD EDITION (WRAT-3)			
Wilkinson, 1993 (Psychological Assessment Resources)	easily administered by well-trained examiner; 15–30 minutes	grade equivalents, standard scores, percentile ranks	3 subtests measuring **Reading, Spelling**, and **Arithmetic** (spelling and arithmetic can be group administered)
INTELLIGENCE TESTS			
WECHSLER ADULT INTELLIGENCE SCALE, THIRD EDITION (WAIS-III)			
Wechsler, 1997a (Psychological Corporation)	individually administered by a highly trained examiner; 50–70 minutes	IQ scores, standard scores for subscales, standard scores for factor indices, percentile ranks	3 **IQ scores** (verbal, performance, full scale); 14 subscales organized into 4 factor indices: **Perceptual Organization** (Picture Completion, Block Design, Picture Arrangement, Object Assembly, Matrix Reasoning), **Processing Speed** (Digit Symbol, Symbol Search), **Memory** (Arithmetic, Digit Span, Letter-Number Sequencing), **Verbal Comprehension** (Vocabulary, Similarities, Information, Comprehension)
STANFORD-BINET INTELLIGENCE SCALE, FOURTH EDITION (SBIS-IV)			
Thorndike, Hagen, & Sattler, 1986a, 1986b (Riverside Publishing Company)	individually administered by a highly trained examiner; 50–70 minutes	IQ-type standard scores, standard scores for subscales, standard scores for factors, percentile ranks	1 composite **intelligence score**, 15 subscales organized into 3 factor scores: **Verbal Comprehension** (Vocabulary, Comprehension, Absurdities, Verbal Relations, Memory for Sentences), **Nonverbal Reasoning** (Pattern Analysis, Copying, Quantitative, Bead Memory, Matrices, Number Series, Equation Building, Paper Folding and Cutting), **Memory** (Memory for Sentences, Memory for Digits, Memory for Objects)

Continued

TABLE 4-3   Overview of Samples of Abilities Tests (continued)

REFERENCE (PUBLISHER)	ADMINISTRATION AND TIME REQUIREMENTS	TYPES OF SCORES	DESCRIPTION OF ITEMS OR SUBSCALES
KAUFMAN ADOLESCENT AND ADULT INTELLIGENCE SCALE (KAIT)			
Kaufman & Kaufman, 1993 (Psychological Assessment Resources, Inc.)	individually administered by a highly trained examiner; 60 minutes for core; 90 minutes for expanded	IQ-type standard scores, standard scores for subscales, standard scores for factors, percentile ranks	1 composite **intelligence score**, 10 subscales (4 of which are optional) organized into 2 core factors and 2 optional factors: **Fluid Intelligence** (Rebus Learning, Mystery Codes, Logical Steps), **Crystallized Intelligence** (Definitions, Auditory Comprehension, Double Meanings), **Memory, Mental Status Information**
WOODCOCK JOHNSON TESTS OF COGNITIVE ABILITY - REVISED (WJ-TCA-R)			
Woodcock & Mather, 1989b (Riverside Publishing Company)	individually administered by a highly trained examiner; 30–60 minutes	standard scores, percentile ranks, discrepancy scores, clinical cut-offs	3 **Broad Ability** scores (Early Development Standard Scale, Extended Scale); 21 subtests organized into 8 **Cognitive Abilities** clusters (Long-term Retrieval, Short-term Memory, Processing Speed, Auditory Processing, Visual processing, Comprehension-Knowledge, Fluid Reasoning, and Oral Language) and 5 **Differential Abilities** clusters (Reading, Mathematics, Written Language, Knowledge, Oral Language)
NEUROPSYCHOLOGICAL BATTERIES AND SCREENING TESTS			
HALSTEAD-REITAN NEUROPSYCHOLOGICAL TEST BATTERY			
Reitan & Wolfson, 1985 Jarvis & Barth, 1994 (Psychological Assessment Resources)	individually administered only by specially trained examiner; 8–12 hours (multiple sessions)	impairment indices; clinical cut-off criteria	battery comprised of 10 separate tests: **Category Test** (learning, problem solving), **Speech Sounds Perception** (verbal/auditory processing), **Seashore Rhythm Test** (sound perception), **Tactual Performance Test** (spatial analytical memory and problem solving), **Finger Oscillation** (motor speed/manipulation), **Trail Making A+B** (sequencing, problem solving), **Sensory-Perceptual Examination, Aphasia Screening**, WAIS-III, MMPI-2;* assessing location and impact of impairment
LURIA-NEBRASKA NEUROPSYCHOLOGICAL BATTERY			
Golden, 1981; Golden, Purish, & Hammeke, 1991 (Western Psychological Services)	individually administered only by specially trained examiner; 2 hours 30 minutes	T-scores, (mean: 60; SD: 10), clinical cut-off criteria	269 items grouped into 11 subsections of functioning: **Motor Functions, Rhythm, Tactile Function, Visual Function, Receptive Speech, Expressive Speech, Writing, Reading, Arithmetic Skills, Memory, Intellectual Processes** assessing localization and impact of pathognomy via 8 localization scales, a pathognomonic scale and right versus left hemisphere scales
REY-OSTERRIETH COMPLEX FIGURE TEST			
Meyers & Meyers, 1996 (Psychological Assessment Resources)	individually administered by well-trained examiner; 45 minutes (including 30 min. recall)	accuracy score to be compared to normative data	one 8x11 stimulus card to be copied, reproduced via Immediate Recall and via Delayed Recall (30 minutes); scored on a 36 point scoring system measuring **visual-spatial dimensions of functioning** (including **visual-spatial memory**)

TABLE 4-3   Overview of Samples of Abilities Tests (continued)

REFERENCE (PUBLISHER)	ADMINISTRATION AND TIME REQUIREMENTS	TYPES OF SCORES	DESCRIPTION OF ITEMS OR SUBSCALES
BENDER GESTALT VISUAL MOTOR TEST			
Bender, 1938; Koppitz, 1975 (Psychological Corporation)	individually administered by well-trained examiner; 5–10 minutes	standard scores, percentile ranks, clinical index scores	9 designs for reproduction with possibility of immediate and delayed recall and background interference administration; measuring visual **motor perception and speed**; screens for **brain pathology**; can score for **emotional indicators**
WECHSLER MEMORY SCALE, THIRD EDITION (WMS-III)			
Wechsler, 1997 (Psychological Corporation)	individually administered by a highly trained examiner; 50–60 minutes	weighted scores, index scores, percentile ranks	9 subscales clustered into an **Index of General Memory** and 6 domains: **Auditory/ Verbal** (Logical Memory, Paired Associates, Word List) Immediate & Recall; **Visual/Nonverbal** (Family Pictures, Memory for Faces, Visual Reproduction) Immediate & Recall; Working **Memory** (Mental Control, Spatial Span, Letter-Number Sequencing); **Delayed Recognition**
WISCONSIN CARD SORTING TEST, REVISED (WCST-R)			
Heaton, Chelune, Talley, Kay, & Curtiss, 1993 (Psychological Assessment Resources)	individually administered by a highly trained examiner; 30–60 minutes	standard scores, percentile ranks, clinical cut-off	128 response cards assessing **perseverative thinking, learning efficiency, abstract reasoning ability**, and **failure to maintain set**

*see Table 4-4 for a description of the MMPI-2

(see Gregory, 1996 and Kaplan & Saccuzzi, 1997, for further information).

Structured Personality Tests. As mentioned above, there are single-trait and multitrait structured personality tests. Single-trait tests are usually very easy to administer, have quick administration times, are straightforward with regard to interpretations, but also tend to be less reliable and less valid than more comprehensive multitrait instruments. Single-trait instruments are often used by providers not specifically trained in testing and may be the only instruments for which such use can be appropriate. In fact, in several chapters in this book, various such instruments are recommended to supplement information gathering about clients. Single-trait instruments can also provide objective data about changes in clients across time. Such data may satisfy criteria of managed care companies for outcome evaluation and assessment. Even the use of these single-trait instruments requires thorough familiarization with administration and interpretation procedures. Workshops or seminars about certain instruments can be extremely helpful in creating more skilled consumers. If clinicians decide to incorporate single-trait instruments in their counseling or therapy practice, they best seek some consultation until they have gained sufficient experience and facility with the test to use it independently.

Multitrait instruments are more complex, generally more difficult to administer, require longer administration times, and are complex with regard to drawing conclusions and making interpretations based on their test results. Many require special scoring instructions; some have to be sent to the publisher for scoring. Multitrait instruments can look deceptively simple to administer and interpret but should never be used by the nontrained clinician. Details about the most commonly used structured personality inventories are summarized in Table 4-4.

Unstructured Personality Assessment. This grouping of tests is the most complex with regard to administration, scoring, and interpretation. Unfortunately, because the tests often look deceptively simple to use, this is also one of the most abused groups of tests. For example, many untrained clinicians ask their clients to engage in projective drawings, a testing procedure that actually requires strict administration standards and complex guidelines for scoring and interpretation. This grouping of tests also has received the greatest criticisms with regard to their perceived lack of reliability and validity. Although these criticisms are appropriate for some instruments in this class of tests, they are overstated with regard to others. The key to the proper use and usefulness of these tests is education and experience of the examiner. A skilled examiner will derive much useful information from a projective test; an unskilled, poorly prepared examiner, on the other hand, will perpetuate the rampant abuse that has occurred with these assessment tools. Some debate has taken place over whether the instruments in this class of tests should be called tests at all, or whether they would be better referred to as clinical tools, thus eliminating or reducing the requirements for reliability and validity studies. Currently, this debate remains unresolved and individual examiners must decide where they stand on this issue (Groth-Marnat, 1997). However, for some of the commonly used projective tests (e.g., the Rorschach Inkblot Test), reliability and validity data do exist,

as do standardized testing, scoring, and interpretation instructions, making these instruments tests, not tools. Responsible examiners will be aware of the standardization requirements of these instruments and will follow them carefully. Facts about the most commonly used projective tests are provided in Table 4-5.

DOCUMENTATION AND RECORD KEEPING

The need for psychological testing is best documented in the client's chart as part of a progress note or as a recommendation in the intake report. The details of the actual discussion, as well as its outcome, with the client about the referral are best placed in a progress note. If a number of referral target names are given, all need to be documented. The name of the examiner finally chosen by the client is then recorded once that information becomes available. Most generally, the referral target will ask for some information from the referral source that is best rendered in writing, often in the form of the intake report (with appropriate signed ROIs). A note is made in the client's record about which documents were released to the examiner. If a special referral report is needed, a copy of this report would be mailed to the examiner; the original retained in the clinician's chart for that client. Some examiners have their own referral forms that need to be completed by a referred client's clinician. If that is the case, a copy of the completed form is best retained in the client's chart.

Once the assessment is complete, the clinician can expect to receive a written report from the examiner. This copy becomes a permanent part of the clinician's file for that client. Most generally, the clinician does not have the right or permission to release this report to other practitioners; if clients want the testing report released to other providers, they will in all likelihood have to direct this request (and release of information) to the examiner directly.

TABLE 4-4   Overview of Samples of Structured Personality Tests

REFERENCE (PUBLISHER)	ADMINISTRATION AND TIME REQUIREMENTS	TYPES OF SCORES	DESCRIPTION OF ITEMS OR SUBSCALES
MULTITRAIT PERSONALITY TESTS			
MINNESOTA MULTIPHASIC PERSONALITY INVENTORY, SECOND EDITION (MMPI-2)			
Butcher, Dahlstrom, Graham, Tellegen, & Kaemmer, 1991 (NCS Assessment)	easily individually administered by a well-trained examiner; 50–60 minutes	*T* scores, clinical cut-off ranges, critical items, profile	567 items grouped into 4 validity scales and 3 indices, 10 clinical scales (**Hypochondriasis, Depression, Hysteria, Psychopathic Deviate, Masculinity/Femininity, Paranoia, Psychasthenia, Schizophrenia, Mania, Social Introversion**), hundreds of supplemental scales, and 14 content scales (e.g., Anxiety, Fears, Anger, Type A, Family Problems)
MILLON CLINICAL MULTIAXIAL INVENTORY, THIRD EDITION (MCMI-III)			
Millon & Davis, 1996 (NCS Assessment)	individually administered by a well-trained examiner; 25 minutes	base-rate scores, *T*-scores, percentile ranks	175 items grouped into 14 personality scales coordinated with **DSM-IV Axis II** and 10 clinical syndromes coordinated with **DSM-IV Axis I** (Anxiety, Somatoform, Bipolar, Dysthymia, Alcohol/Drug Dependence, PTSD, Thought Disorder, Depression, Delusion); 3 modifying indices; and a validity index
SYMPTOM CHECKLIST 90, REVISED (SCL-90-R)			
Derogatis, 1992 (NCS Assessment)	easily individually administered by a well-trained examiner; 12–15 minutes	*T*-scores, percentile ranks, profile	90 items for 9 symptom dimensions (**Anxiety, Somatization, Obsessive-Compulsive, Depression, Interpersonal Sensitivity, Phobic Anxiety, Paranoid Ideation, Hostility, Psychoticism**) and 3 severity indices (**Global Severity Index, Positive Symptom Distress** and **Global Distress Index**)
CALIFORNIA PSYCHOLOGICAL INVENTORY, THIRD EDITION (CPI)			
Gough, 1996 (Consulting Psychologists Press)	easily individually administered by a trained examiner; 60 minutes	*T*-scores, percentile ranks, profiles grids	462 items grouped in 20 vectors (e.g., **Empathy, Capacity for Status, Sociability, Dominance, Social Presence, Self Acceptance, Tolerance, Independence, Responsibility, Socialization, Self Control, Good Impression, Intellectual Efficiency, Sense of Well-Being**)
PERSONALITY ASSESSMENT INVENTORY (PAI)			
Morey, 1991 (Psychological Assessment Resources)	individually administered by a well-trained examiner; 40–50 minutes	*T*-scores, profile, clinical cut-off criteria	344 items in 4 validity scales, 2 interpersonal scales (**Dominance, Warmth**), 5 treatment scales (**Aggression, Suicidal Ideation, Stress, Nonsupport, Reject Treatment**), 11 clinical scales (**Somatic Complaints, Anxiety, Anxiety Related, Depression, Mania, Paranoia, Schizophrenia, Borderline, Antisocial, Alcohol Problem, Drug Problem**)

Continued

TABLE 4-4   Overview of Samples of Structured Personality Tests (continued)

REFERENCE (PUBLISHER)	ADMINISTRATION AND TIME REQUIREMENTS	TYPES OF SCORES	DESCRIPTION OF ITEMS OR SUBSCALES
SINGLE TRAIT PERSONALITY TESTS—EMOTIONAL FUNCTIONING			
TRAUMA SYMPTOM INVENTORY (TSI)			
Briere, 1997 (Psychological Assessment Resources)	easily administered by any examiner; 20 minutes	*T*-scores, profile	10 clinical scales (100 items) grouped into 3 clusters called **Self, Dysphoria, Trauma**: Anxious Arousal, Depression, Dissociation, Anger/Irritability, Intrusive Experiences, Sexual Concerns, Dysfunctional Sexual Behavior, Impaired Self-Reference, Tension Reduction Behavior
BECK DEPRESSION INVENTORY, SECOND EDITION (BDI-II)			
Beck, Steer, & Brown, 1996 (Psychological Corporation)	easily administered by any examiner; 5 minutes	diagnostic ranges, critical items, clinical cut-off criteria	21 items measuring intensity of **depression**; factor-analytic subscales have been reported by some researchers for the BDI-I
STATE-TRAIT ANXIETY INVENTORY (STAI)			
Spielberger, 1995a (Psychological Assessment Resources)	easily administered by any examiner; 10 minutes	total scores to be compared to male and female norms	40 items measuring two types of **anxiety**: State Anxiety (situational anxiety) and Trait Anxiety (characteristics level of anxiety)
STATE-TRAIT ANGER EXPRESSION INVENTORY (STAXI)			
Spielberger, 1995b (Psychological Assessment Resources)	easily administered by any examiner; 15 minutes	*T*-scores, percentile ranks, profile	44 items measuring 5 dimensions of **anger**: State Anger, Trait Anger, Anger In, Anger Out, Anger Control
SINGLE TRAIT PERSONALITY TESTS—PERSONAL FUNCTIONING			
ADULT SELF PERCEPTION PROFILE			
Harter, 1986 (University of Denver)	easily administered by any examiner; 15–20 minutes	total score, subscale scores, clinical cut-off criteria	50 items measuring self-perception (self-esteem) in a **Global Self-Esteem** score and 11 **subscales** (Sociability, Job Competence, Nurturance, Athletic Competence, Physical Appearance, Ability as Provider, Morality, Household Management, Intelligence, Intimate Relationships, Sense of Humor)
COPING RESPONSE INVENTORY			
Moos, 1993 (Psychological Assessment Resources)	easily administered by any examiner; 10–15 minutes	*T*-scores, percentile ranks, profile	8 coping scales clustered into 2 factors: **Approach Coping Style** (Logical Analysis, Positive Reappraisal, Seeking Guidance, Problems Solving); and **Avoidant Coping Style** (Cognitive Avoidance, Acceptance, Seeking Alternative Rewards, Emotional Discharge)

TABLE 4-4   Overview of Samples of Structured Personality Tests (continued)

REFERENCE (PUBLISHER)	ADMINISTRATION AND TIME REQUIREMENTS	TYPES OF SCORES	DESCRIPTION OF ITEMS OR SUBSCALES
DEFENSE MECHANISM INVENTORY			
Gleser & Ihlevich, 1969 (Psychological Assessment Resources)	easily adminis- tered by any examiner; 15–25 minutes	total score for each defense (ipsative)	assesses 5 clusters of defenses via reaction to 10 stories: **Turning against Self, Projection, Turning against Object, Principalization, Reversal**
FIRO-COPE			
Schutz, 1962 (Consulting Psychologists Press)	easily adminis- tered by any examiner; 10–15 minutes	total score for each defense with higher scores indicating greater use	measures 5 clusters of defenses via 5 ques- tions to 6 scenarios: **Denial, Isolation, Projec- tion, Regression, Turning against Self**
SINGLE TRAIT PERSONALITY TESTS—INTERPERSONAL FUNCTIONING			
INTERPERSONAL STYLE INVENTORY			
Lorr & Youniss, 1985 (Western Psychological Services)	easily adminis- tered by any examiner; 30 minutes	standard scores, per- centile ranks; must be computer scored	300 items (150 in short form) and 15 sub- scales grouped into 5 broad band factors: **Interpersonal Involvement** (e.g., Nurtu- rance), **Self Control** (e.g., Orderly, Persistent), **Stability** (e.g., Stable, Approval-Seeking), **Socialization** (e.g., Trusting, Tolerant), **Au- tonomy** (e.g., Directive, Independent)
FIRO-B			
Ryan, 1977 (Consulting Psychologists Press)	easily adminis- tered by any examiner; 5–10 minutes	total scores, profiles	54 items providing a personality profiles for: **Control** (Expressed versus Wanted), **Affec- tion** (Expressed versus Wanted), **Inclusion** (Expressed versus Wanted)
SELF-REPORT FAMILY INSTRUMENT (BEAVERS)			
Beavers, Hampson, & Hulgus, 1985 (W.R. Beavers, M.D., 12532 Nuestra, Dallas, TX 75230)	easily adminis- tered by any examiner; 10–15 minutes	total score, subscale scores, clinical cut off criteria	36 items providing information in 5 dimen- sions of family functioning: **Family Conflict, Family Communication, Family Cohesion, Family Leadership,** and **Family Health** and for overall family functioning
INDEX OF FAMILY RELATIONS (IFR)			
Hudson, 1992 (WALMYR Pub Co, P.O. Box 24779, Tempe, AZ 82285)	easily adminis- tered by any examiner; 10 minutes	total score, two clini- cal cut-off criteria	25 items assessing overall **extent, severity, and magnitude of family problems**

Continued

TABLE 4-4 Overview of Samples of Structured Personality Tests (continued)

REFERENCE (PUBLISHER)	ADMINISTRATION AND TIME REQUIREMENTS	TYPES OF SCORES	DESCRIPTION OF ITEMS OR SUBSCALES
SINGLE TRAIT PERSONALITY TESTS—BEHAVIOR ASSESSMENT			
SUBSTANCE ABUSE SUBTLE SCREENING INVENTORY (SASSI)			
Miller, 1985 (Psychological Corporation)	easily administered by any examiner; 15 minutes	clinical cut-off scores, T-scores	provides 8 subscales assessing **substance use**: Obvious Attributes, Subtle Attributes, Denial, Defensive Abuser vs. Nonabuser, Alcohol vs. Drug Codependency, Risk Prediction Score (Alcohol and Drug)
MICHIGAN ALCOHOL SCREENING TEST (MAST)			
Selzer, 1971 (Melvin Selzer, M.D, 6967 Paseo Laredo, La Jolla, CA 92037)	easily administered by any examiner; 10 minutes	total score, clinical cut-off criteria	25 items that provide rapid, effective screening for **lifetime alcohol-related problems and/or alcoholism**
EATING DISORDERS INVENTORY, SECOND EDITION (EDI-2)			
Garner, 1996 (Psychological Assessment Resources)	easily administered by any examiner; 20 minutes	standard scores, percentile ranks, profile	91 items assessing **disordered eating** via 11 subscales: Drive for Thinness, Body Dissatisfaction, Ineffectiveness, Interpersonal Distrust, Bulimia, Perfectionism, Maturity Fears, Interoceptive Awareness, Impulse Regulation, Social Insecurity, Asceticism
DEROGATIS SEXUAL FUNCTIONING INVENTORY (DSFI)			
Derogatis, 1975 (Johns Hopkins University)	easily administered by any examiner; 30 minutes	total scores, clinical cut-off criteria, some standard scores	247 items assessing 7 domains of **sexual functioning**: Information, Experience, Drive, Attitudes, Affects, Fantasy, Gender Role Definition (some dated language revised by Neves, 1995)
SINGLE TRAIT PERSONALITY TESTS—RISK ASSESSMENT			
ADULT SUICIDAL IDEATION QUESTIONNAIRE			
Reynolds, 1991 (Psychological Assessment Resources)	easily administered by any examiner; 10 minutes	risk cut-off score, T-score, percentile rank	25 items providing one score of **suicide risk potential** and interpretations regarding need for further assessment
BECK SCALE FOR SUICIDE IDEATION			
Beck, 1991 (Psychological Corporation)	easily administered by any examiner; 5–10 minutes	clinical cut-off criteria, diagnostic ranges, critical items	21 items assessing **risk for suicide**; 5 item short-form for crisis situations
SUICIDE PROBABILITY SCALE			
Cull & Gill, 1989 (Western Psychological Services)	easily administered by any examiner; 5–10 minutes	T-scores, weighted scores, clinical cut-off	35 items assessing **suicide risk** via 4 subscales: Hopelessness, Suicidal Ideation, Negative Self-Evaluation, Hostility

TABLE 4-4   Overview of Samples of Structured Personality Tests (continued)

REFERENCE (PUBLISHER)	ADMINISTRATION AND TIME REQUIREMENTS	TYPES OF SCORES	DESCRIPTION OF ITEMS OR SUBSCALES
BECK HOPELESSNESS SCALE			
Beck, 1993 (Psychological Corporation)	easily individually by any examiner; 5 minutes	diagnostic ranges, clinical cut-off criteria	20 items measuring 3 aspects of **hopelessness**: Feelings About Future, Loss of Motivation, Expectation
CHILD ABUSE POTENTIAL INVENTORY (CAP)			
Milner, 1980 (Psytec Inc., P. O. Box 564, DeKalb, IL 60115)	easily administered by any examiner; 20 minutes	total scores, clinical cut-off criteria	77 items providing 6 validity/response distortion indices, 1 overall **Abuse Potential Score** and 6 clinical subscales: Distress Scale, Rigidity, Unhappiness, Problems with Child and Self, Problems with Family, Problems from Others

TABLE 4-5   Overview of Samples of Unstructured (Projective) Personality Tests

PUBLISHER	ADMINISTRATION AND TIME REQUIREMENTS	REFERENCES FOR SCORING SYSTEMS	DESCRIPTION AND/OR AVAILABLE INSTRUMENTS AND VERSIONS
RORSCHACH INKBLOT TEST			
Western Psychological Services	individually administered by a specially trained examiner; 50–60 minutes	Beck, 1950a, 1950b, 1957; Klopfer et al., 1954, 1956; Exner, 1991, 1993	10 inkblots are presented with the directions "What might this be?" to assess a variety of personality dynamics, traits, perceptions, pathology indicators, emotional indicators, cognitive processing, problem-solving style, etc.; multiple scoring and interpretive systems have been developed
PROJECTIVE DRAWINGS			
Charles C. Thomas	individually administered by a specially trained examiner; 5–15 minutes	Goodenough, 1926; Koppitz, 1968; Handler, 1996	many projective drawing tasks exist, ranging from Draw-A-Person, Draw-A-House-Tree-Person, Draw Your Family, and Draw Your Family Doing Something Together to Draw-A-House-Tree-Person with Action Going On; many scoring and interpretive systems exist for assessment of a wide range of personality traits
THEMATIC APPERCEPTION TEST			
Western Psychological Services	individually administered by a specially trained examiner; 30–50 minutes	Murray, 1943; Bellak, 1993	a subset of 31 stimulus cards is presented with the directions to make up a story that includes beginning, middle, end, action, thoughts, and feelings; multiple interpretive systems exist to assess a range of personality traits and characteristics

Continued

TABLE 4-5   Overview of Samples of Unstructured (Projective) Personality Tests (continued)

SENTENCE COMPLETION TESTS

Psychological Corporation	individually administered by a specially trained examiner; 10–20 minutes	Rotter & Rafferty, 1950; Forer, 1950; Holsopple & Miale, 1954	many sentence completion blank versions exist by age, topic area, etc.; multiple scoring systems have been developed to assess a wide range of personality variables; in almost all cases the incomplete sentences are presented with directions to finish each started sentence as fast as possible

SUMMARY AND CONCLUDING THOUGHTS

Psychological testing is only as good as the examiner, and hence skillful selection of the referral target is critical. The entire referral process hinges on a good choice of examiner and is never taken lightly. To be good consumers, counselors and therapists need to explore the background of their referral targets, speaking to them directly and gaining an appreciation of their educational backgrounds, experience, and interpersonal skills. The referral to the examiner will often be a test of the client–clinician relationship; a referral to a poor provider can introduce a component of distrust into this relationship that may continue to interfere with rapport building for some time.

Psychological testing is a complex process that requires specific skills and education. A clinician not trained in psychological testing must never attempt to test a client. Further, psychological testing requires a different relationship between provider and client, introducing the possibility of a confusing, if not dual, relationship between a clinician and a client who attempt to do both therapy and testing together. Of course, these cautions are not exclusive to referrals for psychological testing. Knowing the referral target and making proper referrals when they are indicated by a client's needs is not exclusive to psychological testing referrals; such procedures are crucial to successful case management and treatment of all clients requiring any type of referral.

PART II

Challenges Arising During Treatment

5

The Challenge of Crises:
Assessment and Intervention

The Chinese term for crisis (weiji) is composed of two characters, which signify danger and opportunity occurring at the same time.

KARL SLAIKEU, *CRISIS INTERVENTION*

The Chinese character for crisis (meaning both danger and opportunity) correctly suggests that a crisis, however difficult and painful at the moment of presentation, may open new doors for the client. A crisis can be defined as any event that outstrips a client's resources or coping skills, thereby creating emotional upheaval, cognitive distortion, and behavioral difficulties. Although this definition focuses on a triggering event, "the defining characteristic of a crisis is not the instigating event, but rather the client's inability to cope with a situation" (Zaro, Barach, Nadelman, & Dreiblatt, 1977, p. 126). In other words, the same event that triggers a crisis for one individual may have no effect on another individual. Similarly, a person may experience the same event at different times in life, at one point experiencing a crisis and on the other occasion coping adequately. Thus, a crisis truly is a combination of event, context, and person.

TYPES OF CRISES AND TRIGGERING EVENTS

The types of events that may lead to crises have been categorized in various ways. Most simply, they have been labeled as situational versus developmental crises and have been hypothesized to be upsetting to the client because they are perceived to be blocking a major life goal (Patterson & Welfel, 1994; Slaikeu, 1990). A situational crisis is defined as a crisis that occurs suddenly in response to an unexpected and unpredictable event (Gilliland & James, 1993). A developmental crisis is defined as a crisis that occurs as part of growing up and getting older (Gilliland & James, 1993; Slaikeu, 1990). These categories have been broken down further by others who postulate at least three types of situational and three types of developmental crises. For example, Okun (1997) indicates that situational crises can be dispositional in

133

FIGURE 5-1 Events That May Trigger a Crisis

	Situational Crisis	**Developmental Crisis**
Psychological	public embarrassment	identity confusion
Physical	natural disaster (e.g., earthquake)	illness connected with aging
Social	divorce, death of a friend	adolescent leaving home
Moral	role model convicted of a crime	conflict with peer group behavior

nature (where a crisis is induced because a client does not know how to deal with a particular event, such as deciding on a type of medical treatment for a newly diagnosed illness), induced by traumatic stress (wherein the client was suddenly exposed to a traumatic event, such as a rape, an accident, or a war), or psychiatrically induced (wherein the client has a psychiatric emergency, such as a psychotic break or suicidal ideation). Similarly, developmental crises can be due to life transitions (e.g., getting married, retiring, becoming pregnant), due to psychopathology that developed over the years (such as psychological problems that complicate an event to crisis proportions [e.g., sudden recognition of one's sexual preferences, gradual realization that abuse occurred in childhood]), or they can be maturational in nature (e.g., age-inappropriate developmental needs leading to relationship problems, recognition of lack of life meaning, emptiness or void in one's life). The psychopathological crisis hypothesized by Okun (1997) has also been labeled an intrapsychic crisis; the maturational crisis, an existential crisis (e.g., by Gilliland & James, 1993). These six categories can account for most of the common triggering events which include death of a loved one, health concerns, reproductive issues, relationship disruptions, experience of violence, disruption of work or school, natural disasters, or financial emergencies (Patterson & Welfel, 1994; Slaikeu, 1990).

Another way to categorize triggering events is by their impact on the client's physical, psy-

chological, social, or spiritual well-being (Gilliland & James, 1993). Any perceived transgressions, threats, or losses in these four areas are believed to be capable of setting off a crisis in an individual. A client's physical well-being is threatened when the client is concerned about his or her ability to maintain personal safety with regard to such matters as food and water, shelter, and employment. Psychological threats are events that challenge the client's self-concept, emotional well-being, or identity. A social event that may trigger crises revolves around a perceived threat to the client's family, friends, or coworkers. A client's spiritual or moral realm can be threatened by personal values, beliefs, or integrity. This system of categorization can easily be superimposed upon the situational and developmental dichotomy to result in a grid of sample events (see Figure 5-1).

In addition to categorizing potential crisis events by their content, they can also be categorized by their level of acuteness. This distinction is particularly important because it has implications for treatment or intervention. Namely, an acute (or immediate crisis) needs to be attended to through immediate crisis management, minimally consisting of catharsis, assessment, action plan, and follow-up. It is this type of crisis that will be the focus of this chapter. A nonacute (or chronic) crisis is a crisis that can be dealt with through regular psychotherapy—meaning that assessment can take some time and intervention will be less directive and more client centered. Most commonly, situational crises are acute; developmental crises are chronic. However, this is not always true, and it is important for clinicians to get some sense for the level of acuteness of a crisis regardless of how they may classify it. For example, a diagnosis of Alzheimer's disease, in essence, can be a developmental crisis that affects the person's physical realm because it is connected to a maturational issue. However, the initial diagnosis may act like a traumatic stressor, thus warranting immediate crisis intervention. This type of combined developmental (chronic) and situational (acute) stressor or crisis essentially needs double intervention: first, the counselor or therapist responds with crisis intervention to deal with the acute situation and its consequences (i.e., with the diagnosis and the effects it has on the person's physical and psychological well-being); then, the clinician must attend to the long-term issue through therapeutic intervention (i.e., with the adjustment to a degenerative disease and to aging). Crises can also be evaluated with regard to whether they are easily resolved or difficult to manage, whether they are comprised of a singular event or multiple events, and whether they are genuine or manipulative (i.e., true crises or means to get attention or receive reassurance).

EFFECTS OF CRISES ON INDIVIDUAL FUNCTIONING

Crises can have a number of effects on the client and these effects vary idiosyncratically. Generally, any or all of the three major areas of functioning (affect, behavior, cognition) of a human being are affected; in other words, there is an emotional response, a behavioral response, and a cognitive response. A multitude of emotional reactions is possible, with a range of severity (Gilliland & James, 1993; Slaikeu, 1990; Zaro, Barach, Nadelman, & Dreiblatt, 1977). These affects include the following:

- anxiety (ranging from mild symptoms to disorganizing, disabling panic)
- phobia (ranging from mild fear to disabling phobic reaction)

- depression (ranging from mild symptoms such as withdrawal to severe impairment)
- anger (ranging from mild frustration to aggressive hostility and belligerence)
- helplessness (ranging from mild need for assistance to incapacitation)
- guilt or shame (ranging from mild self-deprecation to severe self-punishment)

The severity of these reactions, as already stated, can range from mild to severe, with specific manifestations being highly idiosyncratic. However, mild affective impairment is defined as arousal that is appropriate to the triggering event or crisis situation. Low impairment implies that the experience of the mood state is more severe and/or prolonged than warranted by the stressful event. Moderate impairment refers to a level of affective arousal that is incongruent with the stressful event in that it is more intense and prolonged than would be expected and may be accompanied by lability of affect. Marked impairment is defined as obviously incongruent affect that has excessive intensity, is accompanied by mood swings, and is perceived by the client as out of control. Severe impairment implies affect that is so intense and prolonged that it results in psychological decompensation and depersonalization (Gilliland & James, 1993).

Cognitive effects of a crisis generally imply that the client does not process information as efficiently and accurately as at other times in life. Common cognitive distortions or problems include the following (Gilliland & James, 1993; Slaikeu, 1990; Zaro, Barach, Nadelman, & Dreiblatt, 1977):

- confusion
- irrational thinking
- distorted belief system
- poor logic
- lack of concentration
- poor or biased attention
- obsessions

The level of severity of cognitive distortions can again be placed on a continuum ranging from mild to severe (Gilliland & James, 1993). Mild cognitive distortion refers to understandable preoccupation with the event, along with adequate interpretation and perception of reality, as well as otherwise adequate problem-solving skills. Low impairment implies diminished control over thinking, with thoughts about the crisis preoccupying the client at the expense of concentration and problem solving, along with mild distortion of reality as it relates to the crisis event. Moderate impairment suggests significant problems with concentration and problem solving due to significant preoccupation with the crisis event that appears to be clearly misperceived and misinterpreted. Marked impairment implies intrusive thoughts about the event that clearly interfere with concentration and problem solving, thoughts that are clearly out of context and out of proportion with the actual event. Severe impairment is defined by total preoccupation with the crisis event at the expense of any productive thought production, implying confusion, absence of problem-solving skills, self-doubt, and severe misunderstanding or misrepresentation of reality as relevant to the crisis event.

Behavioral responses can be equally varied, though all imply a degree of psychological paralysis or poor goal-directedness. The specific nature of this behavioral helplessness can take three forms: approach, avoidance, or immobility (Gilliland & James, 1993). In an approach reaction, the client appears highly motivated and active. Upon closer inspection, however, it becomes clear that the behavior is nonproductive or counterproductive due to its random nature and lack of a clear goal or target. In an avoidance reaction, the client attempts to flee the crisis situation through extreme psychological avoidance or actual physical removal from the situation. Although this may appear to be an appropriate response, it becomes pathological if it is not directed toward a goal and if it does not contribute to the solution of the problem at hand. Finally, if clients are at the point

of total immobilization, they become psychologically paralyzed or trapped. This paralysis does not necessarily mean complete inaction (though it may), but rather is defined by the complete lack of forward movement in terms of resolving the crisis. As Gilliland and James (1993) point out, immobility may be exemplified by the action of a battered wife who keeps returning to the abuser despite attempts by herself and others to remove her from the abusive situation.

Like the emotional and cognitive response, the behavioral response can be graded on a continuum of severity or appropriateness (see Gilliland & James, 1993). Mild behavioral impairment implies occasional ineffective coping, with basic day-to-day living skills intact. Low impairment is defined by behavior that becomes increasingly nonproductive, with some impairment in coping and some neglect of day-to-day tasks. Moderate impairment refers to maladaptive (not just ineffective) coping and compromised ability to engage in routine daily living skills. Marked impairment is defined by behaviors that are attempts at coping but that actually worsen the crisis situation along with inability to perform daily living tasks. Severe impairment results in behavior that is erratic and unpredictable and that is likely to cause harm to self and others.

Clearly, although the three human functioning areas were dealt with separately, the overall state of crisis is best judged by considering how all three areas of functioning interact. A client with severe impairment in affect (e.g., severe depression), who also experiences severe behavioral difficulty (e.g., is stuck in a nonproductive cycle of returning to an abusive spouse) along with cognitive distortions that suggest strong self-doubt, poor problem solving and misjudgment of the crisis situation, is much more impaired than a client who has severe impairment in only one or two of the three realms of functioning. This issue of assessing the client's level of functioning by looking at the emotional, behavioral, and cognitive repercussions of the crisis will later be revisited in the assessment section of this chapter.

A MODEL FOR CRISIS INTERVENTION

To help a client deal with an acute (or situational) crisis, a series of steps is necessary. Most generally, the mental health care provider and client will progress through the following procedures:

Relationship-Building Procedures and Communication of Safety

- catharsis (client vents affective arousal)
- sharing (client outlines a personal perception of crisis)
- support (clinician attempts to anchor client)

Assessment Procedures

- clinician gleans insight into the triggering event
- clinician assesses client's state of current functioning
- clinician evaluates client's resources and coping skills
- clinician assesses client's danger to self or others

Action Plan Procedures

- plan development (client and clinician together develop coping strategies, resources, and social supports)
- implementation (client puts action plan in place and clinician monitors client's success)

Follow-Up Procedures

- client and clinician reevaluate implementation of action plan
- client and clinician plan future treatment

Although these procedures are listed as if they represent distinct stages, several of them are actually carried out simultaneously or repeatedly, depending upon the client's needs. For example, although relationship building is usually the process that precedes all other interventions, it does not really ever end. Throughout the intervention, the counselor continues to give the client opportunity to vent and share, remaining supportive while other procedures are initiated. Similarly, assessment is somewhat flexible in that it may begin during the relationship-building stage and continue in the action-plan phase. Specifically, as the client shares and vents, the therapist can already begin to gather data to assess various aspects of the crisis and consequences for the client. Similarly, even once the action plan process has begun, the therapist continually reevaluates the client's responsiveness to the planning process, adapting the action plan to changes in the client's frame of mind, emotional state, or behavioral capabilities. The only stage that is distinct is the follow-up stage, which, by definition, succeeds the implementation of the crisis plan. Nevertheless, even this step of the intervention process will include relationship building and continued assessment. Sometimes, if the action plan was less than optimal, follow-up may even include new planning for implementation of alternative strategies.

In other words, although each of the procedures will be covered separately and distinctly, well-prepared clinicians will be open to all four processes at any time during the entire intervention. Clinicians will flexibly revert to relationship building whenever necessary, incorporate more assessment whenever new information begins to emerge, and they will plan as early as appears appropriate to the circumstances. Flexibility and quick thinking are truly the most important characteristics of successful crisis intervention. "The challenge of crisis intervention lies, not only in working efficiently and effectively with the client, but also in being flexible enough to mobilize a full range of suprasystem resources (family and community) in working toward crisis resolution for the client" (Slaikeu, 1990, p. 9).

Relationship Building

The goals of the relationship-building portion of the crisis intervention are to help the client feel understood, to reduce the client's emotional arousal through venting, and to reestablish some level of cognitive control on the part of the client (Hoff, 1989; Slaikeu, 1990). These goals can be attained through catharsis and sharing and through the communication of support, caring, respect, and safety. Relationship building is an important component of the crisis intervention process because it ensures that client and counselor establish a bond and make psychological contact. Only if clients feel some attachment to their clinicians (however, tenuous and temporary), will they be responsive to the clinician's interventions. Thus, the relationship-building phase cannot be undervalued. This is also the reason why relationship building and psychological contact remain important throughout the entire intervention.

Catharsis and Sharing. The emotional venting of difficult affects, or catharsis, is often quite successful in diffusing emotionally tense situations and in reducing a person's level of affective arousal. Such reduction in affective arousal in turn tends to restore a sense of affective control to the client that will be perceived as a great relief. Sharing is the equivalent process, involving cognition (and to some extent behavior) rather than emotion. Through sharing of content about the event and the person's reaction to it, the client can diffuse behavioral agitation and can often revise distorted thinking. Merely retelling one's story often leads to emotional improvement, cognitive insight, and improved behavioral control. However, because they seldom occur entirely spontaneously, catharsis and sharing need to be facilitated by the therapist. This is done through the creation of an environment that provides emotional support, cognitive direction, and behavioral control. Emotional support is achieved through various empathic and listening strategies; cognitive direction is achieved through structured questioning that guides the client through a maze of information that needs to be shared; and behavioral control is achieved through implementing a number of strategies that help the client feel safe and protected. All of these strategies are elaborated upon next.

Providing Support and Making Psychological Contact. To help the client feel emotionally supported, cognitively directed, and behaviorally safe, the counselor must begin by communicating comfort, support, understanding, and respect (Hoff, 1989; Okun, 1997). The clinician must "maintain calm confidence and hopeful expectation" (Patterson & Welfel, 1994, p. 215) lest the client will spin further out of control. Calmness can also be communicated and passed along to the client through cautious physical touch (Slaikeu, 1990) that helps the client reestablish some sense of physical boundary and safety. Attentiveness in listening, empathy, and understanding are also critical skills, as is the communication of acceptance. The latter can often be signaled if the clinician takes care not to minimize the client's concern or the situation that triggered the reaction, even if the client is perceived to be overreacting to the event (Patterson & Welfel, 1994).

Calmness on the part of the therapist is important because it coveys reassurance about the possibility of solving the problem and helping the client feel better and because it projects confidence on the part of the counselor about being able to handle the crisis (Okun, 1997). Reassurance is also achieved by emphasizing the strengths of the client without minimizing the severity of the event (Meier & Davis, 1993). The environment created by the mental health care provider must be conducive to psychological safety as well and hence needs to be designed to provide comfort, privacy, and minimal time pressures (Patterson & Welfel, 1994). Additionally, clinicians must be concrete and explicit in their approach to the client, for they cannot count on higher-level cognitive processing by the client at this time of cognitive confusion and potential impairment. The therapist must use leading strategies and must provide as much structure as possible (Patterson & Welfel, 1994), taking charge without being controlling (Meier & Davis, 1993). Structure (and hence safety) is also achieved through directiveness, goal-directedness, and session focus (Bellak & Siegel, 1983).

All of these direct supportive techniques can be applied in a crisis setting without the usual concerns about fostering dependency (Cormier & Cormier, 1999). Nevertheless, all necessary direct action is still best implemented in such a way as not to be perceived by the client as controlling (Okun, 1997). The level of directiveness also needs to be adapted according to the idiosyncratic needs of the client. A client who is in crisis but remains relatively mobile and composed will need less direction than a client in crisis who experiences moderate to severe impairment in all realms of functioning. Thus, the skilled clinician does not misperceive engaging in crisis intervention as simply an opportunity to give advice.

Communicating Safety. To be able to communicate safety, counselors must feel confident that they are capable of dealing with difficult situations, not only in terms of helping clients decide upon an action plan for the future, but also in terms of keeping clients safe while they are with them. Clients in crisis states are often aware that they are spinning out of control emotionally, are cognitively irrational, and are behaviorally ineffective. This recognition often only serves to make their level of impairment worse by adding anxiety about level of functioning to the already stressful situation that triggered the crisis. Clinicians working with clients in crisis (and in any therapy setting for that matter, since a crisis can occur at any time) must have some skills to diffuse potentially explosive or painful in-session events. For this reason, they must feel confident about their ability to reduce a client's anxiety, to control a panic attack, to master a client with aggressive impulses that are being expressed with the clinician, to deal with psychotic behavior, and to intervene if a client depersonalizes or dissociates (Lipkin & Cohen, 1998).

The very nature of the intervention model presented here suggests that there will be times when clients will have intense feelings during their sessions. Intense experience of affect is not in and of itself a dangerous thing nor is it an event during which the clinician must intervene. Most of the time, even the most intense affects are successfully resolved through the usual therapeutic tools, processes, and catalysts that have already been discussed in detail (e.g., catharsis, corrective emotional experience, holding environment). However, occasionally, it is necessary for a therapist to intervene with a particularly intense affect. For example, if clients experience intense emotion over which they appear to be losing control toward the end of a session, a counselor may have to intervene to help clients get ready to leave. Another example that requires the intervention of the clinician is the client whose affective experience is so severe that it could result in a dangerous situation either for the client or the clinician (e.g., hyperventilation during a panic attack; potential aggression acted out during intense experience of anger). In such situations, the clinician needs to know how to restore a sense of safety to the counseling session. One important aspect of providing a holding environment is the clinician's capacity to contain clients' affects if they threaten to become overwhelming. The following helpful hints will assist the mental health care provider to develop the skills necessary to deal with clients' intense affects when they need to be managed.

Dealing with Loss of Control over Emotions. Even the most out-of-control affect cannot last forever; in fact, it usually cannot even sustain itself for a whole session. Consequently, depending on the amount of time left in a session, the first approach to apparent loss of emotional control is to let it play out as long as it is not destructive to the client. The catharsis that comes from allowing emotion to flow freely for a while is often quite helpful for the client as long as it is not accompanied by gravely uncomfortable physical feelings or dangerous behaviors. For example, out-of-control anger may need to be reined in if the client appears to be on the verge of violent action; likewise, a panic attack is always controlled to prevent the client from hyperventilating or experiencing other severe physical consequences. If the emotion is one for which catharsis is appropriate (e.g., anxiety not of panic proportions, depression or sadness, frustration), this is the first choice. The intensity displayed by the client will decrease on its own once the affect has been given free range for a while.

If the length of time elapsed becomes too long (either because the session is almost over or because either client or therapist becomes too uncomfortable), the clinician can calmly begin to take steps to deal with the affect. The clinician starts by asking the client to begin to focus on the counselor (e.g., "I need you to look at me," or "to look up," "to catch my eyes," or "to look at my hand"). Next, any behavior connected to the affect is talked about and stopped (e.g., "I want you to stop pacing now; please sit back down." Or "I need for you to stop bouncing your leg and picking at your hair." Or "Please stop wringing your hands now."). In so doing, clinicians lower their voice significantly and speak much more slowly than usual, but with emphasis. Once clients have stopped the associated behavior and have begun to focus on the clinician, the clinician can ask clients to pattern their breathing after the therapist's (e.g., "Let's get your breathing back to normal. Follow my lead. Slowly breathe in [clinician takes a long, calm breath] and out [clinician releases the breath forcefully]; in . . . and out . . ."). This joint rhythm of breathing is maintained until the client becomes notably calmer. All along, the counselor remains keenly aware of his or her own personal body language, which is most effective if it expresses confidence, calmness, and collectedness. This is one time when the mental health care provider definitely does not want to mirror the client's body language. All demeanors on the clinician's part should exude relaxation and calmness. For clients who lose control over their affects on a regular basis, it may be useful to develop a structured and predictable sequence of interventions for dealing with this situation. Not only will this help clients begin to be aware of how to regain control over affect with the therapist's help in session, but it will also help teach them to use the same, regularly rehearsed strategies outside of the session.

Dealing with Uncontrollable Crying. If uncontrollable affect includes uncontrollable crying and this crying needs to be stopped (again, if feasible, catharsis is the first choice because it may be sufficient to stop the crying in and of itself), a few additional strategies are available. First, handing clients a box of tissues is usually a nonverbal signal that it is time for them to pull themselves together. Once clients have been given this nonverbal message, the clinician asks them to look directly into his or her eyes and continues to talk calmly (e.g., "I am going to help you stop crying now so that we have enough time to talk about what happened before you have to leave today. I need you to look into my eyes—look directly into my eyes."). Once clients are able to maintain firm eye contact, the clinician can initiate the same breathing exercise mentioned previously. The reason for the insistence on firm eye contact is simple: it is physiologically difficult (if not impossible) to cry and focus one's vision at the same time. Thus, if clients are asked to focus on the clinician's eyes, crying will automatically stop in most cases. Occasional clients will respond to increased eye contact with more intense experience of affect. If a clinician encounters such a client, eye contact may not be the best method to stop the client from crying. However, in most cases eye contact is incompatible with crying. Clients can be taught this trick for home use as well—they can simply look at and focus on their own eyes in a mirror and get crying under some control. If the breathing exercise can also be added to such independent practice, calming will almost always be assured.

Dealing with Strong Anxiety Reactions and Panic. A strong anxiety reaction or panic attack requires intervention much sooner than the type of emotion referred to above because it often is self-perpetuating and can be physically dangerous. Highly anxious or panicked clients have physiological responses that further frighten them and often serve to increase the panic. The actual intervention is not very different from the previously outlined procedure. However, in these cases it is even more important for the clinician to appear calm, in control, and capable of setting firm limits (e.g., "I have to stop you from . . . *now*." Or "You need to stop talking about . . . *now*." Or "We need to move on *now* to think about how to get you ready to leave today."). The practitioner uses voice and body language to underline the command nature of the directives, providing verbal and physical structure (e.g., handing the client a tissue box; taking away the pillow the client may be beating; stopping the client from twirling hair, perhaps by physically moving and then holding on to the client's hand). The clinician's voice must be firm but caring, as well as calm and controlled. Once the structure has been set, the clinician asks the client to find a focal point and to place total attention on it. When the client has established eye contact with the focal point (e.g., a picture in the therapy room), the client is asked to describe it (e.g., "I need you to look at . . . right there across from you. Okay, now tell me what you see. Describe it in detail."). This simple task serves to shift the client's focus of attention away from the distressing thoughts that fuel the anxiety or panic. All the while, the clinician also needs to pay attention to the client's breathing. If the client is hyperventilating, intervention with regard to breathing is imminently important. This can be accomplished through the previously outlined breathing exercise, in which the client is asked to model breathing frequency and intensity after the therapist's example. In extreme cases, the counselor may need to ask the client to breathe into a paper bag or through a straw (or to engage in any similar strategy that prevents overbreathing). As was explained with regard to intervening with other out-of-control affects, it is important to slow all interactions with the client to a calm level. This is accomplished through lowering and slowing one's voice. All along, the clinician gives reassurance about the

safety of the room and setting, doing so with calming firmness. If the client is still panicking, it is often helpful to ask the person to pick up a pillow and hug it tightly to the body. This action helps reestablish some body boundaries and may help with any beginning symptoms of depersonalization.

Any and all of these strategies can be combined in any order, depending upon what the clinician believes is most likely to work. If the client is hyperventilating, the first step of intervention would be the breathing exercise; if the client appears to have a sense of loss of boundaries and self, the pillow intervention is best (if no pillow is available in the room, clients can hug a purse or bag; or as a last resort, they can hug themselves firmly around the torso). If the client appears extremely cognitively preoccupied, the focal point exercise is best. Combining strategies can be easily accomplished as well, since the client can hug a pillow, do slow breathing, and focus on a specified object (including the clinician's eyes) all at the same time, while the clinician calmly talks about the safety of the room and the setting. Certainly, whenever anxiety is high, the client may need special help with transitions of any sort, including beginning and ending sessions or changing topics within sessions. Specific instructions, more time, clear directions, and similar structuring events may assist clients in moving more successfully through the treatment process even while highly anxious or agitated.

Dealing with Thought Racing and Pressured Speech. If the primary problem appears to be that the client's thoughts are racing and resulting in pressured speech, the mental health care provider again intervenes with calming strategies that can include the breathing exercise previously outlined. The clinician begins by asking the client to stop talking altogether and to begin breathing and relaxing. The client is asked to model breathing according to the therapist's example and to talk during this exercise. The simple breathing can be kept up for some time if the clinician believes that it is sufficient in slowing down the client's thought process. If the client's thoughts appear to continue to race (as perhaps suggested by a difficulty in slowing down the breathing or fidgety psychomotor behavior), the focal point exercise previously described can be added. It will serve to distract the client's focus of attention away from the racing thoughts. In this case, the client may need some assistance, because he or she may be more preoccupied with thoughts than is the simply anxious client. If so, the counselor may need to do some modeling of description (e.g., "Okay—what do you see? I see a brown picture frame—do you see it? Okay. What shape is it? . . . Yes, that's how I see it: it's almost square. . . . What else?"). Overall the clinician focuses on helping the client stop thinking about the obsessive thoughts by slowing cognitive processing and by modeling calmness and relaxation through calm breathing, controlled body language, low and slow voice, and firm directives about what the client needs to do next. The focal attention exercise is generally very successful with these clients.

Dealing with Depersonalization and Dissociation. During an episode of depersonalization, a client loses a clear sense of self and feels detached from the self. Sometimes this is manifested by perceptions of being out of the body, being an automaton, or otherwise being a spectator of the self as opposed to being in control of the self. Depersonalization reactions are often closely tied to anxiety and a sense of lost personal boundaries. Not surprisingly, the same techniques that have been outlined so far (for emotional control, for anxiety and panic, and for racing thoughts and pressured speech) all are useful to some extent. The breathing exercise may be the most important and effective exercise because depersonalization can be due to overbreathing (Bellak & Siegel, 1983) or underbreathing, where clients breathe so minimally that they fall into a trance state (L. Olson-Webber, personal communication, August 14, 1997). In severe cases of overbreathing, asking the client to breathe into a paper bag or through

a straw may be helpful. The pillow hugging exercise is very useful in that it provides the client with some definite body boundaries that can be experienced concretely. If the hugging is not enough in and of itself, the clinician can ask the client to explain what it feels like to hug the pillow, focusing on establishing a clear perception of body boundaries (e.g., "Tell me how the pillow feels. Is it soft? Hard?" And "On which parts of your body can you feel the pillow? Your arms? Where else?" And "Tell me how your stomach feels with the pillow pressing on it?" Or "Squeeze the pillow a little tighter. What changes can you feel? Where can you feel the additional pressure in your body? In your arm muscles? On your stomach?"). Using the client's name can be helpful in getting the client's attention and beginning all verbalizations by repeating the client's name is good routine practice.

For clients prone to dissociation (i.e., clients traumatized as children or adults by events such as chronic and inconsistent abuse, combat, natural disasters, and so forth [Pope & Brown, 1996]), the therapist must learn to pace sessions carefully and must prepare signals for the client to become alertly oriented to the present and to return to the safety of an unaltered state of consciousness. Dissociation-prone clients need help in learning to recognize the feelings and thoughts they have that might signal the onset of a dissociative experience so that they may learn to prevent the episode from occurring both in and out of sessions (Sanderson, 1996). Careful pacing of clinical material is critical to avoid retrieval of difficult memories that are too quick and painful that they trigger a dissociative episode. As explained by Gil (1988), it is most helpful to determine when dissociation occurs (i.e., in what setting or under what circumstances), its precipitants (i.e., the specific events that lead to the flight response), and emotions associated with it. Further, the client is helped to understand dissociation as an adaptive strategy developed for purposes of psychic or emotional survival. Once clients have gained this understanding, they can be taught alternative strategies of coping or defense for circumstances that usually would trigger a dissociative response (e.g., relaxation exercises, activities for purposes of distraction, conversations with supportive others, and so forth). If dissociation occurs during a session, the clinician needs to be prepared to assist the client in regaining a normal alert state of consciousness. This is best achieved by first developing a bridge between the dissociative, or trauma-related, state of consciousness and the present, or nondissociative, state of consciousness (Dolan, 1991). One such bridge is called symbol for the present, wherein the client is asked to identify an item in the client's possession that can be used as a reminder of the here-and-now. Should a dissociative event threaten or occur, the symbol of the present can be used by client and clinician to bring the client back to a normal alert state (Dolan, 1996). A similar bridge, also recommended by Dolan (1996), is called the first session formula task in which the client is asked to make a list of events or activities that are currently ongoing in the client's life to which the person has a strong positive commitment. The list is used during dissociative periods to remind the client of current resources that were not available to the client when he or she developed the dissociative defense. A third bridging or grounding technique developed by Dolan (1991, 1996) is called the older, wiser self. This technique, which can be invoked to prevent a dissociative episode, involves seeking advice from an older, wiser version of the client's self during stressful periods. This older, wiser self is described to the client as follows (Dolan, 1996, p. 406):

> Imagine that you have grown to be a healthy, wise, nurturing, old woman (or man) and you are looking back in this time in your life in which you were integrating, processing, and overcoming the effects of the past experience of sexual abuse [or other traumatic event]. What do you think this wonderful, old, nurturing, wiser you would suggest to you to help you get through this

current phase of your life? What would she/he tell you to remember? What would she/he suggest that would be most helpful in helping you heal from the past? What would she/he say to comfort you? And does she/he have any advice about how therapy could be most helpful and useful?

All of these centering or grounding techniques can be used to return the client to the here-and-now. More explicit techniques that involve imagery have also been described (e.g., Sanderson, 1996). These grounding techniques encourage a client to imagine himself or herself as a tree with a strong root system that is anchored in a safe setting and is indestructible even by the most powerful forces. Such visualization exercises can help the client regain equilibrium after a dissociative episode and can also be useful in preventing dissociation during stressful periods (Sanderson, 1996).

Dealing with Psychotic Breaks, Hallucinations, and Delusions. The key to successful intervention with a client who is experiencing a psychotic break is to reestablish psychological contact with the person. Counselors should use the client's name frequently and calmly to get the client's attention. They also should respond with calmness to the client's hallucinations or delusions to make psychological contact and avoid getting caught up in the psychotic content of the delusions or hallucinations the client is describing. It is best to respond to delusions and hallucinations by listening and expressing caring and concern without encouraging or validating the psychotic content. In other words, delusions are neither agreed with nor argued with; instead, the clinician focuses on giving the client a sense of being heard and understood. It is very likely that the client is used to becoming the target of ridicule, challenge, and harassment when voicing delusional thinking or while talking about specific hallucinations, and it is important for the clinician to provide an alternative experience. An understanding attitude by the clinician is very useful in these instances and greatly facilities psychological contact.

The next goal is to reestablish the client's contact with reality. This is accomplished, not by challenging the delusional content of the client's verbalizations or by challenging the reality of the perceptions, but rather by redirecting the client toward a here-and-now concern (e.g., "I understand what you are saying. Tell me, how did you deal with your son when you heard this voice telling you to kill yourself?" And then, "And what do you do to get your son to daycare on time when this happens?" and similar interventions to get the client refocused on a real, but related problem). It is also important to explore and then allay any fears stemming from the hallucination or delusion by asserting the clinician's awareness of reality (e.g., "I understand your fear, but please let me assure you that I can guarantee you that Satan is not in this room with us."). If there is a kernel of truth to a client's delusion (and there usually is), it is important to find it and to respond to it (e.g., "I believe that you have been followed, especially that one time you told me about when . . . Can you tell me more about *that* incident?").

While reestablishing psychological contact and some degree of touch with reality, counselors remain calm and focused themselves, being careful not to become frightened of the client and not to be persuaded to reinforce the client's delusions. It is quite possible to express understanding as to why the client may have certain thoughts or beliefs without suggesting that the counselor share them. It is also quite possible to acknowledge that the client is hearing or seeing things and that this is distressing, while remaining clear that others do not hear or see these same things. It is important to maintain an empathic stance that provides understanding and guidance. In other words, it is not enough to keep the client happy by giving emotional support; it is also important to set firm limits on the client's behavior to ensure the client's and the clinician's safety (e.g., "I need you to do . . . before we can go on." Or "I understand you

feel . . . ; however, right now we need to do . . . to keep you safe."). A final note is necessary here: despite their portrayal in the media, psychotic individuals are no more dangerous and no more aggressive than the general population. They certainly can be aggressive (especially if their delusions involve paranoia and clients experience a need to defend themselves from a threat). However, more often than not, the clinician does not have to fear aggression from the psychotic client. As a rule of thumb, it is more important for the clinician to get the client reoriented than to worry about personal physical safety. Obviously, if the client cannot be reoriented, the in-session intervention may need to end with the institutionalization of the client. A client is typically not allowed to leave the clinician's office alone and without follow-up plan while flagrantly psychotic.

Dealing with Anger and Hostility. If the counselor or therapist is faced with an angry, agitated, or hostile client, the primary concerns become the diffusion of the affect and maintenance of behavioral safety for client and therapist. It is best not to challenge the accuracy or truthfulness of an agitatedly angry client, and this is increasingly true as the levels of anger and agitation rise. Rather than challenging, a counselor best acknowledges the client's feeling and validates it (e.g., "I certainly understand that you are very angry right now. And I can certainly see why. After all, what happened to you when . . . was very upsetting."). It is important to remain calm and not to get defensive, even if the client's affect and behavior become a personal attack against the clinician (e.g., "And it's all your fault. If you were a better therapist you would have helped . . . by now."). In other words, the clinician should avoid getting caught up in the client's affect. If, for example, clients accuse clinicians of various transgressions or misdeeds, clinicians best do not defend themselves but rather acknowledge the clients' experiences For instance, a clinician would not say, "I think you are wrong there. I really have been doing my best, but you have not been following the advice or recommendations I have made." Similarly, the clinician would not say, "If you had listened you would know that that is not what I said. What I really said was . . ." Instead the clinician would say, "I understand that you feel as though I have let you down. Please tell me what I could have done differently to help you better." Or "I really appreciate that you are disappointed in therapy. We have not progressed as much as you wanted. . . . " Or "I am sorry that you heard me as so critical of you. How could I have said things differently that would not have been so hard for you to hear?"

While acknowledging clients' affects and validating their right to them, therapists insist on basic safety rules. Clients are allowed to get angry, but they do not have permission to act out this anger physically in aggressive or hostile ways. Behavioral boundaries on clients' actions have to be very clear (e.g., yelling is fine, so is hitting a pillow; but acting out physically against the counselor or breaking therapy room furniture is not). The mental health care provider may choose to avoid too much direct eye contact if the client escalates and should generally avoid touching the agitated client. Providing extra interpersonal space can be helpful. Especially when dealing with angry affect that appears to have the potential to be acted out, clinicians best remember the caution never to be heroes. If a client becomes too agitated or openly aggressive, it may be time to end the session or call for help. It is generally clear to a therapist when a situation reaches a danger zone of potential physical aggression. The clinician will begin to feel unsafe and will sense a loss of control on the client's part that involves not merely affective but behavioral control. The strength of a client's voice alone is generally not the best predictor. Better predictors are a client's eyes and physical movements. Specific physical symptoms that signal increasingly angry affect include muscle twitching or restlessness, getting up and pacing, pantomiming aggression (such as pounding, choking someone, beating), staring or lack of eye contact, shallow breathing, quivering or loud voice, clenched fists, and angry words. When counselors begin to notice these symptoms, quick diffusion of the affect is important or the session may need to be discontinued. This resolve is communicated to the client directly as it may often serve to diffuse the behavioral reaction (e.g., "Unless you can calm down a little bit, we will not be able to keep working today."). Similarly, if the clinician feels the need to end the session or call in a helper, an explanation is given (e.g., "I believe we are no longer safe in this room because I am sensing that you are about to blow up. Let's stop for today and continue our work next time." Or "Let's stop for a moment and call in one of my colleagues to get her/his perspective on this issue.").

If the clinician is working with a client who regularly and predictably becomes uncontrollably angry, it is a good idea to plan ahead and have supervision and consultation available. If the mental health care provider is lucky enough to have videotaping facilities or one-way mirrors, it is best to have a colleague watching the session from another room. Thus, if the clinician misjudges the intensity of the client's affect and behavior, help is immediately available. If this route is chosen, the clinician needs to make sure that the observer does not overreact. It is best for the clinician and the observer to agree on a (nonverbal) sign that tells the observer when to intervene. If nothing else, knowing that backup is available will help the therapist feel less nervous and concerned and thus more emotionally accessible and available to the client.

Even with the knowledge of how to handle in-session affective crises, the counselor must never try to be heroic and must know when it is time to seek assistance (Bellak & Siegel, 1983). If clients are imminently dangerous due to out-of-control anger and hostility, clinicians need to get help; if therapists are anxious and unsure about their ability to manage a crisis, they need to seek input and immediate consultation or supervision. If mental health care providers anticipate working with dangerous clients on a regular basis (e.g., due to a caseload of court-referred clients), it is best to be prepared in terms of having special equipment as well. Such clinicians may consider installing an alarm system, as well as having in place a backup support system of staff who are present in the clinic when (certain) clients are being seen. If a client is extremely dangerous, but the counselor would like to attempt to manage the crisis on an outpatient and immediate basis, the clinician may have to weigh confidentiality against safety and arrange for someone else to monitor the session through video or one-way mirrors.

Assessment

The basic goal of crisis assessment is to "understand the impact of the contemporary crisis in terms of the patient's life situation" (Bellak & Siegel, 1983, p. 83). Minimally, crisis assessment allows the clinician to get an accurate picture of the nature of the event that triggered the crisis, with an exact definition of the problem, and the consequences it had for the client in terms of functioning in the realms of thoughts, behaviors, and feelings. Assessment is a relatively directive process that is focused and goal directed without ignoring the client's ongoing needs for catharsis and sharing. The client often perceives the focus and goal-directedness of the process as supportive because it provides structure and clarity to a situation that overwhelms the client. It must be remembered throughout that the assessment is about the client's current crisis situation (and about past behaviors, thoughts, and feelings only as they are relevant to the current situation). In other words, the clinician needs to focus on the traumatized aspect of the client, not the long-term picture. Crisis assessment is not to be confused with intake assessment; it is a much narrower and often much quicker process. It does not make use of forms or tests but rather relies on simple and direct questions that are to the point (cf., Gilliland & James, 1993).

Defining the Problem and Severity of the Crisis Event. The therapist asks clarifying and direct questions about the current crisis and

probes the client for as much specificity and detail as he or she can tolerate in this state of crisis. This is important because therapists can only assist with planning a safe course of action if they have a complete and accurate picture of the situation that triggered the problem (Patterson & Welfel, 1994). Counselors need to get a clear sense of the precipitating events, the occurrences during the crisis, and the consequences involved for the client and for others in the client's network. To get an accurate picture, the clinician explores what happened before, during, and after the triggering event; what the gravity of the stressor is from the client's perspective; what the event means subjectively to the client; what the event means to others in the client's network from the client's perspective; who and what contributed to the crisis (e.g., whether there is a family member who feeds the fire [Slaikeu, 1990]); and how and if the event involves any of the client's significant others. Specific questions that assess these components of the problem may include the following possibilities (the generic "the event" in these questions should be replaced by a rephrasing of the stressful or triggering occurrence):

Samples of Problem Definition Questions

- What exactly happened?
- When did it happen?
- Tell me the first thing that happened that was upsetting.
- Tell me everything that happened in the order that it happened.
- Tell me more about that.
- How did it all start?
- When did it start?
- What happened right before that?
- What happened next?
- Then what?
- What happened after that?
- How do you think it happened?
- What was the sequence of events?
- I need to know everything that went on (at that moment, right after that, right before that).

- I want to know exactly what went on in your head.

- Who else was there? Did you know them?

- How were they involved? Were they part of the event or were they witnesses? Did they encourage the event?

- What was your role in the event? What did you do? How did you react?

- How did you feel when . . . ?

- How does/will this affect you?

- What has this done to you?

- Does this have any effects for (family, friends, colleagues)?

- How did you react at that moment?

- How do you feel about it now?

- Is there anything we have left out?

- What else do you remember that I haven't asked about?

A clinician would not ask all of these questions, but would select a certain approach. The questions listed are merely provided as examples of how to probe for more detail in the areas of investigation previously outlined (i.e., precipitants, the event itself, consequences, etc.). They are intended to demonstrate the level of specificity that is being sought, as well as the directive nature of this aspect of crisis management. They certainly need to be adjusted to the client's frame of mind. For example, if a client is not ready to talk about a traumatic event in detail, this needs to be accommodated by keeping the questions more generic and shorter, or by not asking them at all and moving directly to the assessment of the client's current emotional state. The level of detail that is obtained from the client also depends somewhat on the circumstances that triggered the crisis. If an earthquake precipitated the crisis, for example, many of these questions would be nonsensical; on the other hand, if the crisis was precipitated by a fight with a significant other, most would apply.

It is also helpful to round out the discussion of the event by asking about any plans the client may have considered for reaction or resolution.

This can provide an important and efficient transition into the exploration of possible solutions, alternatives, and action plans. Because the level of the plan's appropriateness can be evaluated, it also can help give the therapist an appreciation of the client's level of impairment. Finally, any actions the client considered can be evaluated with regard to risk of danger to self or others. If clients indicate at this time that they have considered suicide or aggressive action against a perceived offending third party, a suicide or aggression assessment and intervention may need to be initiated. In fact, any time a client threatens suicide or aggression, crisis intervention becomes very specific to these threats and the counselor must be prepared for a specific set of questions and interventions. Because of the great specificity of the legal, assessment, and intervention issues, the topics of child abuse and neglect, suicide, and aggression will be dealt with in Chapters 6–8. All of the crisis skills discussed in this chapter, however, still apply and provide the mental health care provider with a solid basis of skills and knowledge that will assist in these specific circumstances as well.

Assessing the Client's Current Level of Functioning. This portion of the assessment looks at the impact of the triggering event on the client's level of functioning, with emphasis on short-term adjustment but also with an awareness of potential consequences for the client's long-term adaptation (Patterson & Welfel, 1994). The three areas of human functioning that are evaluated in this portion of the assessment are emotions (or affective arousal and impairment), thoughts (or cognitive distortion and impairment), and behavior (or behavioral paralysis and appropriateness). Although these areas of functioning are treated separately below, the evaluation may proceed by mixing questions from different realms of functioning if this makes the interview more logical and efficient. The main point is to ascertain that the therapist can evaluate the level of impairment or strength in each area to have data that can assist with planning an intervention.

In exploring affective arousal and impairment, the counselor asks questions that assess the client's current emotional state, identify the primary affect (e.g., anxiety, depression, anger), and evaluate the level of impairment experienced in the affective realm. The levels of impairment were previously outlined and need to be kept in mind as the clinician assesses this arena of functioning. Further, the clinician checks for emotional issues that are commonly evaluated in mental status exams. As such, control over affect (overly emotional versus overly controlled), emotional expressiveness (flat affect versus hysteria), affect appropriateness (congruent versus incongruent affect), chronicity of the affective disturbance (entirely explained by the crisis situation versus chronic impairment), and similar issues are to be considered. It is a good idea for the clinician to reread the mental status section in their basic psychotherapy text (e.g., Brems, 1999a) to be prepared for this aspect of the crisis interview. Finally, it is helpful for the clinician to gain an appreciation of the client's reservoir of emotional strength, which may range from helplessness and hopelessness to optimism and emotional resourcefulness.

With regard to exploring cognition and thought impairment, the clinician attempts to gain an appreciation of the client's level of cognitive distortion and rigidity. Issues such as concentration, attention, logic, and complexity can be used to evaluate cognitive functioning and generally do not have to be asked about but are gleaned from answers to questions probing other areas of assessment. In other words, many of the client's answers to the clinician's questions about the situation surrounding the crisis will provide insight into current level of impairment in reasoning. The clinician can listen for distortions, irrational beliefs, obsessive thoughts, preoccupations, and illogical conclusions, while also assessing concentration, attention, problem-solving skills, and ability to deal with day-to-day tasks.

The clinician assesses the behavioral realm by evaluating the client's level of psychomotor activity and behavioral impairment. The clinician looks for psychomotor retardation that signals behavioral paralysis or ineffectiveness; for psychomotor agitation that may signal clumsiness or impulsivity; and for any other behavioral indices that the client may be caught in either an approach, avoidance, or paralysis approach to the triggering event. The clinician will attempt to make a judgment regarding the client's overall behavioral equilibrium and mobility. Behavioral equilibrium suggests that the client will not engage in any impulsive or inappropriate action that may cause any further harm. It also includes cognitive and emotional components, because the clinician essentially probes for information that provides reassurance about the client's ability to maintain emotional stability, cognitive balance, and behavioral poise. The clinician also evaluates capability for autonomous, flexible, and adaptive coping responses to assess the client's capacity for mobility.

Finally, the therapist or counselor considers the client's immediate needs, which must be addressed through the action plan or through the relationship in the session, and juxtaposes them to the client's long-term needs, that can wait until follow-up or later therapy (Slaikeu, 1990). Some questions the therapist can keep in mind while attempting to clarify the client's current level of functioning are listed below (also see Gilliland & James, 1993). These listed questions are obviously never asked directly of the client, but merely guide the counselor's inquiry and observation.

Emotional Realm Level of Functioning Questions

- What is the primary emotion displayed?

- Is the emotional response in proportion with the event?

- Is the emotional response congruent with the event?

- Is the affective response typical of people in this type of circumstance?

- Is the level of emotional expressiveness appropriate?

- What level of control does the client have over the affective response?

- How would I rate the client's overall affective impairment?

Cognitive Realm Level of Functioning Questions

- Is the client's thinking realistic given the circumstances?

- Is the client distorting the event (e.g., rationalizing, exaggerating, believing part-truths, overanalyzing)?

- Is the client overly preoccupied with the event (i.e., are there obsessions)?

- Is the client's thinking overly concrete or rigid (e.g., does the client seem to have blinders on)?

- Is the client confused?

- Can the client concentrate and attend appropriately?

- Is the client delusional?

- How would I rate the client's overall cognitive impairment?

Behavioral Realm Level of Functioning Questions

- Is the client immobilized?

- What approach is the client taking to the crisis (approach versus avoidance)?

- How is the client's psychomotor activity (retarded versus agitated)?

- Is the client's behavior goal directed?

- Is the client's behavior maladaptive, erratic, or unpredictable?

- Is the client's behavior a threat to self or others?

- Are the client's general day-to-day living skills intact?

- How would I rate the client's overall behavioral impairment?

To stress an important point one more time: many of the answers to the questions in this portion of the assessment can be gleaned through observation and can be answered from having gone through the definition-of-the-problem assessment. Direct inquiry about the client's emo-

tional, cognitive, and behavioral state can generally be kept to a minimum, saving precious time during the session for other assessments and intervention. The assessment of the client's level of functioning and perception of the triggering event is best understood as a facilitative assessment. In other words, the "data gleaned about the client are used as part of an ongoing helping process, not simply filed away or kept in the [clinician's] head. The key to facilitative assessment [of this type] is to focus on the client's inner emotional world—not the [clinician's] analysis of that world" (Gilliland & James, 1993, p. 37).

Assessing the Client's Resources and Coping Skills. In this portion of the assessment, the clinician attempts to evaluate several coping and problem-solving client skills and resources. First, the clinician assesses the client's internal resources (which include not only current skills but also past ability to cope and solve problems). Then, the clinician assesses the availability of a social support network to the client. The clinician attempts to identify the client's current ability to solve problems efficiently and effectively. A variety of client skills are explored in this regard. Namely, the clinician attends to the client's:

- perceptual skills
- cognitive skills
- wellness and stress management skills
- emotional skills
- problem-solving skills

The clinician assesses perceptual skills by considering distortions in the client's current perception of reality (such as, at worst, hallucinations or illusions), the client's orientation to reality (such as orientation to time, place, and person), and the client's ability to screen out sensory stimulation to maintain a healthy level of arousal. Cognitive skills refer to the client's ability to reason logically, think efficiently, and contemplate the situation rationally. These cognitive skills are critical to problem-solving ability, and the mental health care provider must assess whether clients currently trust their own ability to cope. Wellness and stress management skills refer to the client's level of self-care. Issues to be explored in this regard are sleep, appetite, and day-to-day functioning, such as being on time for work, taking care of daily hygiene, attending to health care needs (e.g., taking medications as prescribed), maintaining a healthy diet and nutrition, and exercising. Leisure skills may be assessed as well, though they may not be as critically important to the current crisis as they would be in an overall evaluation of the client. Emotional skills refer to the client's ability to experience, identify, and express emotions. If these critical skills are not experienced consciously, they are generally neither expressed fully nor appropriately, often leading to physical or psychological symptoms that further impede the client's functioning during a crisis. Assessment of emotional skills begins with figuring out whether the client can consciously (i.e., without prompting and without help) identify affects experienced at the current time. Emotional expressiveness can be gleaned through observation. Emotional appropriateness has been dealt with in the previous section of assessment.

All of these skills (perceptual, cognitive, wellness, and emotional), in combination, will lead to a reasonable assessment of the client's overall problem-solving ability. If clients have poor cognitive skills, misperceive reality, are emotionally over- or underexpressive, and are unable to sleep, they are unlikely to be able to cope very well and will need to receive intensive intervention to become fully functional again. On the other hand, clients who have highly distorted and irrational thoughts, but who have preserved their wellness skills, are aware of and express their emotional arousal, and have clear perception, are likely to respond well to minimal intervention, which focuses on realigning their cognitive reasoning skills. These latter clients, with some guidance, will be able to begin to generate ideas and potential solutions to their crisis. The level of intervention (i.e., the level of directiveness) chosen by the clinician in a crisis situation is directly related to the client's ability to cope.

Coping skills can also be indirectly assessed by taking a look at the client's past ability to solve problems, especially in situations similar to the current crisis. This process is not only helpful in assessing the client's coping ability in general, but it is often a helpful transition process into the action plan phase of the crisis intervention. Specifically, in assessing the client's past coping skills, the counselor will ask questions about how the client has coped with similar situations in the past. Through these questions, client and therapist essentially go over possible solutions for the current crisis, in that they are taking a look together at what has worked in the past. The best predictor of future behavior is past behavior. If this portion of the assessment reveals that clients coped well in the past and were able to pull themselves out of crisis mode, then this information can be taken as a positive prognostic indicator. If the client has never been able to solve a crisis independently, prognosis is much poorer, and the clinician will take a more active and directive stance. Questions for the counselor to keep in mind while exploring past and present coping are as follows:

- How did the client approach/cope with similar situations in the past?
- What have been the client's past coping and problem-solving approaches in general?
- What was the client's problem-solving style (e.g., internalized versus externalized, active versus passive, proactive versus reactive, methodical versus haphazard, trial-and-error versus planning)?
- How well or successfully has the client coped in the past?
- How flexible and creative has the client been in solving problems in the past?
- Has the client been able to mobilize personal strengths and internal resources independently or only with assistance?
- Has the client been able to mobilize external (i.e., social support) resources?
- Who have been the client's past external supports (names, phone numbers, addresses, relationships)?

- What were the client's past actions (behavioral history) that appear to have been most helpful during crises (e.g., increased exercise, music, self-help books, socializing)?

- What have been past successful solutions the client developed and implemented?

- How well did the client function before the crisis?

Another component of this assessment is the consideration of the client's interpersonal resources. The clinician first questions the client about his or her current living situation. If the client lives alone and appears relatively isolated, this portion of the assessment will take more time and effort than if the client lives with or has easy access to a supportive adult. The therapist needs to explore whether the client can identify friends, family, neighbors, or colleagues who may be available, willing, and able to help the client during this crisis state. The more external resources that are available, the less the clinician has to rely on agency and institutional supports. This is, of course, preferable, because it represents the least restrictive treatment alternative. In identifying resources and social supports, it is important to assess how realistic the client is in evaluating and describing this network. Some people in crisis will claim that there is no one who can help them. They indicate that their family is unavailable or unwilling to help, that they have no friends, that they do not know their neighbors, and that they cannot trust their colleagues. At the other extreme, some clients may claim that they have lots of friends and that no specific arrangements need to be made for social support because someone is always available. Both extremes are unlikely and uncommon and should be approached cautiously by the clinician. A realistic assessment of external resources includes the client's description of some of the helpful traits of identified resource persons, their addresses and phone numbers, and the date of last contact. A client who claims to have no resources is asked specific questions about when he or she last talked to a variety of people, especially those relevant to the client's history (i.e., "When did you last speak to your . . . "). The mental

health care provider can also ask for addresses and phone numbers of these individuals.

A list of individuals who can be inquired about is provided below. In going over this list with the client, the clinician can help assess individuals with regard to their availability, willingness, and ability to help. The clinician must keep in mind that availability is not the same thing as the ability to help. The fact that the client has a spouse does not immediately imply that this spouse will be a usable support resource. In fact, it is entirely possible that a particular person is actually part of the problem, not only ruling that individual out as a resource person, but also suggesting that the client may actually need to keep distance from this individual. Therefore in exploring resources, the therapist must also ask questions about how well the client and a particular person get along, how they tend to interact, if there is a certain level of trust, if and how the person has supported the client in the past, if this person has ever had a problematic relationship with the client, and whether this person is a potential asset or potential liability.

Potential Social Network Resources

- parents
- siblings
- grandparents
- aunts and uncles
- cousins
- other relatives or in-laws
- significant others
- close friends
- distant friends
- acquaintances
- neighbors
- colleagues
- church members
- club members
- members of other groups to which the client belongs
- classmates

- internet acquaintances

- people the client's friends, acquaintances, etc. have mentioned who have experienced a similar crisis

If the client is unable to reveal the support of anyone on this list of potential resources, it will be up to the clinician to generate ideas about community resources that can take the place of social resources. The clinician attempts to connect the client to resources such as self-help groups in the community, crisis lines, health clinics, mental health agencies that have groups related to the triggering event, support organizations relevant to the client's problem (Alzheimer or cancer support groups; support groups of family members of crime victims, etc.), hot lines for traumatic events such as the client's (e.g., rape hot lines, hot lines for runaway adolescents), shelters (e.g., for the homeless or for battered women), and so forth. If a client has used any of these resources in the past, the clinician needs to find out how well they worked and how willing the client is to make use of them again.

Once they have assessed the client's internal, past, and external resources, practitioners make a mental list of the client's strengths. The more resources of any type, the longer the list of strengths. Once therapists have a list in mind, it is good practice to review it briefly with the clients to allow them to recognize that they have resources with which to work. Clients in crisis often have the self-perception of complete inadequacy, isolation, and weakness and are surprised when counselors begin to summarize traits and resources that will be of assistance during the crisis. This summary often returns some sense of self-efficacy to clients, greatly facilitating their active collaboration during the action plan stage.

Action Plan

"There is no substitute for a good plan for crisis resolution. Without careful planning and direction, a helper can only add to the confusion already experienced by the person in crisis" (Hoff,

1989, p. 118). The action plan phase of crisis intervention has two goals. First, it restores the client to a level of functioning that includes emotional equilibrium and coping skills, which provide order and predictability. Second, it identifies and implements one or more solutions to address the immediate needs of the client (Meier & Davis, 1993; Slaikeu, 1990). To achieve the first goal, throughout this entire action plan process, the therapist needs to stay focused on the problem at hand and not to get derailed by the client (Patterson & Welfel, 1994). Clients in crisis tend to jump from topic to topic, often involving many aspects of their lives that are actually unrelated to the crisis. It is the clinician's role to discourage this type of generalization to keep the situation manageable. In line with this task, the clinician also helps clients form an accurate cognitive understanding of their situation, assisting them in reducing or eliminating cognitive distortions and irrational beliefs that exaggerate the stressful event or blow its consequences out of proportion (Patterson & Welfel, 1994). To achieve the second goal, the clinician leads the client through three stages: exploring alternatives and developing possible solutions, committing to a set of solutions that are to be attempted, and implementing the action plan.

Exploring Alternatives and Developing Possible Solutions. At this point in the crisis intervention process the clinician coaches the client into generating ideas and possible solutions to the crisis situation. Coaching is accomplished through questions that help the client focus on the problem in small pieces, rather than focussing on the entire situation all at once. Once the problem situation has been broken down into manageable bits, coaching helps the client generate ideas about how to approach each aspect of the problem individually and then, after each component of the problem has received attention, integrate these aspects into the entire plan. By breaking the problem situation into manageable chunks, the client focuses, becomes less overwhelmed, and can begin to organize thoughts more clearly. It also helps reduce hopelessness and self-defeating thoughts by increasing the client's cognitive flexibility and by infusing realism into how the client perceives the problem.

It is during this step in particular, that the counselor suspends some of the usual counseling rules and becomes quite concrete, almost to the point of giving advice by following a step-by-step problem-solving approach. The clinician helps the client generate alternatives, evaluating each with regard to its appropriateness and likelihood of success for resolving the moment of crisis and only stopping when several acceptable solutions have been generated. The strategies that will be generated will be selected and developed only with the crisis in mind; more growth-oriented alternatives can be developed later, after the crisis has been resolved and the normal therapeutic process is reentered.

A large number, in fact an almost infinite number, of strategies can be considered for crisis moments. The strategies generated with the client should, however, include community resources for additional referrals and supports, which will clarify for the client that, in a moment of crisis, going it alone is not the best option. For each individual client, a plethora of strategies may exist that reflect a unique support network, personal strengths and skills, and idiosyncratic emotional, cognitive, and behavioral needs and desires. Strategy choices need to consider what is necessary to ensure the client's safety. In some situations, it may be important to remove the client from a traumatic situation (e.g., removing a battered woman from her domestic violence situation into a shelter), whereas in other situations it may be important for the client to confront the stressor (e.g., encouraging appropriate assertiveness skills with a friend who owes money to a client who is in a financial crisis). Finally, although it may seem obvious, any action plan should be tailored to the client's cognitive resources, developmental or functional level, dependency needs, and financial resources. Additionally, the plan will have a greater likelihood of success if it is consistent with and

respectful of the client's lifestyle and cultural background (Hoff, 1989). A few generic options are listed below.

Strategies to Consider

- catharsis
- removal of the stressor
- cognitive restructuring
- behavioral contracting
- assertiveness training
- systematic desensitization
- relaxation strategies
- psychoeducation
- skill building
- referrals
- recruitment of a network of family, friends, neighbors, colleagues
- placement in a shelter
- institutionalization

Solutions are generated for each aspect of the crisis; once several acceptable solutions have been identified for a particular aspect, client and mental health care provider move to the committing-to-action phase for that aspect. Client and clinician do not attempt to generate ideas for *all* problem components before moving to the committing-to-action stage. Instead, they cycle between the idea-generating phase and the committing-to-action phase repeatedly for each problem component until all aspects of the crisis have been addressed.

All of these issues are reflected in the sample transcript that follows the "Committing to an Action Plan" section below. This transcript is a replay of a small portion of a solutions-generating and action-planning conversation between a clinician and a client who is in a mild state of crisis.

Committing to an Action Plan. In this phase of action planning, client and clinician work jointly to select the strategies that are most agreeable to the client and most likely to result in success (from the therapist's perspective). Depending on the level of cooperation expected from the client, the clinician may choose to formalize the action plan by making a written behavioral contract with the client. This is particularly important if the action plan is needed to prevent clients from harming themselves or from harming another person (see contracting in Chapter 6 about suicide and Chapter 7 about violence). If suicide or violence against others is a possibility, the counselor must also be certain to take all necessary actions to remain within the bounds of laws and ethical obligations to the client (see Slaikeu, 1990). These issues are addressed in great detail in the relevant chapters of this book.

For each aspect of the problem, a specific plan is developed that includes all the necessary steps to ensure its success. In identifying the specific steps in the plan of attack, the clinician needs to be as concrete as possible while still being respectful and constructive. It is important to understand that the overall goal is not always to resolve the problem; sometimes, the goal has to be to get the client to adapt to a new situation that is perceived as traumatic (e.g., if a partner left, it is unrealistic to try to resolve the problems between the two individuals; it may be more realistic to try to help the client adapt to the idea of being single again [Patterson & Welfel, 1994]). Additionally, it may be necessary to begin the process by assuring client safety, which may involve removing the client from the pathogenic situation (e.g., encourage a battered woman to seek out a shelter) or from the unhealthy setting (e.g., encourage a client to ask for a transfer away from an abusive boss [Bellak & Siegel, 1983]).

If the client needs support from within a personal support network, the clinician and client should identify specific individuals who are needed to help. It is not enough to decide that the client will rely on friends for support (Meier & Davis, 1993). Instead, the names of the friends should be mentioned, and if possible the client will be encouraged to call them during the ses-

sion (see "Implementing an Action Plan" below). If the client needs to learn new coping skills, a plan needs to be made as to how this is best accomplished. Again, it is not enough, for example, to conclude that the client needs to learn how to relax; a specific plan of action is needed that outlines how the client will learn this new skill or accomplish this task (e.g., the client may be sent home with one or two relaxation tapes while also being scheduled for another session with this or another clinician to learn more about relaxation in the future).

If the client is to make use of some existing skills that might help in difficult situations, it is not enough to identify these skills, rather the clinician should work out a schedule for when the client will engage in them. For example, if the client copes better after having had a good aerobic workout, clinician and client should agree on a plan of action that includes the amount of time to be spent in this activity, when the client will engage in it, and (if desired) with whom. In this way, the client may agree to do an aerobic workout at a gym at a specific time for a certain length of time. If the client does not know the schedule for aerobic workouts at the gym, the client can call the gym right then and there to obtain this information. It is important in the commitment phase to make sure that the client does just that: *commits* to a plan of action. It is not enough to talk about possibilities (as was done in the idea-generation phase). In this phase it is also a good idea to build contingencies into the plan. Thus, if a planned action does not work out for whatever reason, the client has a fallback position. For example, if it is decided that the client should take a certain length walk after waking up in the morning, a contingency plan needs to be made for extremely cold or inclement weather. For example, in this case, the client may engage in a specified indoor aerobic activity instead.

What follows is a sample of a portion of an idea-generating and commitment-to-action cycle with a client in mild crisis. The sample portrays a mild crisis to prevent the transcript from being overly lengthy. In more severe crises, when the client is even less functional, this aspect of the intervention may take more time and may require more creativity. The sample demonstrates how the counselor narrows the conversation to one particular aspect of the problem and then proceeds through a series of questions with the client to generate several possible solutions. Once these solutions have been generated, the therapist moves to the next phase (committing-to-action) for that particular aspect of the problem. Once that aspect of the situation is decided upon, the mental health care provider moves to the next component of the problem and back to the idea-generating stage. This process of cycling from idea generation to the commitment-to-action phase will repeat itself until all critical pieces of the problem situation have been addressed through a plan for action. Once this is accomplished, the clinician pulls all the components together into one cohesive overall action plan with clear directions for the client about how to proceed with the overall implementation of the plan in its entirety

Sample Transcript of the Solution-Generating Phase of Crisis Intervention

(This sample is based on the following client crisis: The client's home was burglarized three days ago; she is now in a state of crisis that has unbalanced her emotionally, made her hypervigilant and overly fearful for her physical safety, and left her behaviorally immobilized in that she does not know what to do next; her impairment in the emotional and cognitive arenas is mild to low; in the behavioral area it is low to moderate.)

Clinician: Okay, I now have a really good handle on what happened and how you have been feeling since the break-in. What strikes me is that you have not only experienced quite a trauma, but that you also have not been able to take care of yourself very well ever since. We now need to figure out what we can get you to do to help you pull yourself together to overcome the trauma and to

start taking care of yourself again. Let's start by figuring out what you need to do about taking care of yourself. Okay?

Client: Okay, but I have no idea what to do. . . . *(still close to tears)*

Clinician: We will figure it out together. We'll take it one step at a time. So many things have happened that by trying to focus on all of it at once I sure could understand why you would feel overwhelmed and unable to figure out any solutions. Let's start with the fact that you haven't eaten a thing since the burglary. What can you do about that?

Client: Well, I just have felt too upset about it to have much of an appetite. I really haven't thought about food—but now that you mention it, you're right, I haven't eaten at all. . . . No wonder I feel so weak.

Clinician: So, the first thing we can do is to get you to start thinking about food!

Client: Yeah—I guess that's a start. . . . *(smiles weakly)*

Clinician: Do you have any food in the house?

Client: No. They even took the fridge and everything in it. Can you believe that? Who would do such a thing and why to me? I live in an apartment for goodness sakes—why not break into a house where there is some real stuff? *(gets agitated and derails from the topic at hand)*

Clinician: Yes, I know you don't understand why this happened; you may never know, but right now we need to figure out how to get you back on your feet. So let's get back to food. You have no food in your house. What can you do about that?

Client: I don't know. . . . *(tearful; still somewhat preoccupied with her derailed thoughts)*

Clinician: Okay—I need you to focus now. Think food. You had your purse with you at the time of the burglary and there was no cash in your apartment. So, as we talked about, at least you don't have to worry about cash for now. Given the fact that you have money and credit cards, what's a logical next step in terms of getting you something to eat?

Client: I could stop at McDonald's on my way home. . . .

Clinician: That's a start. Where does that leave you tonight and tomorrow morning? You'll need food in the house.

Client: I guess maybe I should stop at the grocery store instead. . . . *(looks for reassurance)*

Clinician: That's a great idea as well. Remember—right now we are just generating ideas; in a little while we'll decide which way you'll go. What other options do you have?

Client: Well, several friends have called me since the burglary and asked if I need anything—I couldn't think of anything; I guess I could ask them for help. . . . Oh yeah, and one of my neighbors said I could stay with her for a while; till I get some new stuff. I can't believe they cleaned me out completely—there is nothing left. I slept on the floor last night—well, I didn't really sleep. I was so worried that they might come back and the two nights before that I didn't sleep much at all either. *(getting agitated; speech is getting pressured)*

Clinician: *(interrupts the flow of thought and directs client back to the topic)* Okay, so you have several food options. You can go to a restaurant, you can go to the grocery store, you can go to a friend's house for food, you can ask friends to shop for you and drop things by your place, you can go stay and eat with your neighbor. All of these are good options. Tell me, who are the friends who offered help?

Client: Oh, they are a couple of people I have known for a long time. . . . *(does not offer any elaboration)*

Clinician: What are their names?

Client: Well, there was Sylvia; she called yesterday. She heard about it at work—we work together.

Clinician: You said there was more than one friend who called?

Client: Jeffrey called, too. He's engaged now—he was my boyfriend a long time ago, but we're still friends. . . .

Clinician: How well do you get along with Sylvia?

Client: We really like each other but we haven't done a lot of stuff together. She is new at work and I only met her a couple of months ago. We keep meaning to go to lunch or something. But we've both been pretty busy.

Clinician: So you haven't ever asked her for help with anything else?

Client: Except at work. She has helped me with some projects and we worked well together.

Clinician: That's terrific. Sounds like Sylvia may be someone who could really help you out. You like her and you know you can at least work with her. How about Jeffrey? How well do you get along with him?

Client: He is one of my best friends! We keep in touch and we see each other once a week or so for a run—we've been running partners for years. We do really well together as friends. We were lousy as boyfriend and girlfriend though. . . .

Clinician: You mentioned Jeffrey is engaged now. Has that changed your relationship?

Client: No not really; his girlfriend knows about me. We've done stuff together and get along okay.

Clinician: Has Jeffrey ever helped you out with anything before?

Client: Yeah, he is a physical therapist. When I broke my ankle a couple of years ago, he helped me get back on my feet *(client grins at her own pun—a good sign)* and he's done other stuff too.

Clinician: *(smiles too)* . . . How is it getting help from Jeffrey? Did you feel okay about it?

Client: *(nods "yes")* . . . I help him out with his taxes every year, so we feel like there is a give-and-take. . . .

Clinician: Perfect. So would you say that Jeffrey is someone you can still rely on for help?

Client: Oh yeah.

Clinician: Okay—that's great. So we have two people for sure that want to help out and that you feel okay about asking for help: Sylvia and Jeffrey. Do you have their phone numbers with you?

Client: I have them memorized.

Clinician: Great! Now tell me about the neighbor who offered help. . . . How well do you know her?

Client: Really well. We look out for each other's apartments when we're gone—pick up each other's newspapers and mail and stuff. We've had dinner and gone for walks. We talk almost every day. I guess maybe Carol is more than a neighbor, though I've never thought of it like that before.

Clinician: Certainly Carol is someone you can rely on!

Client: Yes, definitely.

Clinician: Okay—so let's review one more time: You can go to a restaurant, you can go shopping, you can call Sylvia or Jeffrey, or you can stay with Carol. Which one of these options sounds best to you?

Client: The easiest thing would be to stay with Carol. . . .

Clinician: Okay, so that's settled. Before you leave today we'll call Carol and make the necessary arrangements. If we can't get a hold of her, we'll talk about a backup plan then. Let's move on to the next thing. I've also been really concerned about your not sleeping. How about you?

From here client and clinician tackle the next component of the problem, progressing through all components until each aspect of the critical situation has been covered and an action plan has been made and committed to for each. Once that is accomplished, it is time to move on to the implementation stage.

Implementing an Action Plan. During the implementation stage, the counselor reaffirms the client's ability to carry out the plans that they committed to in the previous stage of the intervention. It is crucial that the clinician pay attention to reestablishing a sense of control and competence in the client. This is important from here on out because clients will need to begin to function increasingly independently as they are getting ready to leave the session to tackle the plan that was made. The therapist can offer additional support to the client at this stage by building contacts with the therapist into the action plan. For example, extra sessions can be scheduled after critical time intervals, phone contacts can be scheduled, crisis numbers can be made available to clients so that they know there is always someone who can be reached (e.g., numbers of the local crisis hot line), and specific instructions can be left with clients about what to do if they need to schedule another appointment with the clinician. Clients should be equipped with an appointment card for the counselor and a brochure of the clinic (if available) because these are good transitional objects that can reassure clients about the existence of a place and a person from whom help is available.

The implementation phase begins with the clinician's review of all the problem components that were identified and the action plans that were made for each. During a crisis intervention, the clinician may want to take notes (a practice not usually encouraged in therapy settings) because it is important for the clinician to remember all of the problem aspects and plans when doing the implementation summary. In fact, it will be helpful to give clients these written lists, since they may not be in the cognitive frame of mind to remember everything after leaving the session. This written documentation is essentially a cheat sheet for the practitioner and the client. A copy of it needs to be retained in the client file because it provides good documentation of what was done on behalf of the client. The cheat sheet copy in the client chart will accompany the progress note written about the crisis situation.

Appropriate charting of the crisis situation outlines the process clinician and client went through to assess the crisis and its effects on the client, as well as the process of arriving at the action plan contained in the cheat sheet. A sample cheat sheet is included at the end of this chapter.

Once the mental health care provider has listed all the problem components and associated action plans, client and counselor begin with the work that can be done in session to start putting the plan in action. They make phone calls to any individuals who were identified in any of the action plans, ensuring their availability, ability, and willingness to help. If anyone who was written into the plan as a resource person is not available, a backup plan is made for clients to use until they can get a hold of the original social support person and put the preferred plan in action. If the client is too immobilized or emotional, the therapist can make the phone calls; however, it is preferable for the client to make the calls. This avoids issues of confidentiality and restores a sense of control and competence to the client. If the counselor must make the calls, it is best to get the client to sign a release of information form so the clinician does not need to worry about breach of confidentiality. The clinician will only divulge enough information to resource persons to get them to understand the urgency of the situation and the role the clinician would like for them to play.

For any actions that require the client to function independently, it is often a good idea to spell out all of these actions in a behavioral contract. This has two advantages. First, a written contract generally increases a client's commitment to follow through with the agreement. Second, clients are in a state of cognitive impairment and may not remember everything they agreed to do. The behavioral contract spells all of this out for them and hence is an excellent reminder of what needs to be done. Samples of behavioral contracts are contained in Chapters 6 and 7.

If referrals to other practitioners are made, the client needs to leave the current session with phone numbers, names, and appointment times. In other words, it is best to make calls for appointments while the client is with the clinician so the clinician can be confident that the client will follow through. If arrangements are to be made for a shelter visit or for institutionalization, these need to be initiated at this time. How to institutionalize a client is covered in detail in Chapter 6 (which deals with suicide), since this is the occurrence that most often leads to hospitalization.

Essentially, by the time clients leave the session, they should feel ready to face their crisis situation and strengthened through a number of interventions and resources to the point of feeling capable of dealing with whatever comes next. Clients should have a clear idea about what they will do next, who will be involved, and how they should react if a certain plan does not work. Clients will be thinking more clearly and are now relatively stable emotionally. That does not mean clients will not have relapses once they have returned to their situation. However, they now have a behavioral plan that can be put into action even if a relapse occurs. If nothing else, they have phone numbers of resources who can be called should they once again become behaviorally immobilized. If the crisis is severe and the reaction to it strong, the plan should include a resource person who phones the client on a regular basis to check in on progress or regression. A follow-up session with the clinician needs to be scheduled as well so clinicians can assure themselves that the action plan is working and so that they can initiate any revisions that may be necessary. This follow-up appointment will be scheduled at an interval that is directly related to the client's state of crisis. In other words, severe crises results in short follow-up intervals; mild crises may extend the period of time elapsed between the current session and the follow-up appointment.

Follow-Up

It is the goal of the follow-up appointment to receive feedback about the success of the chosen interventions and to revise any interventions that were not optimal. Further, the client can be reevaluated informally (through behavioral observation rather than direct questioning) to assess whether the client's state of mind and emotion has improved. Depending on the frame of mind of the client and the success of the action plan, time may be spent almost exclusively on revising and/or adding action plans or on beginning to ensure proper therapeutic follow-up. The crisis can be used to therapeutic advantage in that it may point the way for long-term treatment planning. If the bulk of the follow-up needs to be spent on revision and addition of action plans, the same procedure would be followed that was previously outlined in the Action Plan section. If the client's state of crisis is greatly reduced and the person is capable of beginning to think about long-term issues, a therapy plan can be made.

If the crisis occurred within a therapy setting, client and clinician can return to their therapeutic work. However, they should jointly assess whether the crisis needs to precipitate changes in the treatment goals and plan that they had agreed upon previously. If the crisis occurred during an intake session, counselor and client would now begin to plan the client's overall treatment, keeping in mind all the issues that emerged for the client during the crisis.

DOCUMENTATION AND RECORD KEEPING

At the end of this chapter is a sample of a cheat sheet that contains the problem components of the crisis situation that was used in the earlier sample transcript. For each problem component, a brief description of the problem is provided. It is then juxtaposed with the action plan and time line that was established. Wherever appropriate, resource persons are named and their phone numbers are noted. This cheat sheet would be copied for the client's chart, and the client would be allowed to retain the original. The copy in the client's chart would accompany the progress note

written for the session. Progress notes are important pieces of data that keep the therapist safe from a legal perspective. In cases of harm to self or others, progress notes are absolutely essential for possible legal purposes. However, even if no legal issues are involved, progress notes of crises are important because so much happens during a crisis assessment and intervention that the counselor cannot possibly later remember everything. Having a thorough progress note that can be reread before the follow-up session is often critical to the success of that session. Samples of crisis progress notes are contained in Chapter 6 for the example of a suicidal client and in Chapter 8 for a case of reportable child abuse.

SUMMARY AND CONCLUDING THOUGHTS

Dealing with client crises is not as difficult as it may sound. The key to successful resolution is a clinician who can keep a clear head and who does not get caught up in the emotional state of the client. The clinician must be able to keep a therapeutic distance and must not overidentify with the client. From this distance, solutions to the problems faced by the client are usually relatively obvious, although clients themselves may be unable to see them. If clinicians can keep a clear head, can stay rational, and can keep their own emotional response under control (regardless of the type and severity of the trauma recalled by the client), they will become an excellent resource and source of support for the client. The ability to keep things slowed down, to maintain control of the topic, to stay calm and confident, and to stay focused, without allowing clients to derail, is essential for clinicians doing a crisis intervention. Focus can be obtained through a clear action plan for crisis situations with clients. This chapter outlined just such an action plan, and familiarity with it will greatly facilitate a clinician's successful crisis intervention.

Sample Crisis Intervention Cheat Sheet

Client Name: Josephine H.

Therapist: Chris B.

Date: 23 August 1997

Session Number: 1 (Intake Interview)

Summary of the Crisis Situation: The client was burglarized two days ago; the burglars took all of her belongings in the apartment, but no cash or credit cards, which the client had on her person. The client has shown poor self-care skills since the burglary and has not mobilized her internal or external coping resources.

PROBLEM	FAVORED SOLUTION	BACKUP PLAN	TIME LINE	RESOURCE PERSON
client has not eaten since the crisis event	client will stay with neighbor Carol	client will contact her friend Jeffrey to make food shopping arrangements with him	for one week	Carol (555-1234) Jeffrey (555-2345)
	client agrees to eat a minimum of three meals per day with proper attention to nutritional needs	client may choose to ask Carol or Jeffrey to monitor her food intake if she finds herself cheating	for one week	Carol Jeffrey
client has not slept properly since the crisis event	client will stay with her neighbor	client will make arrangements to stay with Sandra	for one week	Carol Sandra (555-9876)
	client will go furniture shopping with Jeffrey and with Sandra (alternating) to purchase a new bed	client will go furniture shopping by herself when no one else is available to accompany her	over the next month	Jeffrey Sandra
client has not gone to work since the crisis event	client will call her boss to ask for a week off; she will attempt to use comp time but is willing to use vacation time if needed	client will take a one-week vacation from work	one week	boss (555-0987)
	client will give herself permission not to feel guilty about taking time off work	client will call Jeffrey or Sandra for reality checks if she becomes worried or guilty about work	one week	Jeffrey Sandra
	client will return to work after one week	client will extend her vacation or leave as needed given her emotional state	open	boss

Sample Crisis Intervention Cheat Sheet (continued)

PROBLEM	FAVORED SOLUTION	BACKUP PLAN	TIME LINE	RESOURCE PERSON
client has not left the home except to come to her crisis intake interview since the event	client will resume her daily exercise program at the local gym; she will call the two women who usually join her there to ask them for help in getting her motivated to go	client will resume her exercise program at the gym by herself if neither of the two women are available to go	always	Gwen (555-0854) Beth (555-9874)
	client will seek out other opportunities to leave home for short periods each day (e.g., to go get her mail, to do additional shopping)		always	optional
client has not initiated any interaction with friends or acquaintances since the event	client will go running with friends at the same intervals she used to maintain	client will go running by herself when usual running partners are not available	always	Jeffrey Beth
	client will visit with one friend or acquaintance per week	client will go to a museum or other community event by herself	for one month	Beth, Gwen, Carol, Sandra, Jeffrey
	client will call one person per day	client will go to a shopping mall	for one month	Beth, Gwen, Carol, Sandra, Jeffrey
client must deal with the insurance company	client will make an appointment with her insurance agent for next week		in two weeks	optional
	client will secure the necessary police reports to file her insurance claim		in one week	optional
client shows some symptoms of PTSD	client will return for a complete intake assessment to plan a course of treatment		to be decided	Chris B. (555-4312)
	client will consider joining a support group for victims		immediately	Victims for Justice (555-1289)
	client will learn basic relaxation techniques to assist with reduction of immediate symptoms		next appointment	Chris B.

## 6

The Challenge of Threats of Suicide: The Duty to Protect

Suicide represents the final submission to self-destructive machinations. Negative reactions against the self are an integral part of each person's psyche, ranging from critical attitudes and mild self-attacks to severe assaults on the self. The latter includes feelings and attitudes that predispose physical injury to the self and eventually the complete obliteration of the self. No one reaches maturity completely unscathed by their personal experiences during the developmental years and no person is completely exempt from a suicidal process that leaves its mark on every life.

ROBERT FIRESTONE, *SUICIDE AND THE INNER VOICE*

According to one of the most well-known suicide researchers in the United States, Edwin Shneidman, "suicide is chiefly a drama of the mind" (1997, p. 23). Shneidman believes that suicide as an act is motivated by two primary factors: psychological pain and unmet or distorted psychological needs. The causes for psychological pain and the sources of unmet psychological needs are of course multifold and highly pervasive in modern society, as the following discussion of risk factors will point out. Not surprisingly, the suicide rate for the general U.S. population is approximately 12 per 100,000. It is the ninth leading cause of death overall and accounts for about 1% of (or 30,000) total annual deaths, with one suicide occurring every 17 minutes (Clark, 1995; Firestone, 1997; Meyer & Deitsch, 1996). It is the third leading cause of death for young people between the ages of 15 and 24 and the fifth leading cause for those between 25 and 64 (Cooper-Patrick, Crum, &

Ford, 1994). For several population groups (e.g., at-risk adolescents, substance-using clients, emotionally stressed individuals), suicide risk is greatly increased—these often happen to be the very people a clinician is likely to encounter in mental health treatment (Meyer & Deitsch, 1996). Suicidal threats and attempts occur with even greater frequency, although there are "no reliable national studies elucidating the incidence and prevalence of suicide attempts" (Silverman, 1997, p. 13). Not surprisingly, suicidal ideation and threats are the most commonly encountered crisis situation faced by mental health service providers (Clark, 1995; Firestone, 1997). Perhaps it is best to start a discussion about suicide with several common myths that have been identified by clinicians experienced in researching and treating suicidal clients. In fact, the argument has been made that dispelling myths about suicide is the single most important step in the societal prevention of suicide (Shneidman, 1985). Myths

1–5 are taken from Nystul (1993; p. 348); Myths 6–8 are taken from Grollman (1988; pp. 41–42); Myths 9 and 10 are taken from Gilliland and James (1993; p. 132).

Myth 1: "Suicide is only committed by people with severe psychological problems."

Myth 2: "Suicide usually occurs without warning."

Myth 3: "People who are suicidal will always be prone to suicide."

Myth 4: "Discussing suicide may cause the client to want to carry out the act."

Myth 5: "When a person has attempted suicide and pulls out of it, the danger is over."

Myth 6: "The tendency toward suicide is inherited."

Myth 7: "Suicidal youths are mentally ill."

Myth 8: "Nothing could have stopped her once she decided to take her life."

Myth 9: "A suicidal person who begins to show generosity and share personal possessions is showing signs of renewal and recovery."

Myth 10: "Suicide is always an impulsive act."

As the following discussion will show in great detail, these statements, despite their common acceptance, are incorrect and, in fact, are often virtually the opposite of what statistics about suicide reveal. Briefly, a good proportion of people who successfully committed suicide had not received a prior psychiatric diagnosis (see Capuzzi & Nystul, 1986). Most suicide attempts are preceded by a verbal threat or warning; in fact, at least two-thirds of clients who attempt to kill themselves tell someone about their intent (Faiver, Eisengart, & Colonna, 1995). Suicide can be induced by a temporary crisis; once resolved, the client may not present suicidal ideation again (Slaikeu, 1990). Talking about suicide, quite contrary to commonly held beliefs, may actually decrease a person's risk for carrying out the act; the threat represents a cry for help and discussion permits a means of validating the client's emotional state (Hood & Johnson, 1997). The greatest period of danger may actually occur after a person has made an unsuccessful attempt or after depressive or anxiety symptoms may appear to have been resolved (Nystul, 1993). Despite the finding that familial suicide increases suicide risk, there is only limited evidence that there is a genetic link to suicide (Meyer & Deitsch, 1996); instead, it is more likely that factors leading to suicide (such as affective disorder) may be inherited or that suicidal behavior is modeled (Stelmachers, 1995). Similarly, although mental illness increases risk, it cannot be presumed as a contributing factor, especially among young people (Grollman, 1988). Often suicide threats and gestures are cries for help that, if heeded, can lead to significant improvement in mood as well as the prevention of the action (Fujimura, Weis, & Cochran, 1985). If suicidal persons begin to give away their possessions, this is usually a signal that they have made a final decision to follow through with their intent (Stelmachers, 1995). Suicide is not always impulsive, instead, it is often carefully considered and deliberately planned (Gilliland & James, 1993).

Given that these common beliefs are mere myths, it is important to explore what the true risk factors are that signal that suicide assessment and intervention may be necessary. Only the clinician familiar with risk factors will know how to conduct a suicide assessment and how to make appropriate intervention decisions. This is a critical skill: the client's life may be at stake. It is also important from a legal perspective, because the counselor has some professional responsibility to the client based on ethical codes that apply to the clinician's profession.

LEGAL AND ETHICAL ISSUES FACING THE CLINICIAN OF THE SUICIDAL CLIENT

According to the ethical standards for psychologists established by the American Psychological Association (1992), "[p]sychologists disclose confidential information only as mandated by law or where permitted by law for a valid purpose, such as . . . to protect the patient or client or others from harm" (p. 1606). This confidentiality exemption was written with suicidal clients in mind, clarifying that the psychologist and similar mental health professionals have a duty to protect clients from harm. Similarly, the American Counseling Association, American Medical Association, American Psychiatric Association, National Association of Social Workers, and other professional groups who provide mental health services have endorsed ethical guidelines. Reading these ethical standards serves to remind the mental health service provider that the serious threat of harm supersedes the rules about confidentiality. The mental health care provider's first concern during times of serious threat is to protect the client—this protection in fact is perceived as a duty and generally receives little debate among associations who set guidelines for mental health professionals (see Freemouw, Perczel, & Ellis, 1990).

However, individual providers have argued that any forced treatment (such as commitment to a hospital or placing the client under the constant care and surveillance of a significant other) violates many of the implicit and explicit contracts and understandings between a client and a counselor. In one survey of psychotherapists, 81% indicated that they believe in rational (i.e., defensible) suicides, leading the authors to suggest that current professional standards may need to be revised to incorporate less absolutist standards of care (Werth & Liddle, 1994). Clinicians who believe in the right to suicide argue that forced treatment most obviously violates confidentiality; it also results in the clinician taking over for the client and making major decisions for the client, thus robbing the person of autonomy and willfulness (Lemma, 1996). Although absolutist standards of care that impose the duty to intervene on the clinician present a moral dilemma for some (e.g., Auld & Hyman, 1991), no *legal* gray area exists: according to the law, the clinician has a duty to protect (Swenson, 1997).

One way to reframe this experience of taking over the responsibility for a client's care is to recognize that while clients are in a suicidal crisis (i.e., feel depressed, helpless, or hopeless; or are under the influence of substances or voices), they are not making the best possible judgments and decisions. Thus, suicide prevention at such times is best viewed as a means of assuring that clients do not make decisions they would not have made had other interfering factors not been present. Once clients have come through the suicidal crisis, they may be better equipped to make rational and logical choices about life and death.

Although a significant amount of legal and ethical attention has been paid to suicide threats, attempts, and completion, no single set of standards of care has been identified as the answer to all suicide work (Silverman, Berman, Bongar, Litman, & Maris, 1994). In fact, even if an ideal or optimal standard of care existed, some clients probably would still end up killing themselves despite the best efforts of mental health care providers to prevent this action. This chapter is not meant to provide the definitive answer to suicide assessment or prevention. It can only provide guidelines that may be helpful to the clinician who works with the suicidal patient. Even following these guidelines, however, is no guarantee of successful prevention. The field of suicide assessment and prevention is not yet, and in all likelihood never will be, at a point where it can predict (and thus prevent) all suicides (Stelmachers, 1995). The good clinician must be prepared clinically and knowledgeable ethically and legally; this chapter will assist only with this preparation and knowledge.

Clinicians are held to a legal standard of care, which says they will provide "reasonable care and skill (the degree of learning, skill, and experience which ordinarily is possessed by others of the same profession)" (Stelmachers, 1995, p. 367). This standard assumes and implies that therapists or counselors have the skills necessary to do an adequate risk assessment, and that they will not just simply ask whether a client feels suicidal. Indeed, the skilled clinician must be capable of conducting an

TABLE 7-1   Gradation of Risk Factors for Violence

RISK FACTOR	HIGH RISK	MODERATE RISK	LOW RISK
PERSONALITY FACTORS			
Anger/Range Hostility	strong	moderate	minimal
Impulse Control	absent to poor	variable	adequate
FAMILIAL FACTORS			
Parenting	authoritarian	inconsistent	adequate
Violence	high levels present	some violence	no violence
SOCIAL AND ENVIRONMENTAL FACTORS			
Role Models/Peers	violent acting-out	endorse aggression	reject aggression
Isolation	isolated and rejected	some social contact	regular social contacts
Environment	conducive to violence	conducive to violence	not conducive to violence
PSYCHOLOGICAL FACTORS			
Stressors	multiple	some	none
Recency	within days; chronic	within last month	none or long ago
Sensation-Seeking	strong	moderate	none to low
Fascination with Weapons	strong	moderate	none
MEDICAL AND PHYSICAL FACTORS			
Relevant Illnesses	present	present	absent
Motoric Clues	several present	some present	none present
PSYCHIATRIC HISTORY			
Substance Use	severe	moderate	low
Type of Diagnosis	Antisocial Personality, Conduct Disorder, Intermittent Explosive	Bipolar Disorder, Schizophrenia	none
# of Diagnoses	multiple diagnoses	one diagnosis	no diagnosis
Severity	severe	moderate	none
IMMEDIATE PREDICTORS			
Victim	identified individual	identified target group	none identified
Access	present	present	absent
Method	decided	decided	undecided
Means	in possession	easy access	still to be secured
Time and Place	definitely chosen	tentatively chosen	not chosen
Lethality	high	moderate	low
History of Violence	present	present	absent

ahan, 1993). In making decisions about the risk of harm, the clinician needs to revisit the issue of lethality of the plan as well as the identity and whereabouts of the potential victim. Identification of the threatened person is considered a crucial issue because the clinician must know who is threatened to be able to make an appropriate warning should risk for follow-through and lethality be high (Meyer & Deitsch, 1996). Slaikeu (1990) also suggests that former attempts should be investigated, much as one investigates prior suicide attempts. Specifically, he suggests that the clinician needs to ask whether the client has ever attempted to harm the identified victim before, and, if so, how the attempt was made. The exact circumstances of the event need to be explored to assess how related they may be to the currently planned attempt. The level of success of the prior attempt also needs to be assessed, asking questions about level of harm done to the victim at that time, as well as level of satisfaction perceived by the perpetrator after the action.

A final issue that needs to be considered in the assessment of risk for action and level of harm is the presence of current clues to violence in the client. Although it is entirely possible for clients to act out even if they fail to demonstrate aggression or anger during the session with the clinician, at times, actual physical clues are available in session. These clues, if present, suggest increased risk. They include signs such as muscle tension or twitching, darting eye movements, bulging eyes, paranoid eyes, defensive body posture, pacing, and rapid or confused speech. As noted earlier, these signs are also to be taken seriously by the clinician with regard to potential in-session violence. If clients become extremely agitated during a risk assessment, clinicians may have to consider the possibility that they will not only present a threat to the potential victim identified in their plan for aggression but also to the clinicians who may try to stand in clients' way of carrying out the plan. Thus, any clinician conducting a risk assessment for aggression also needs to be prepared to deal with in-session aggression. Under extreme circumstances, security personnel may need to be involved to protect the

clinician. For a thorough discussion of in-session violence, the reader is referred to Chapter 5.

To assist with risk assessment, the previously presented risk factors and their associated level of danger are summarized briefly in Table 7-1. This table follows the same format that was used for the corresponding table in Chapter 6. The information in this table can be used in the same way as it was used with regard to suicide assessment. Specifically, following (but expanding) Freemouw, Perczel, and Ellis's (1990) model, aggression and homicide prediction may best be based on several issues: imminence of the behavior (ranging from likely never to occur to likely to occur immediately); clarity of danger (ranging from vague or no time, place, method, and opportunity to very specific time, place, method, and opportunity); specificity in the identification of the victim (ranging from threat to the general public to specific victim identified); intent (ranging from no wish to harm the potential victim to strong wish to do harm); lethality of behavior (ranging from nonlethal or minimally harmful method to highly lethal or harmful method); and probability estimate of the occurrence of the aggressive act (i.e., reaction potential; ranging from not likely to occur to highly probable to occur). Each of these decisions is based on the level of risk expressed by the content of each of the risk factors outlined earlier and summarized in Table 7-1.

Once risk for follow-through by the client and level of harm or lethality to the potential victim has been estimated, the therapist plans intervention based on the clinical judgment of the seriousness of the situation. Fortunately for therapists, the main decision from a legal perspective is whether the therapist acted prudently and reasonably (i.e., not negligently); counselors who did a thorough assessment and planned interventions based on the subsequent clinical judgment are generally not penalized by the courts should they have made an honest and unpredictable mistake. This is of little consolation, however, to clinicians who are faced with their mistake in the form of a client who killed or harmed an innocent victim. Hence, intervention must be carefully planned to reduce the risk of harm to a potentially innocent victim while objective assessment and consequent implementation of an intervention plan that can withstand the legal investigation that may ensue if clients do end up killing themselves (Bongar, Maris, Berman, & Litman, 1992). The essential questions with regard to such legal liability of the clinician in the event of a client suicide are related to malpractice or negligence (consisting of an evaluation of foreseeability and causation/control) and confidentiality (Ahia & Martin, 1993; Fujimura, Weis, & Cochran, 1985; Silverman, Berman, Bonger, Litman, & Maris, 1994; Swenson, 1997). Questions about negligence include the following:

Regarding Foreseeability (Appropriate Diagnosis and Prediction of Risk Behavior):

1. Was the clinician aware or should the therapist have been aware of the risk?

2. Was the clinician thorough in her or his assessment of the client's suicide risk?

3. Did the clinician make "reasonable and prudent efforts" to collect sufficient and necessary data to assess risk (Silverman, Berman, Bonger, Litman, & Maris, 1994, p. 155)?

4. Were the assessment data misused, thus leading to a misdiagnosis where the same data would have resulted in appropriate diagnosis by another mental health professional?

5. Did the clinician mismanage the case, being either "unavailable or unresponsive to the client's emergency situation" (Ahia & Martin, 1993, p. 36)?

Regarding Causation (Protection against Harm, Control of Behavior, and Provision of Safe Environment):

6. Was the clinician negligent in the way she or he designed her or his intervention with the client after assessing risk?

7. Did the clinician make adequate attempts to keep the client safe (i.e., set up a plan of contingencies with appropriate resources, phone numbers, etc.)?

8. Did the clinician remove the means to be used by the client in the suicide attempt?

The issue of negligence or malpractice in cases of client suicide is a significant one, representing the fourth most common cause for claims of malpractice against mental health practitioners (Swenson, 1997), although it has been estimated that less than 10% of these claims actually go to trial (Bongar, Maris, Berman, & Litman, 1992). Lawsuits initiated due to suicide are usually tort lawsuits attempting to establish professional malpractice on the part of the care provider. A tort is a civil, as opposed to criminal, wrongful act perpetrated by a person or institution against another person. To win a tort lawsuit, the plaintiff has to prove that the defendant had a legal or special duty to the plaintiff, breached that duty, and injured the plaintiff through this breach of duty (Swenson, 1997). Considerable legal precedent has been established to rule that mental health care providers and clients (even after as little as one intake contact) have formed a relationship that creates a special duty on the part of the therapist or counselor toward the client. This legal duty requires therapists to exercise their services and care for the client in such a manner as to protect clients from themselves, as well as to protect others from the client. However, for the defendant to become legally liable, the plaintiff has to prove actual and proximal cause. This means that the accuser has to prove that the injury to self or other would not have occurred had it not been for the clinician's action (or lack thereof). Not surprisingly, as long as mental health and health professionals have been able to show prudent and responsible care (through assessment of risk and tailored intervention planning), the courts have tended to rule in favor of the practitioner (Firestone, 1997). Given that most suicide-related malpractice claims never go to trial, and that those that do are often found in favor of the defendant, some writers have concluded that "while lawsuits against mental health practitioners that go to trial are traumatic experiences, they remain a relatively rare occurrence. Thus, the fear of being sued probably has more widespread and deleterious effects on clinicians than do actual lawsuits" (Bongar, Maris, Berman, & Litman, 1992, p. 455).

Despite this conclusion, it appears prudent and extremely important that all mental health clinicians be thoroughly familiar with risk factors, procedures for assessment, and guidelines for intervention. Further, proper documentation in case notes and reports that are placed in the client record are essential should a malpractice suit be initiated by surviving family members of the suicidal client. "Any therapist, regardless of how competent, successful, and skilled, may lose a client through suicide. What will distinguish this therapist from another who was clearly negligent, careless, and indifferent to her or his client's suicidal state is the presence of a well-documented, thorough client record" (Freemouw, Perczel, & Ellis, 1990, p. 10).

Questions about confidentiality may be raised by the surviving client or by the client's family members if the counselor activates the client's support network or the police in an attempt to protect the client from harm. However, the mental health practitioner is protected from legal prosecution as long as the risk-benefit to the client can be properly documented. Indeed, most, if not all, states in the United States have laws that protect the practitioner who discloses confidential client information in an attempt to save a client's life during a suicidal crisis. Therapists need to be familiar with the mental health statutes of the state in which they practice, so that they know the limits and boundaries of confidentiality not only as defined by their profession's code of conduct (ethical guidelines) but also by their state's laws.

HELPFUL TRAITS DURING SUICIDE ASSESSMENT AND INTERVENTION

This section is not to imply that particularly extraordinary or special skills are needed above and beyond the usual therapeutic skills necessary for work in the mental health field. However, it is intended to highlight a few of the therapeutic skills possessed by all counselors that are of particular importance and assistance in suicidal crisis situations. It is also intended to clarify the importance of clinicians building a certain knowledge base about suicide assessment and intervention before beginning to see clients.

First of all, the counselor must remain calm and collected, must not panic, and has to be able to keep listening even if the session reaches crisis dimensions (Grollman, 1988). Therapists need to be reassuring and soothing while remaining supportive, but firm (Boylan, Malley, & Scott, 1995; Firestone, 1997). Mental health care providers also pay attention to their reaction to the client, keeping negative reactions under control (Firestone, 1997). If the counselor is unable to differentiate between genuine and manipulative suicidal ideation and behavior (Meyer & Deitsch, 1996) (Shneidman [1985] calls the latter parasuicide, in which the suicidal gesture is undertaken to change others or the environment without the true intent for death), it is best to err on the side of caution and believe the threat (unless there has been a long history of borderline type behavior [Clark, 1995; Grollman, 1988]). This is also a good idea because accidental success among parasuicidal clients may actually lead to death.

Regardless of therapists' own religious, moral, or cultural beliefs, they avoid being moralistic or preachy. It is necessary to maintain an impartial stance toward the client's intent. However, impartiality is not to be confused with indifference (Grollman, 1988); clinicians have a responsibility to protect the client from self-harm. However, they can and must do so without resorting to moral, religious, or cultural arguments against the client's desired action. Similarly, clinicians accept the client's suicidal thoughts; they do not disavow or deny clients their ideation and affect (Boylan, Malley, & Scott, 1995). Further, therapists are willing to talk about the topic of suicide without fear or hesitation. Skilled clinicians do not avoid the topic of suicide. Despite common fears to the contrary, counselors will not induce suicidal behavior by asking about it (Faiver, Eisengart, & Colonna, 1995; Hood & Johnson,

1997). Another reason why mental health care providers may avoid the topic of suicide is that it forces them to face their own fears or attitudes about death (Lemma, 1996). Thus, some self-awareness with regard to attitudes about death and dying and an exploration of personal death anxiety will help prepare a clinician to face and deal with a suicidal client without imposing personal emotions and needs.

As the topic of suicidality is explored, the clinician must not use euphemisms or metaphysical terminology, such as "hurting yourself," "seeking relief," or "wanting to pass away." Suicide is a cold reality and it is best to use terms and labels that reflect this, such as "killing yourself," "dying," or "death" (see Hood & Johnson, 1997). Direct questioning is essential (Firestone, 1997). Exploration of the topic will go a lot more smoothly and will be much more relevant if the therapist or counselor is fully knowledgeable about risk factors, knows how to assess and evaluate them, and is able to recognize the clues for possible suicidal action (see below; Grollman, 1988). Before the clinician sees the client, he or she must be prepared to ask the right questions. A clinician cannot wait until a suicidal client presents to learn about suicide, how to assess its risk, and when to intervene. Instead, this knowledge, much like the ability to listen actively and to empathize, must be part of the clinician's bag of skills. Questions about suicidal ideation must be asked if the client indicates severe depressive symptomatology, as evidenced by periods of significant sleep disturbance, severely dysthymic mood with loss of interest and pleasure in things usually enjoyed by the client, feelings of worthlessness and/or guilt, and hopelessness (Cooper-Patrick, Crum, & Ford, 1994).

In addition to being knowledgeable about risk factors and how to inquire about them, the clinician must also be familiar with the spectrum of intervention procedures that are available. These skills (discussed in detail later in the chapter) include knowledge about the difference between therapy and the crisis intervention mode; understanding of outpatient management options; comprehension of the techniques and components of suicide contracts; familiarity with local resources, such as numbers for crisis lines, hospitals, and so forth; knowledge of the voluntary referral process and quick access to phone numbers; and familiarity with involuntary commitment procedures, and preferably a working relationship with local police officers. Finally, throughout the intervention process, the counselor should try to instill hope while trying to facilitate help-seeking behaviors (Grollman, 1988).

Either during or after the crisis, the clinician must seek supervision and/or consultation, as appropriate to the level of training. Most generally, a therapist who still works under supervision should make some contact with a supervisor before the severely suicidal client leaves the premises to confirm judgment about the level of lethality involved in the case. A trainee must seek supervision when faced with a suicidal client, but should do so after having gathered all the necessary assessment information; then the supervisor should be involved in the decision-making process about level of intervention. Being prepared for consultation by having asked relevant questions that can be discussed with the supervisor (Boylan, Malley, & Scott, 1995) is the best course of action for all levels of practitioners.

RISK FACTORS FOR SUICIDE

"Although it is not possible to prevent suicide, it is possible to recognize the existence of common crises that may precipitate a suicide attempt and reach out to the people who are facing them" (Fujimura, Weis, & Cochran, 1985). Being cognizant of relevant risk factors for and commonly associated variables of suicide will help mental health care providers with such recognition. A variety of risk factors are associated with suicide, ranging from long-term predictors to immediate precursors of the action. All need to be considered in making final judgments about risk or lethality, as will be discussed later in the assessment section.

Demographic Characteristics

The demographic characteristics associated with suicide risk have remained very stable over the decades and have been reported with great consistency by many writers (e.g., Cooper-Patrick, Crum, & Ford, 1994; Clark, 1995; Forster, 1994; Firestone, 1997; Fremouw, Perczel, & Ellis, 1990; Grollman, 1988; Maris, Berman, Maltsberger, & Yufit, 1992; Meyer & Deitsch, 1996; Slaikeu, 1990; Sommers-Flanagan & Sommers-Flanagan, 1995; Stelmachers, 1995). Given this consistency, references are not provided for every single demographic factor. When no reference is provided, the reader can assume that all authors cited above have reported the pattern; only for those demographics for which there is a less clear-cut pattern will references be included below.

Despite the finding that women make more suicide attempts than men, the male to female ratio for completed suicide remains three to one, suggesting significantly higher risk for men than women. This is at least partially related to patterns that indicate that men employ more lethal methods than women do. For example, 65.6% of men who commit suicide do so by using firearms or explosives, as compared to 42% of women; on the other hand, 26.7% of women commit suicide by using poisons as compared to 6.2% of men (Firestone, 1997). In general terms, risk for suicide increases with age. White men who are over 45 present a greater risk than White men who are under 45; White women between 35 and 65 are at greater risk than women under 35 or women over 66. For people of other ethnic groups, risk is highest between the ages of 25 and 34, with lower risk for those aged 24 or younger and those aged 35 and older. There are some exceptions to these ethnic patterns for Native American and Alaska Native groups, with excessive suicide rates for teenage boys (Brems, 1996). Some recent statistics suggest that age is becoming less of a predictor, with other factors such as race and marital status mediating the importance of age (Clark, 1995).

Overall, Whites are more likely than other ethnic groups to complete suicide, again with the notable exception of adolescent boys who are Native Alaskans (Brems, 1996). At particular risk, according to some, are White males aged 15 to 19 (Grollman, 1988). If gender and age are disregarded, the two highest risk groups are those between ages 15 and 24 and those over the age of 65 (Silverman, 1997). According to Silverman (1997), suicide rates for this younger risk group increased steadily during the 70s and 80s, but have now reached a plateau. For the older group, however, rates continue to climb.

There are many social factors that apparently indicate a person's risk for suicide. Among suicide completers, unmarried people outnumber married people four to one. This finding may be related to the finding that solitary living arrangements appear to increase risk as compared to any other living situations (Fremouw, Perczel, & Ellis, 1990). Further, urban dwellers seem to be at higher risk than people from rural populations (Fremouw, Perczel, & Ellis, 1990; Lemma, 1996), again with the possible exception of Alaska Native peoples (Brems, 1996). Risk is higher among the unemployed than the employed, though some types of occupations have higher suicide rates than the general population (e.g., physician's suicide rate is twice the national average; psychiatrist's four times [Grollman, 1988]).

Beyond demographic factors, statistics show that suicide may be a cyclical phenomenon with regard to time of year. Specifically, the suicide rate for April is 120% percent above the average rate for the rest of the year and is rivaled only by the suicide rate around the winter holidays (Grollman, 1988). It must be noted, however, that this finding could be an artifact of situational stress or trauma (accounting for high holiday rates) or of depression associated with seasonal variation (accounting for high April rates: seasonal affective disorder symptoms peak approximately six weeks after winter solstice and clients may take additional time to recover enough energy to carry out the suicide attempt).

Coping and Personal Style Factors

Another set of risk factors has to do with coping or daily functioning skills, in which risk increases with level of impairment and risk decreases with stability in lifestyle or routine (Shneidman, 1985). Risk increases when internal and external coping resources decline (Clark, 1995; Fremouw, Perczel, & Ellis, 1990) In these cases, internal resources refer to factors such as good problem-solving skills, variables that were identified as having kept the client from committing suicide in the past, plans for the future, skills demonstrated in the past to deal with emotional crises; external resources refers to social and familial supports (see below).

A number of personality variables have also been identified as risk factors. For example, hostility has been reported to increase risk with severity, as has the client's inability to express emotions. Perfectionism or overly responsible behavior, which can lead to self-blame and guilt, also increases risk, as do pessimism, dependency, rigidity, impulsivity, and sensitivity or tender-mindedness (Fremouw, Perczel, & Ellis, 1990; Stelmachers, 1995). The level of impulsivity is a particularly important personality factor indicator because clients who easily lose self-control present a higher risk (Sommers-Flanagan & Sommers-Flanagan, 1995).

Social Support and Social Pressure Factors

The more social support resources clients can claim, the lower the risk that they will successfully complete suicide. Support resources include family members, confidants, close friends, neighbors, coworkers, as well as external social supports that consist of professionals with crisis management or therapeutic skills (e.g., community availability of a crisis line or emergency treatment center [Fremouw, Perczel, & Ellis, 1990; Slaikeu; 1990; Stelmachers, 1995]). Both lack of a social support network (Kleinke, 1994; Meyer & Deitsch, 1996) and lack of a therapeu-

tic alliance (Meyer & Deitsch, 1996) increase risk.

Risk increases as the number of reported significant others decreases. If the client can report no (or only one) significant social resource person, risk is considered high; similarly, risk increases as regularity and number of contacts with others decreases (regardless of whether they are significant or intimate others [Fremouw, Perczel, & Ellis, 1990]). For adolescents, there may be a social factor that has not so much to do with social isolation as with social attachment to others or pressure from a peer group (Velting & Gould, 1997). Specifically, some evidence has suggested a clustering effect: if one young person commits suicide, other teen suicides follow in the community, a finding that has implications for community-level suicide prevention (Grollman, 1988; Velting & Gould, 1997). However, it must be noted that in the overall picture of suicide rates, cluster suicides are relatively rare; even among adolescents, where the rates are highest, cluster suicides account for only 1 to 135 of all suicides (Velting & Gould, 1997).

Additionally, it is not enough to explore availability of social supports, rather it may be more critical to determine whether clients are willing to use these resources. The client who cuts off or refuses all communication is at a significantly increased risk (Firestone, 1997; Boylan, Malley, & Scott, 1995). Thus, clinicians must be careful to differentiate very clearly between availability of resources versus the client's willingness to access resources. Numerous people may exist in the person's social network who are willing to lend support; however, if the client is too depressed, helpless, hopeless, or listless to ask for help, these apparent resources may become meaningless (Slaikeu, 1990).

Absence of social support is not the only social factor that appears to increase risk for suicide. Silverman (1997) notes that suicide rates have been increasing among gay and lesbian individuals as well as among those inflicted with HIV/AIDS. He concludes that these increases are due, not to the experience of stress secondary to a person's sexual orientation or health status, but rather to the lack of social acceptance experienced by these individuals.

Family Variables

It has been noted that individuals who come from families with a history of rejection or instability are at higher risk for suicide than individuals without such a history (Lemma, 1996; Meyer & Deitsch, 1996). Family history of suicide has implications for likelihood of suicide completion, with risk increasing from no family history of suicide to family history of parasuicide (also called suicide gestures) to completed suicide by family members (Grollman, 1988; Meyer & Deitsch, 1996). Risk is particularly high if a family suicide occurred recently or if an anniversary of a family member's completed suicide is impending (Faiver, Eisengart, & Colonna, 1995).

Psychiatric and Medical History

Other important risk factors are associated with the client's history of psychiatric and medical illness. Of those persons who successfully commit suicide, 38% have a known history of psychiatric disorder, and of those persons with psychiatric disorders 4.75% end up killing themselves (Stelmachers, 1995). Thus, taking a psychiatric history is critical to proper risk assessment (Firestone, 1997; Forster, 1994). This includes, not only the assessment of prior or current psychiatric diagnosis, but also the client's adaptation to prior psychological treatment. The more negative the client's response was to previous treatment, the greater the risk. With regard to psychiatric diagnosis, several diagnostic groupings have been identified that significantly increase risk of suicide completion. Not surprisingly, affective disorders are one of these categories. Risk is higher for clients with a history of diagnosed depression or bipolar disorder, especially disorders with highly cyclical mood swings (Clark, 1995; Meyer & Deitsch, 1996).

Clients with a diagnosis of schizophrenia also represent a significant risk, especially if voices are present that challenge them to kill themselves. In such a case, risk is always sufficiently severe to warrant hospitalization (Hood & Johnson, 1997). The schizophrenia factor may be related to the finding that disorientation and disorganization correspond to suicide risk, with risk increasing as the severity of these symptoms increases (Gilliland & James, 1993). Another diagnostic category commonly associated with high suicide completion risk is panic disorder/attacks (Bellak & Siegel, 1983; Clark, 1995; Meyer & Deitsch, 1996) and other anxiety disorders (Faiver, Eisengart, & Colonna, 1995). Patients with personality disorders are also at high risk, though they are more likely to make gestures or attempts, rather than successfully completing suicide (Clark, 1995). For all of these disorders, risk increases with severity (Faiver, Eisengart, & Colonna, 1995; Grollman, 1988; Hood & Johnson, 1997; Meyer & Deitsch, 1996).

Cooper-Patrick, Crum, and Ford (1994) reported odds ratios for a variety of psychiatric disorders with which participants in their sample of over 6,000 general medical patients had been diagnosed. These odds ratios clearly demonstrate the highly increased risk for clients with psychiatric disorder; namely, patients with bipolar disorder were 65.3 times as likely as those without it to report suicidal ideation. Odds ratios for panic disorder were 47.3; for major depression 40.3; for schizophrenia 25.3; for drug use (other than alcohol) 11.9; for dysthymia 9.8; for alcoholism 9.5; and for phobias 6.3.

Clients with psychiatric histories are particularly at risk for suicide completion during the three months immediately after discharge from a psychiatric hospital (Clark, 1995). Over 80% of psychiatric patients who commit suicide were recently discharged from inpatient to outpatient care, a fact that points "to the need to provide aggressive follow-up care after hospital discharge" (Stelmachers, 1995, p. 377). Risk also increases if there are unanticipated signs of improvement or a sudden disappearance of symptoms with the outpatient (or even inpatient) client suddenly becoming calm and resolved (Faiver, Eisengart, & Colonna, 1995; Grollman, 1988; Hood & Johnson, 1997; Meyer & Deitsch, 1996). This sudden calmness may be related to the client's decision to end life and may reflect an inner peacefulness that was reached in the face of finally having made such a decision. It signals grave danger with regard to the person's resolve and the risk of suicide completion. Risk may be increased, not only if sudden improvement was noted, but also if there are any changes in clinical features or any other abrupt behavior changes (Gilliland & James, 1993; Meyer & Deitsch, 1996).

Last but not least in the category of psychiatric illness, both drug and alcohol use are highly associated with suicide risk, with risk being highest among chronic abusers (Clark, 1995; Forster, 1994; Grollman, 1988; Lemma, 1996). Drug use not only precipitates emotional or physical states (such as depression and anxiety; or intoxication and withdrawal) that may lead to suicide, but also provides the client with the means to complete the act. In fact, it has been pointed out that "drugs are currently a major means of committing suicide" (Grollman, 1988, p. 10). Drug use in conjunction with other psychiatric illness, especially with affective disorders, increases risk even further (Brems & Johnson, 1997; Clark, 1995). Further, drug and alcohol use is perceived as a pattern or evidence of self-destructive behavior that eventually leads to death, sometimes referred to as "subintentioned suicide" (Stelmachers, 1995, p. 376).

Finally, with regard to physical illness and its relationship with risk for suicide, the strongest link has been established for chronic, incurable, and painful conditions such as cancer, peptic ulcers, spinal cord injuries, multiple sclerosis, head injury, and Huntington's Chorea (Grollman, 1988; Stelmachers, 1995). The relationship between physical illness and suicide appears to be stronger among men than women and for depressed versus nondepressed medical patients (Stelmachers, 1995). The latter suggests that depression may be a mediating factor; the physical

condition in and of itself may not cause suicide, but rather it may lead to depression, which in turn may establish the link to suicidality. One final comment must be made in the context of physical illness: the information in this chapter does not deal with the issue of assisted suicide among the chronically, severely, or painfully medically ill patient who chooses to die with dignity. This chapter deals with suicide as a psychological issue only, a common approach to the topic of suicide assessment and intervention by mental health practitioners (see Fremouw, Perczel, & Ellis, 1990).

Psychological Factors

Not surprisingly, a large number of psychological variables need to be factored into the suicide prediction and assessment equation. Recent losses such as a financial loss, the loss of a job, a relationship, or a dream can be related to suicidal ideation (Lemma, 1996). Generally, the highest risk exists within days of the loss, especially if clients blame themselves for the event (Gilliland & James, 1993; Hood & Johnson, 1997). From a psychological standpoint, the experience of trauma is quite similar to the experience of loss. Recent trauma (Kleinke, 1994) and other environmental stressors (e.g., promotion, illness, academic failure [Gilliland & James, 1993; Grollman, 1988]) make the client more vulnerable to suicidal ideation and completion. However, this is true, not only for recent trauma, but also for those time periods that signify the anniversary of a trauma experienced some time ago.

"[T]he most powerful antecedent" of suicide is hopelessness (Stelmachers, 1995, p. 374) and increases in its level are directly related to increases in risk. Similarly, despair increases risk (Seligman, 1996), as does helplessness (Kleinke, 1994; Shneidman, 1985). Cooper-Patrick, Crum, and Ford (1994) found that the most profound relationship of suicidal ideation was with symptoms of hopelessness, guilt, depressed mood, panic attacks, loss of libido, and insomnia (in that order). Related to the concept of hopelessness, the client's inability to articulate reasons for living is also considered to be a strong danger sign (Firestone, 1997). Hopelessness is, not only the best predictor of immediate suicide, but also of long-term suicide. In fact, it is the only variable that has been identified in the literature as being significantly related to the prediction or postdiction of both forms of suicide (Lemma, 1996; Stelmachers, 1995).

Immediate Suicide Predictors or Factors

Hopelessness was previously cited as the most powerful predictor of immediate suicide. In addition to hopelessness, there are several other factors that predict immediate (as opposed to long-term) suicide. All of these factors tend to be directly related to suicidal thinking or action. The first of these factors is the verbal suicide threat. Clients who have openly admitted to thinking about suicide or about killing themselves are at greater immediate risk for committing suicide than clients who have not made such threats (over the long run, however, this is not a good predictor of which clients may end up killing themselves). It has been found that 66–80% (depending on source) of people who kill themselves first informed someone of their intent (Faiver, Eisengart, & Colonna, 1995; Grollman, 1988; Meyer & Deitsch, 1996). Shneidman (1985) has pointed out that most suicidal clients have some ambivalence about the choice and the act of suicide. This ambivalence is expressed when the client chooses to make the threat of suicide to the clinician. Although this threat may hence be perceived as a request for help (Firestone, 1997; Fujimura, Weis, & Cochran, 1985), it nevertheless serves to increase the client's risk.

Another immediate risk factor is the client's history of suicidal behavior and intent. Immediate risk increases with the number of previous attempts, with lower risk with no previous attempt, moderately higher risk with one attempt, and significantly higher risk with multiple attempts. Of those persons who actually end up

killing themselves, 30–40% had made previous attempts (Clark, 1995; Maris, 1992); additionally, 20–25% of chronically suicidal clients ultimately succeed in killing themselves (Litman, 1992). Immediate risk increases further if the client has evidenced impulsive behaviors in the past that suggest a loss of control over behavior that may result in a suicidal act (Sommers-Flanagan & Sommers-Flanagan, 1995).

Another immediate risk factor has received many labels ranging from closure behaviors (Grollman, 1988) to preparations (Stelmachers, 1995) to final arrangements (Meyer & Deitsch, 1996). These behaviors include actions such as making wills or otherwise getting one's affairs in order, calling friends or family members to discuss life events or plans for funerals and memorial services, giving away prized possessions, reminiscing with friends and family, and writing suicide notes. Risk is considered highest if a written note was left and possessions were given away; risk is deemed moderately high with other types of preparations (Hood & Johnson, 1987; Seligman, 1996; Stelmachers, 1995).

Perhaps the most important set of immediate suicide risks or predictors, is related to the suicide plan, about which the therapist or counselor assessing suicide threat must ask very specific questions (more detail about the actual assessment will be provided later). Four aspects about the suicide plan are important (Faiver, Eisengart, & Colonna, 1995; Hood & Johnson, 1997; Seligman, 1996; Sommers-Flanagan & Sommers-Flanagan, 1995): (1) the method, (2) the availability of the means, (3) decisions about time and place, and (4) the lethality of the plan. With regard to the method, clinicians check whether clients who have indicated suicidal ideation have thought about how to kill themselves. Difficult as this question may be for the novice counselor to ask, its answer is critical to the decision-making process about how to intervene. Methods for suicide are wide ranging and include too many possibilities to cover here. Some of the most common forms of planning or committing suicide appear to include drug overdoses (using prescription, over-the-counter, and/or illegal substances), hanging, shooting or stabbing, self-initiated car accidents, jumping off of bridges or high buildings, carbon monoxide inhalation (ovens or cars), and poisoning (Lemma, 1996). Methods appear to vary across cultures and gender. For example, the most common method among men in the United States is the use of guns; women, on the other hand, are most likely to use drug overdoses of prescription psychotropics (Lemma, 1996). Risk increases significantly if clients report that they have chosen a method and increases further if clients have engaged in mental rehearsal of the action (Clark, 1995).

Strongly related to method is the availability to the client of the means chosen. Not surprisingly, if clients indicate that they have chosen a particular method and have the means to carry through the plan, their immediate risk for committing suicide is significantly higher than if the means must yet be secured (Meyer & Deitsch, 1996). For example, a client who has indicated that she plans to shoot herself in the head but neither owns a gun, nor knows anyone who does, represents less of an immediate risk than the same client who has access to a gun owned by a friend or significant other, who in turn presents less of a risk than the client who upon hearing the question opens her bag to show the gun to the therapist! In the latter case, which is not as extraordinary as one may think or hope, it is critical to find out about the availability of ammunition. How to deal with such a situation will be covered in the intervention section later in this chapter.

Related to method and availability, but a separate issue and therefore assessed independently or additionally, is whether the client has decided upon a time and place for the action. If a method has been chosen and the means are available, the fact that the client has chosen a time and place significantly increases risk. However, even if the client has not chosen the means, a decision about time and place suggests a great seriousness of the threat (Freemouw, Perczel, & Ellis, 1990; Gilliland & James, 1990).

The final consideration about the suicide plan is the level of its lethality. This refers to the risk of method itself, not the overall risk of the client carrying out the threat. For example, if clients plan to shoot themselves, this represents a highly lethal method, because shootings are more likely to result in death than other methods, such as a drug overdose with sleeping pills. Similarly, if the plan involves hanging oneself, lethality of method is higher than it would be if the plan involved holding one's breath until asphyxiation (Firestone, 1997; Gilliland & James, 1993; Hood & Johnson, 1997). A client who has decided to commit suicide at a certain time and in a certain place (perhaps during an upcoming family event), who has chosen the method (e.g., driving a car off a known cliff on the way to the event), and who has the means available (i.e., owns a vehicle and has sufficient money to purchase gasoline) is at very high risk. The method is relatively lethal (survival, in other words, is unlikely), a definite plan exits, the means are available, and the client has made a decision about where and when. If the client has mentally rehearsed the method and has positive fantasies about it (see below), risk increases further.

One final possible piece of information to collect regarding immediate risk factors is a client self-rating with regard to risk. Self-reported risk may be assessed with a question such as "How likely do you think it is that you will act on your thoughts of suicide?" (Hood & Johnson, 1997, p. 256). This question is strongly related to, though not identical with, an assessment of strength of suicide intent (Sommers-Flanagan & Sommers-Flanagan, 1995). A similar question may be posed to the client in this regard: "How strongly do you feel that you really want to die?" Although no literature exists that gives scientific support to the predictive value of these questions, they may nevertheless be useful, at least by helping the clinician get a feel for how much ambivalence the client is experiencing about the possible suicidal act.

ASSESSMENT OF SUICIDE RISK

Once clinicians are theoretically familiar with or knowledgeable about all the risk factors that have been covered so far, they are ready to assess a client's suicidal ideation and to make a reasonable judgment about risk and lethality. Risk assessment is accomplished through careful listening and questioning (Gilliland & James, 1990), touching on all of the risk factors outlined above (Firestone, 1997). If mental health care providers already know the client well, they may already have several of these pieces of data, especially most of the general risk factors. More than likely, the client who has been with the therapist for several weeks has already disclosed much about family background, has demonstrated the level of social and coping skills, has given the counselor an idea of the level of social isolation versus connectedness, has discussed psychiatric history (including substance use and prior suicide attempts), and has revealed much about personal psychological state, including the degree of feelings of guilt, helplessness, and hopelessness. As such, with regard to risk factors for such a familiar client, all that may remain to be assessed are the immediate predictors that are suicide related, as well as a reassessment of any of the background factors that may have changed since the last time they were discussed in treatment. However, the therapist or counselor who meets with a new client who expresses suicidal ideation faces the challenge of securing all of the background information as well as the immediate predictors. Although this may appear to be a daunting task, there are no shortcuts and no omission. To be able to predict risk and to plan an appropriate intervention, the therapist has to work with a full repertoire of information. It is such complete data collection that, not only protects clients from self-harm, but also protects counselors from being ruled negligent by a court of law, should clients succeed in killing themselves despite a clinician's best efforts.

In addition to assessing the general background and immediate risk factors used to assess risk and lethality, several other pieces of information are useful in the final decision-making process about the risk presented by the client and the intervention that needs to be planned. These data include the client's current level of reality testing, illogical thinking, or cognitive bias; the presence of a triggering or precipitating event; the presence of inhibitory factors; and an assessment of the balance between inhibitory and triggering or facilitating factors.

Current Level of Distortion of Reality, Illogical Thinking, and/or Cognitive Bias

As previously alluded to, it is important to assess clients' fantasies associated with suicide. Clients who have positive fantasies about their death may be more at risk than clients who have painful or difficult fantasies or no fantasies at all. For example, a mother may have some painful fantasies that involve her children: she may fear for their welfare after her death or their guilt about her action. This fantasy may serve to inhibit her action if she is invested in not hurting her children. Conversely, an adolescent may have fantasies of how his parents and friends might react with guilt and shame after his death, feeling some vindication about his behavior. Such a fantasy may be a facilitating factor because the teen may fantasize about the pain he thus inflicts on people whom he would like to hurt. Fantasies about the suicide and its aftermath often give the clinician insight into the motivation for the suicide. No matter how irrational the fantasy or motivation may be, it is its force (i.e., triggering versus inhibiting potential) that must be considered (Seligman, 1996).

Another reality distortion or cognitive bias may be explored by assessing the client's perceptual accuracy with regard to current emotional state, perceived behavioral choices, and evaluation of others. Clients who "view their psychic pain as inescapable, intolerable, and interminable" (Meyer & Deitsch, 1996, p. 378) are at greater risk than clients who have a more realistic perception of their emotional state. Similarly, clients who believe that suicide is the only option left to change the path their life has taken show a great deal of distortion that may trigger the action. Finally, clients who believe that no one is available to them for help, either because they have isolated themselves intentionally, or because it is their perception that no one cares about them, are also at greater risk than those who can point to an available social support network and are willing to use it.

Distorted thinking is another common trait shared by suicidal clients (Grollman, 1988; Nystul, 1993). Clients may have distorted beliefs about the act of suicide as well as about the act of remaining alive. For example, some clients may view death as the only available "catalytic agent for the cessation of distress" (Meyer & Deitsch, 1996, p. 378); others may believe that they have nothing left to live for. They believe they have been abandoned by family and friends and have distorted the evidence around them to fit this perception of the world. For example, a client may report that his children no longer care about him because they have not called in a month whereas they used to call at least every other week. This suicidal client may not be willing to consider a reality alternative to his explanation of such (random) behavior (he hasn't been home to receive his children's calls). The more distorted clients' thinking becomes, and the less they respond to challenges to their distorted cognitive process, the higher the risk that this distortion may contribute to triggering or facilitating (rather than inhibitory) factors.

Suicidal clients also often evidence cognitive rigidity or constriction (Shneidman, 1985) with no ability to perceive options (Meyer & Deitsch, 1996). Such tunnel vision (Bellak & Siegel, 1983; Boylan, Malley, & Scott, 1995) significantly increases risk as the client inhibits the capacity for creative and flexible problem solving. Cognitive constriction interferes with healthy decision making and generally has a poor effect on coping ability. Tunnel vision results in the refusal or inability to explore alternative options, and suicide comes to be seen as the only possible act that will achieve a desired goal (e.g., the cessation of pain, the punishment of others; see fantasy and motivation about suicide later in this chapter). In this type of thought bias, suicide is perceived by the client as a solution, not a problem (Shneidman, 1985).

Occasionally, the clinician is faced with the client who denies suicidal intent and appears to have repressed all conscious knowledge or motivation for suicide. However, there may be many other signals and factors, such as hopelessness and social isolation, that clearly point toward the client's true intent and the client's risk. These types of cognitive distortion or misperception of reality may be just as dangerous as the other cognitive biases discussed so far, because they may lull the clinician into perceiving low risk, when, in reality, risk is quite high. Clients in denial about their intent despite other obvious signs of suicidal motivation need to be dealt with as if they had openly expressed intent.

Presence of a Triggering or Precipitating Event

The presence of a triggering or precipitating event is sometimes difficult to assess because such events may have occurred at other times in the client's life without resulting in suicidal ideation or threat (Bellak & Siegel, 1983). The fact that the event occurred may be less important than the fact that it occurred at a time of great emotional or psychological vulnerability. Thus, in assessing a possible precipitant, the clinician must take into consideration the mental status of the client at the time of the occurrence of the event. Another possibility is that several precipitating events exist, any one of which would not have been sufficient to trigger suicidal ideation. However, their cooccurrence in time may have resulted in emotional or cognitive overload, which in turn became unbearable for the client. In fact, the trigger for suicidal action is often identified by the client as unendurable psychological pain (Shneidman, 1985).

Not surprisingly, some researchers have argued that there is no true triggering event since the factors that seem to precede a suicide attempt appear to be indistinguishable from chronic stressors in the client's life (Maris, 1992). In other words, chronic and acute stressors are similar in content, and the critical event preceding the suicide becomes critical through its occurrence at a time of high vulnerability and sensitivity (Stelmachers, 1995). The clinician needs to take care not to misjudge the importance of a prior event described by the client. Just because the person was able to cope with a similar or identical life event at another time, the same event at this time may be sufficient to contribute to the client's hopelessness and to an attempt at suicide. Conversely, the mere presence of some of the critical events listed below is not sufficient to make a judgment of grave lethality.

Common potential precipitating or triggering events include recent trauma, recent loss, recent fights with significant others, or recent major changes in a person's life circumstances. It is definitely worthwhile (from an intervention perspective) to explore whether there is a problem to which the client ties the suicidal ideation (Clark, 1995). In addition to discrete triggering events or precipitants, the clinician must also consider other facilitating events. Most of these have been discussed in the context of risk factors. Most importantly, the therapist or counselor must assess whether the client presents with the situation (e.g., solitary lifestyle; poor support network), circumstances (e.g., psychiatric history, family history of suicide, psychological vulnerability expressed via guilt or hopelessness), plans (i.e., method, time, and place), and means (i.e., tools to carry out the plan) to create an opportunity for follow-through. In other words, the mental health care provider must be sensitive, not only to the concept of a major trigger or stressor, but also to facilitating life circumstances that may push the client in the direction of committing suicide.

Presence of Inhibitory Factors

Quite to the contrary of precipitating or facilitating events, there may be circumstances in the life of a client that, if activated or discussed, may actually inhibit the client from engaging in the ac-

tion. For example, religious or cultural beliefs may keep a client from self-harm despite very strong ideation, plan, and means, if these beliefs happen to portray suicide as an unacceptable alternative (Fremouw, Perczel, & Ellis, 1990; Stack, 1992). In many religious systems (e.g., Islam, Catholicism), suicide is considered one of the gravest sins committed by a human being, whereas other religions take less severe stands on the issue. For example, in Judaism suicide is considered a violation of the Ten Commandments but has been recognized as a human behavior that cannot be legislated against. Instead, according to Judaism, "suicide must not simply be condemned. It must be understood and prevented" (Grollman, 1988, p. 20). In some cultures, suicide is not only accepted but endorsed as a means of protest or redemption. For example, in traditional Japanese culture, suicide was "embedded in religious and national tradition" (Grollman, 1988, p. 23). Thus, exploration of cultural and religious beliefs is always in order to assess whether some help could be gleaned from activating these beliefs again in the client (or whether for some clients these issue may actually become risk factors).

Strongly felt and perceived family commitments act as another inhibitory factor. A client with young children may be extremely suicidal, may have a plan and may have fantasized about her method, but may refrain from acting out on the fantasy due to a strong commitment to seeing her children grow up safely under her care. However, the mere presence of children cannot ever be taken as an inhibitory factor. In some circumstances, children may be perceived as a stressful factor in clients' lives, making them feel even more overwhelmed or inadequate. In such situations, clients may perceive themselves as hindrances to the proper care of the children, fantasizing that the children would be better off under another adult's care.

A strong inhibiting factor may be present if the client expresses concrete and detailed plans for the future (Fremouw, Perczel, & Ellis, 1990) or a history of being able to cope with stressful events and losses (Stelmachers, 1995). Resilience, or the ability to bounce back from trauma and adapt to change (Yufit & Bongar, 1992), is a good indicator, as is strong social support that includes a sense of belonging and connectedness (Yufit & Bongar, 1992). Finally, a client's willingness to sign a suicide contract can be taken as evidence of some inhibiting potential. However, as with all inhibiting factors, the clinician must be alert to the possibility of denial and lack of complete truthfulness on the part of the client. Thus, although inhibitory factors need to be explored carefully, they need to be approached with skepticism and can be judged best if the clinician knows the client well. With new clients, the clinician is more likely to be fooled by the client and may need to exercise some additional caution in balancing triggers versus inhibitors.

Interaction between Triggering Versus Inhibiting Factors

The final aspect of the risk assessment has already been alluded to: the balancing of inhibiting versus facilitating factors. In the attempt to begin to make a decision about risk (i.e., in attempting to predict the client's behavior, a very difficult task indeed [Stelmachers, 1995]), the clinician evaluates whether inhibiting factors outweigh facilitating factors or vice versa. Further, the therapist must consider whether this balance is stable or whether it may change in the future. For example, a client may have indicated that concern for children fully inhibits his suicidal actions, and has done so on several occasions in the past. However, one of the client's children may be approaching adulthood and may have plans to leave the home in the next few months. In this case, although the presence of children had been a sufficient inhibitor in the past, it may not continue to be so a few months down the road. Similarly, clients who express that they would never follow through with their desire to commit suicide because of religious beliefs may show evidence of strong inhibition that overrules the desire to commit suicide. Should such clients, however, suddenly develop doubts about their religion, this balance of inhibitor versus trigger may shift and the clients may follow through with their suicide plans.

Decision Making Based on Risk Assessment

Once the previously listed factors and their relative balance have been assessed, the counselor is ready to define the dimension of the problem, an important action step in any crisis intervention process (Slaikeu, 1990). In the case of suicidal ideation, the problem dimension is defined most importantly in terms of the level of perceived lethality. Most simply, the more factors or variables that are present suggesting increased risk, the higher the lethality of the client's possible suicide (Clark, 1995). However, risk assessment may not be quite this easy; instead, a multitude of issues may need to be considered. Stelmachers (1995) suggests that "risk should be judged on the basis of the following components: (1) the degree of psychological disturbance, often referred to as 'perturbation'; (2) suicidal intent by self-report and objective evaluation; (3) the particular suicidal behavior exhibited; (4) lethality of plan/method selected or used; and (5) the final outcome" (p. 371). Given that this chapter is primarily concerned with risk assessment in an immediate clinical situation, the fifth factor noted by Stelmachers (for research purposes of postdiction) will not be further considered. The other four factors, however, in addition to client impulsivity, need to be weighed carefully by the therapist who has to decide about intervention planning for a suicidal client. There is no clarity or rule that can be spelled out to guide this process in a fail-safe manner. Instead, each clinician must evaluate each client idiosyncratically on all four dimensions. For each individual person, these factors may interact differently.

For example, for one client, the particularly high psychological disturbance (e.g., schizophrenia with voices suggesting self-harm) may outweigh the relatively benign plan expressed by the client (e.g., holding my breath till I die). For another client, the relatively low amount of psychological distress coupled with a high-lethality plan and poor impulse control may result in a similar action plan. Only the most extreme cases (e.g., when all factors clearly point in the direction of high risk) are easily judged and acted upon. In fact reliability estimates across seasoned clinicians in postmortem studies have confirmed the relatively low agreement about lethality assessment (Stelmachers, 1995).

To assist with risk assessment, Freemouw, Perczel, and Ellis (1990) identified risk gradations for some of the most common risk factors for suicide. Following these authors' example, the risk factors presented above and their associated level of danger are summarized briefly in Table 6-1. These authors also suggest that the following issues be considered when attempting to predict risk: imminence of the behavior (ranging from likely never to occur to likely to occur immediately); clarity of danger (ranging from vague or no time, place, method, and opportunity to very specific time, place, method, and opportunity), intent (ranging from no wish to die to strong wish to die); lethality of behavior (ranging from nonlethal method to highly lethal method); and probability estimate of the occurrence of the suicide (ranging from not likely to occur to highly probable to occur). Each of these decisions is based on the level of risk expressed by the specific content of each of the risk factors outlined above and summarized in Table 6-1.

Finally, possible assistance in risk assessment may be gleaned from several suicide tests or scales that have been developed for use with suicidal clients (also see Table 4-4 in Chapter 4). It is arguable whether the administration of a psychological screening test is appropriate during a risk assessment interview. However, some clients (who may be less acutely stressed) may be amenable to testing. With such clients the tools presented below may be of some assistance to the clinician's decision-making process. Further, suicide screening tests provide the clinician with a concrete record of the client's state of mind that can become a permanent aspect of documentation of suicide risk in the client's file. Nevertheless, before a suicide test is administered, the mental health care provider must weigh the potential risks and benefits of such administration. In other words, the routine use of suicide screening tests is not endorsed here. However, the oc-

casional use, when appropriate to the client's circumstances, can be helpful. The most commonly used tests are the Suicide Probability Scale (SPS) (Cull & Gill, 1989), Beck Scale for Suicide Ideation (BSSI) (Beck, 1991), Reynolds Adult Suicidal Ideation Questionnaire (SIQ) (Reynolds, 1991), and the Suicidal Behavior History Form (SBHF) (Reynolds & Mazza, 1992), though others exist.

The SPS is a 36-item scale that measures four aspects of suicidality via four subscales labeled hopelessness, hostility, suicide ideation, and negative self-evaluation. It provides *T*-scores, weighted scores, and clinical cut-off criteria for suicide risk. Its administration is brief, requiring only 5–10 minutes. The SPS is available from the Western Psychological Service. The BSSI is a 21-item scale that provides clinical cut-off criteria for suicide risk, diagnostic ranges, and critical item information. It requires 5–10 minutes of administration time; however, for acutely suicidal clients a short version of five items exists that can be orally administered as part of the standard suicide interview. The BSSI is available from the Psychological Corporation. The SIQ is a 25-item scale that converts one total score into a *T*-score and percentile rank. It provides risk cut-off information and recommendations for need for further assessment. Its administration requires approximately 10 minutes. The SIQ is available from Psychological Assessment Resources. The SBHF is a four-page interview guide that can be administered in 10–15 minutes (though some greatly distraught clients may require up to 30 minutes). It provides data regarding suicide risk, psychosocial history, and past suicidal behavior. Given its interview form, it can be seamlessly inserted in a suicide risk interview as outlined in this chapter. However, this instrument provides no formal scoring; it provides clinical interpretation only and as such is less of a test than a tool. The SBHF is available from Psychological Assessment Resources.

Once risk or lethality has been estimated, the intervention is planned based on the clinical judgment of the seriousness of the situation. Fortunately for clinicians, the main decision from a

TABLE 6-1 Gradation of Risk Factors for Suicide

| | RISK FACTOR | | |
	HIGH RISK	MODERATE RISK	LOW RISK
COPING			
Internal Resources	no resources	limited resources	adequate resources
Daily Functioning	impaired	adequate	generally good
SOCIAL SUPPORT			
Number Available	no resources	limited resources	adequate resources
Willingness to Use	unwilling to access	limited willingness	willing to access
Isolation	isolated and withdrawn	some social contact	regular social contacts
FAMILY HISTORY			
Connectedness	no or one significant other	one significant other	more than two significant others
Suicide History	multiple members	one member	no member
PSYCHIATRIC HISTORY			
# of Diagnoses	multiple diagnoses	one diagnosis	no diagnosis
Severity	severe	moderate	none
Response to TRT	negative	varied	good or none needed
Discharge	within 3 months ago	more than 3 less than 12 months ago	more than 12 months ago
MEDICAL HISTORY			
Chronicity	chronic	chronic	acute or none
Pain	severe	moderate	mild to none
PSYCHOLOGICAL FACTORS			
Trauma Presence	several traumas	one trauma	none
Trauma Severity	severe trauma	moderate trauma/stress	mild stress to none
Loss	multiple losses	one loss	no or unimportant loss
Recency	within days; anniversary	within last month	none
Hopelessness	severe	moderate	mild
Prior Attempts	yes	yes	no
Impulsivity	high	moderate	low
IMMEDIATE PREDICTORS			
Method	decided	decided	undecided
Means	in possession	easy access	still to be secured
Time and Place	definitely chosen	tentatively chosen	not chosen
Lethality	high	moderate	low
Preparation	notes written, will made, possessions given away or similar behaviors	some planning	none made

legal perspective is whether the clinician acted prudently and reasonably (i.e., not negligently); therapists or counselors who did a thorough assessment and planned interventions based on the subsequent clinical judgment are generally not penalized by the courts should they have made an honest and unpredictable mistake. This is of little consolation, however, to clinicians who are faced with their mistake in the form of clients who successfully killed themselves. It is therefore not surprising that suicidal clientele are one of the highest burnout factors among therapists (Kleinke, 1994).

INTERVENTION IN CASE OF SUICIDE RISK

Some writers suggest that no person is 100% suicidal and that therefore intervention may always have a chance (Shneidman, 1985). "Assessment of warning signs . . . can be translated into life-saving actions by crisis workers or anyone else in the physical or emotional proximity of suicidal persons. . . . However, if the risk factors, clues, or cries for help go unnoticed or unrecognized, the chances for effective intervention are greatly reduced" (Gilliland & James, 1993, p. 135).

The main thrust of intervention should be to prevent an irreversible decision that is made while the person is not fully rational or aware of all options and alternatives (Slaikeu, 1990). Such a "goal of saving a human life supersedes total allegiance to confidentiality" (Slaikeu, 1990, p. 120). Intervention, according to Slaikeu (1990) consists of at least three steps: (1) generation of possible solutions, (2) concrete action, and (3) follow-up.

If a client threatens suicide, this is best understood as an expression of some ambivalence about the act; one of the most important suicide prevention skills that a mental health care provider can have is the ability to recognize a client's communication about suicide (Firestone, 1997; Fujimura, Weis, & Cochran, 1985). Threats or communications about suicide are not always

verbal, but rather may consist of a set or pattern of behavior that must be recognized by the therapist as preparations for suicide or contemplation of the act. Examples of nonverbal communications about suicide can be gleaned from the previously discussed risk factors. (For example, a depressed client who suddenly becomes calm and stops complaining of psychic pain, who makes a will, writes letters to family members, and has increased energy needs to be approached about suicide.) It is worth repeating that talking to clients about suicide does not cause them to do it. In fact, as pointed out by Fujimura, Weis, and Cochran (1985), "[p]eople are not driven to suicide by a caring person who inquires as to whether or not they are suicidal. People may, however, be driven to suicide by an avoidance of the topic on the part of the listener, from whom they need a concerned response" (p. 613).

Counselor characteristics were already discussed at some length earlier in this chapter. However, it is worth reiterating that clinicians best remain supportive and calm to facilitate rapport and trust (Gilliland & James, 1993). They must not panic or get frightened, lest they prove to be of no help to suicidal clients who are panicked or frightened themselves. A panicked mental health care provider is not capable of making "every effort to guarantee that a suicidal person reaches appropriate professional help" (Meyer & Deitsch, 1996, p. 385), and it is this aspect that many authors feel is the most essential to successful suicide intervention.

When suicide intervention is necessary, that is, if the assessment showed that significant risk is present and that the therapist must take steps to protect the client from harm, the clinician must shift to a crisis intervention mode of interacting with the client, abandoning traditional therapeutic neutrality and interactions (Bellak & Siegel, 1983; Fujimura, Weis, & Cochran, 1985). This means that the counselor becomes more active, directive, problem-solving oriented, and less concerned with fostering dependency (Seligman, 1996). However, the clinician does so respectfully, as collaboratively with the client as possible.

Throughout assessment and intervention, the therapist needs to try to manage the situation at the calmest and lowest level possible. In other words, not every client who voices suicidal ideation must be hospitalized. The clinician will attempt to take several steps first to deal with the client on an outpatient confidential basis, then may move to outpatient management that involves others (i.e., loosens the confidentiality of the matter), and only moves to voluntary and finally involuntary treatment if no other options worked.

While still attempting to manage the situation in the least restrictive, confidential outpatient setting, the clinician defines clear goals for the session and imposes these goals on the client, even if the client is not in total agreement (Meyer & Deitsch, 1996). These goals include helping the client increase the recognition of options and choices beyond suicide (Stelmachers, 1995); helping the client reduce tunnel vision and improve cognitive processing and reality testing (Nystul, 1993); alleviating acute symptoms of agitation or anxiety (Clark, 1995); and focusing on delaying the suicidal client's impulse and action (Slaikeu, 1990). The latter may be accomplished by pointing out the irreversibility of suicide, the client's ability to stand the pain, and the client's coping ability. Further, the mental health care provider needs to support the client's attempts at coping by helping the client "(1) separate thought from action, (2) reinforce expression of affect, (3) anticipate consequences of action, and (4) focus on precipitating events and constructive alternatives" (Gilliland & James, 1993, p. 150). The clinician may also point out that suicide can always be engaged in later once the client is no longer in crisis and can make a more level-headed decision.

If these relatively nonintrusive interventions do not suffice to reduce the client's risk significantly and sufficiently, additional actions need to be taken. The next step is often the attempt to make a suicide contract with the client (Boylan, Malley, & Scott, 1995; Meyer & Deitsch, 1996). Such a suicide contract maintains that, even

though the therapist or counselor is aware that the client really wants to pursue the right to commit suicide, the therapist want to make a pact with a given client that the client not do so. The suicide contract relies heavily on the client's truthfulness and is not an option if the clinician believes that the client would agree to anything to keep the clinician from preventing the client from committing the act. A suicide contract spells out client and therapist name, an agreement about a time frame during which the client will not induce any harm, the time of the next appointment with the clinician, an agreement to discard all means necessary to carry out the suicide plan, and sources of support should the client's suicidal ideation increase. A sample is shown in Figure 6-1.

All interactions around the establishment of the suicide contract are best designed in such a manner as to retain as much decision making on the part of the client as possible. As such a suicide contract is negotiated, not imposed (Gilliland & James, 1993). A suicide contract has all of the same components that are present in any other therapeutic contract (see Hutchins & Vaught, 1997). In this way, it has specificity with regard to the targeted behavior (in this case the suicidal acting out), clarifies the time frame for which it is in effect, delays gratification or action, is personalized, leaves control with the client to largest extent possible, and spells out consequences of noncompliance (Hutchins & Vaught, 1997). The client's input is sought during the specification of the contract's time period; the client can also collaborate in the preparation of the calling card for emergencies (e.g., can add phone numbers of friends or family members who could be called instead of or in addition to the crisis line); and the client gives feedback about any contingency plan that is included (e.g., what will happen if the client does not call the mental health care provider at a specified time). If the accessible means necessary to carry out the plan are with the client in the session (e.g., if the client has the gun along), the counselor confiscates these means immediately and includes this information in the

---

FIGURE 6-1 Sample Suicide Contract

I, _____ (*insert client's name*) agree not to kill myself, harm myself, or attempt to kill myself for the time period beginning now, _____ (*insert current date and time*) until _____ (*insert agreed upon next meeting or contact time*).

I agree to call the crisis numbers on the crisis card given to me today by my clinician, _____ (*insert clinician's name*), should I have the urge to kill or harm myself at any moment during this time period.

I agree to come to my next appointment on _____ (*insert day and date*) at _____ a.m./p.m. (*insert exact appointment time*) with _____ (*insert clinician's name*).

I agree to get rid of all things that I have thought about using to kill myself. Specifically, I will discard _____ (*insert means identified as necessary to the client's method*). I will contact my clinician at _____ (*insert phone number where clinician can be reached*) as soon as I have discarded the means to confirm that I did this. If I don't call by _____ (*insert a time later today*), my clinician will alert my family and the police.

I realize that this contract is part of my therapy contract with my clinician at _____ (*insert name of clinic/hospital/etc.*). I am aware that my clinician can break the agreement of confidentiality if I do not comply with this suicide contract.

_____   _____
Signature of Client   Signature of Clinician

_____
Date and Time

---

suicide contract. Further, information is included about when and how these possessions will be returned to the client. Sometimes, a suicide contract may involve family members of the client. If this is the case, these individuals are to be contacted by client and clinician during the session so that the clinician can be assured of these people's availability and responsibility. However, usually the involvement of family and friends signals that the clinician has assessed the risk presented by the client to be greater; that is, the intervention has moved to another level of restriction or restraint.

If a suicide contract involving only the client does not appear to suffice, the therapist or counselor can consider a suicide contract that includes involvement of family. Such involvement of family can range greatly. Whenever the clinician considers mobilizing family and community resources or support networks, it is best (if at all possible) to get clients' permission to do so by having them sign appropriate releases of information. This leaves control with the client and is less intrusive than the unilateral decision of the clinician to breach confidentiality. However, if family or others must be involved and the client refuses releases of information, the mental health care provider can overrule the client and can contact family or other support persons regardless. Contacting others for support and help may be presented to the client as a trade off: that is, the client can either agree to involve support people or can agree to seek hospitalization.

If outsiders are involved (whether with or without the client's consent), this is done to enlist, not only support, but also awareness and understanding from the family members or friends who are being contacted (Clark, 1995). If the client is sufficiently severely suicidal that the clinician sees no option but to involve family or hospitalize, it is important to involve family members and close friends anyway because they are the ones who can prevent, not only the present, but also future suicide attempts. It is family and friends after all who are around the client more often than the clinician. It is important to teach family members of chronically suicidal clients how to assess and intervene in emergency situations when the client voices suicidal ideation or makes behavioral threats. Family members and significant others need to learn to identify the clues given by the client and to recognize these cries for help to ensure effective and early intervention and prevention (Fujimura, Weis, & Cochran, 1985; Gilliland & James, 1993).

When family or friends are involved, they can assist with several functions. They can be engaged in removing the means that were identified; they can be with the client so that there is no unsupervised time; they can provide a temporary home; they can contribute support services, such as food, shopping assistance, child care, transportation; they can provide additional support by calling at regular intervals; and so forth. The level of their involvement is clearly dictated by the level of risk presented by the client. The higher the risk, the more intense the involvement. The clinician must seek careful guidance from the client with regard to the choice of whom to involve. A clinician should not jump to conclusions in this regard; although there may be a spouse, it is entirely possible that this is not the

right person to draw into this situation. It is important to make sure that the person(s) involved are not people who contribute to the client's stress or worse are triggers for the client (e.g., nagging and judgmental parents, abusive spouses, helpless children, depressed and suicidal friends). In other words, this is not a time when people who have a history of conflict with the client should be engaged. If there are no family members or friends with whom the client has no conflict, the clinician faces a decision: is it possible to mediate the personal conflict and use the family member or friend as a resource, or should the client be hospitalized? Finally, in involving others, it is important to balance the client's confidentiality needs with safety needs (Gilliland & James, 1993). Thus, involving a boss or supervisor might need to be a last resort so as not to jeopardize the client's employment situation (which would only serve to add stress to the client's life).

As clinicians move along the lethality continuum, assessing and reassessing risk as interventions are considered in collaboration and cooperation with the client, they also make decisions regarding disposition once the client leaves the office. This decision refers to the continuum of care that ranges from intensified outpatient treatment, to voluntary hospitalization, to involuntary hospitalization or commitment. Intensified outpatient treatment generally consists of an increased number of scheduled contacts with the clinician as well as possible phone check-ins between sessions. It is also necessary to provide the client with a card that has local information about resources that have 24-hour hot line numbers for crisis circumstances. If such intensified outpatient work is chosen as the best avenue for intervention, a suicide contract must be made and serious consideration must be given to involving family members. The client must make a strong commitment to utilize the emergency contacts should the suicide crisis recur between scheduled sessions or contacts.

If the client is deemed incapable of inhibiting the suicidal impulses, voluntary hospitalization needs to be considered and initiated with the client's collaboration. Voluntary hospitalization is obviously much preferable to involuntary commitment because it retains the choice for treatment with the client and is generally seen by client and clinician as a collaborative endeavor on behalf of the client, not an intrusion on the client or a stripping of control from the client. If the client rejects voluntary inpatient treatment, this refusal may be more a reflection of a transient state of hopelessness than a true rejection of help (Clark, 1995).

If the client agrees to voluntary hospitalization, necessary steps need to be taken right there and then to facilitate the process. The mental health care provider must be sufficiently familiar with local resources to know which facilities are available to the client and must have phone numbers handy. Client and therapist then collaborate in calling these facilities to secure a treatment spot for the client and to let the facility know that the client is on the way. Once a treatment site has been identified and secured, client and clinician must turn to the issue of transportation of the client to the facility. It is best not to rely on clients to transport themselves, even if their car is in the parking lot. It is best to enlist the assistance of a trusted family member or friend to meet the client at the clinician's office and to transport the client to the facility. If no family members or friends are available, or if the client refuses to involve them, a taxi may need to be called, with the counselor delivering the client to the cab and instructing the driver about the destination and the need to go there without detour. In some communities, it is possible to enlist the assistance of local law enforcement officers or security guards to provide transportation, even in cases of voluntary hospitalization. However, in many places, these officials will not assist unless a commitment is initiated. It is best for therapists to know which options exist before they begin practice in a certain setting or community so that they do not have to learn about these resources while the crisis is in progress.

If clients' situations are considered highly lethal and risky, but they refuse to check themselves into an inpatient facility on a voluntary basis, the clinician needs to bring up involuntary hospitalization. It is not uncommon for clients who have refused voluntary hospitalization to consent after all, once confronted with the reality that otherwise they will be committed to inpatient treatment involuntarily. Often the decision between voluntary and involuntary hospitalization influences where the client will be hospitalized and, generally, voluntary placements are more desirable from a client's perspective. Commitments often lead to hospitalization in the local nonprofit, state-administered psychiatric hospital, an option that many clients would rather avoid (given the lesser quality of the surroundings and greater social stigma). Voluntary hospitalization, on the other hand, is generally to private psychiatric hospitals or to psychiatric units in general hospitals or teaching hospitals, a much-preferred setting for most clients. When faced with these choices, clients often change their perspective and choose voluntary placement over commitment.

If commitment is necessary, and sometimes it is, especially if the client has severe psychiatric symptoms that reduce reality testing even further, the clinician must take care to have sufficient information to initiate commitment proceedings. Most specifically, the clinician generally must have evidence that the client has a diagnosable mental illness (i.e., a DSM-IV diagnosis) and clearly presents a danger to self. For this reason, it is important to have organized the information that was collected from the client so far and to have made at least a preliminary DSM-IV diagnosis. The clinician must be prepared for questions about the assessment of the client's level of dangerousness. As with voluntary hospitalization, commitment is initiated in the client's presence from the clinician's office. The therapist or counselor needs to work closely with the client to try to maintain a therapeutic relationship despite the involuntary nature of the commitment proceedings. Involving the client in all steps as much as possible will facilitate the maintenance of the therapeutic alliance and must be maximized within the client's level of cooperation (Clark, 1995). The exact nature of the process of involuntary commitment varies greatly from state to state and across professional credentials of the mental health care provider. In most states the process is easier for practitioners who have a medical degree (e.g., psychiatrist with M.D.) or a Ph.D. and a license to practice independently (e.g., a licensed psychologist). The commitment process usually begins with a petition to a judge or magistrate for an ex parte order. If this petition is found by the official to be appropriate and warranted, this person then issues a custody warrant that essentially declares that the client will be placed in custody of the psychiatric facility for further evaluation. This evaluation needs to take place within a specified period of time (usually 72 hours, but anywhere from 2–10 days depending on the state) and findings will be presented at a probable cause hearing at the end of this time period. During the probable cause hearing a judge makes a ruling with regard to whether due process has been followed on behalf of the client as well as to the client's need be committed.

In some states, physicians and licensed psychologists are able to sign the initial petition form to the court that paves the way for a custody warrant. This form, called a Peace Officer's Application in the state of Alaska (the name of this form may vary from state to state), is usually signed by either a peace officer or an emergency room physician. A client with a signed Peace Officer's Application indicating immediate suicide risk and need for commitment will be accepted for transportation to the state psychiatric facility by local police officers. Once at the facility, the attending staff conduct their own evaluation, contact the court, and secure the custody warrant. Then the hospital conducts the probable cause hearing to assess whether the client needs to remain beyond the initial commitment period.

The nondoctoral level clinician cannot sign a Peace Officer's Application and, to function independently of a supervisor with the necessary credentials, would have to go through probate court (a subsection of the state's superior court) to receive an ex parte order from a judge. To do this, the clinician would have to file the petition in person to receive the commitment order or custody warrant for the client. Clearly this would not be an efficient or even possible process if the suicidal client were in the clinician's office at that moment. Thus, the master's level practitioner may want to seek assistance from law enforcement officials who can complete a Peace Officer's Application, take clients into custody, and then transport them to the psychiatric facility. The potential problem with this process is that peace officers called to the clinic must make their own assessment of the client's mental status, and, if the client denies suicidal ideation to the peace officers, the latter may decide not to complete the Peace Officer's Application. Once the client has been transported to the facility, another assessment will be conducted by hospital staff and a commitment proceeding will be initiated wherein the hospital seeks a petition from a judge who will issue a custody warrant for the client. Initial commitment usually lasts only 72 hours; after this time period, the client must be given a hearing and a judge must decide whether the client is to remain within the facility (i.e., continues to represent a threat to personal safety) or is to be released.

Involuntary commitment is greatly facilitated if the client's family cooperates and agrees; however, at times commitment proceedings may have to be initiated, not only against the client's, but also against the family's, wishes. If clinicians judge that clients present an unacceptable risk, they must initiate involuntary commitment even under such hostile family circumstances. In fact, this must be done, even if the client's family threatens legal action against the mental health care provider, to protect the client from harm and the clinician from legal ramifications (Clark, 1995). The clinician is not legally liable for violations of confidentiality in this case, because the client signed an informed consent before treatment began that clearly outlined the limitations of confidentiality. There are no states or set of ethical guidelines that do not have an exclusion for self-harm, which allows the counselor to disclose critical client information if the client threatens suicide and is considered at high risk for carrying out the action. In fact, it is more likely that clinicians will face legal action should they choose not to hospitalize and the client attempts suicide, than if they choose to break confidentiality and hospitalize the client and the client is discharged by the facility after the 72-hour initial commitment period.

Either during or immediately after a suicidal crisis with a client, the therapist or counselor needs to seek consultation and/or appropriate supervision, depending on level of training. The clinician also needs to make appropriate referrals for the client, regardless of level of intervention (i.e., outpatient versus inpatient). A psychiatric referral may be in order to assess need for medication (Hood & Johnson, 1997). This is particularly true if the suicidal ideation and gestures are part of a regular manifestation of the client's psychopathology (Fujimura, Weis, & Cochran, 1985). Once the suicidal crisis is over, client and clinician can settle back down to the therapeutic work they had started or were about to begin when the suicidal crisis was identified. This therapy work will follow the same procedures and processes outlined elsewhere in this book. However, the clinician who works with a client who has weathered a suicidal crisis needs to remain alert and careful not to miss signs of recurrence. This is particularly true during the few weeks after the suicidal crisis, especially if the client was hospitalized. This is a high-risk period for many clients, and the clinician must remain aware and should address the issue of suicidal ideation regularly (Fujimura, Weis, & Cochran, 1985; Meyer & Deitsch, 1996; Stelmachers, 1995).

DOCUMENTATION AND RECORD KEEPING

"Cynical malpractice attorneys say that doctors' notes are more important than their clinical practice. The lawyers exaggerate but the point is essentially correct. If suit comes, the defense must rely on the clinician's notes written at the time. After the fact explanations, written or oral, are of little use" (Beck, 1987, p. 698). This caution clearly points to the importance of good written documentation after a suicide assessment and intervention has been conducted. Such documentation must occur immediately, must be thorough, and must reflect the entire thought process the clinician went through in working with the client. Verbatim quotes are particularly helpful, as are citations of standards of care that were followed by the care provider.

Provided at the end of this chapter is a sample of a progress note that would be written after a suicide crisis with a client. This sample (using the same sample client used in Brems [1999a]) outlines for the reader how appropriate suicide charting is done and covers all of the cautions that I have previously mentioned. Also attached to the progress note would be a copy of the suicide contract and of the crisis numbers that were presented to the client. Any other written documents prepared with the client during the session should also be copied and attached. Any phone calls made in addition to the ones included in the sample should be recorded in the same way. Any message slips received about the case of the client (e.g., messages from family members or other professionals returning calls about the client) also need to be attached (with time and date of receipt).

SUMMARY AND CONCLUDING THOUGHTS

A helpful mnemonic device for guiding assessment of the suicidal client is offered by Patterson, Dohn, Bird, and Patterson (1983) to help clinicians remember the most crucial aspects of a suicide assessment. This mnemonic, SAD PERSONS, draws attention to the most pertinent risk factors that have been identified in the literature, and is as follows:

S	stands for	**S**ex (males>females)
A	stands for	**A**ge (older>younger)
D	stands for	**D**epression
P	stands for	**P**revious Attempt
E	stands for	**E**thanol (alcohol) Abuse
R	stands for	**R**ational Thinking Loss
S	stands for	**S**ocial Support Lacking
O	stands for	**O**rganized Plan
N	stands for	**N**o Spouse and
S	stands for	**S**ickness (especially chronic or terminal illness)

This mnemonic device is also used as a scale that can be scored from 0 to 10. Specifically, 1 point is given for the presence of any one of the dimensions tapped by the scale. The scale developers recommend that clients must be hospitalized if they manifest 7 to 10 of the symptoms; clients are encouraged to consider voluntary hospitalization if they acknowledge 3 to 6 of these symptoms; and clients are sent home with appropriate follow-up if none or only up to 2 symptoms are reported (Patterson, Dohn, Bird, & Patterson, 1983).

Another good resource for the novice clinician is the checklist provided by Sommers-Flanagan and Sommers-Flanagan (1995) and the suicide consultation form provided by Boylan, Malley, and Scott (1995) in their book for counseling interns. The reader has permission to copy this form for use with clients. This form essentially represents a guide for a structured interview that can be used to complete the assessment interview with the client and to record findings in a comprehensive

and organized manner. The form can be used also to ensure that a record of care is placed in the client's chart to protect the clinician legally should a client commit suicide despite the clinician's best efforts to intervene (Boylan, Malley, & Scott, 1995). A checklist is provided in Figure 6-2 that can be used to guide the assessment interview with the client as well as the intervention plan.

It is very important to follow the steps of assessment and intervention outlined in this chapter (including the seeking of consultation and supervision) because they will increase the likelihood that the mental health care provider can document reasonable care. Documentation of all steps taken is critical should the case go to the courts, either because of a successful suicide attempt or a claim of breach of confidentiality. Only information contained in the client's record will be useful to the clinician defending an action in such cases. Recollections are generally not accepted as evidence that proper procedure was followed. Also, all documentation must happen quickly and should reach the client's chart on the same day as the crisis. Trainees and other clinicians working under supervision must seek signatures from supervisors and also need to document their supervision of the case. Psychologists are best served to seek out consultation with a colleague, even if they no longer are required to work under supervision and should document this consultation in the client's chart. In their case, however, no signatures from consultants are needed.

FIGURE 6-2 Suicide Assessment Guide for Interview and Intervention Planning

Immediate Risk Factors

Suicide Ideation: 7 ---------- 6 ---------- 5 ---------- 4 ---------- 3 ---------- 2 ---------- 1
                      extremely strong   extremely weak

Suicide Intent: 7 ---------- 6 ---------- 5 ---------- 4 ---------- 3 ---------- 2 ---------- 1
                      extremely high   extremely low

Suicide Plan: 7 ---------- 6 ---------- 5 ---------- 4 ---------- 3 ---------- 2 ---------- 1
                      extremely specific   extremely vague/none

   Details of Plan   _____
   Method   _____
   Availability   _____
   Place   _____
   Time   _____
   Lethality of Method   _____
   Preparations   _____

Hopelessness: 7 ---------- 6 ---------- 5 ---------- 4 ---------- 3 ---------- 2 ---------- 1
                      extremely hopeless   hopeful

Impulsivity: 7 ---------- 6 ---------- 5 ---------- 4 ---------- 3 ---------- 2 ---------- 1
                      extremely impulsive   good self control

Closure Behaviors: 7 ---------- 6 ---------- 5 ---------- 4 ---------- 3 ---------- 2 ---------- 1
                      several present   none present

Other Risk Factors

Demographics: high risk   moderate risk   low risk
   Age   _____
   Gender   _____
   Marital Status   _____
   Life Style   _____

Coping: poor   moderate   good
   Internal Resources   _____
   Daily Functioning   _____

FIGURE 6-2 Suicide Assessment Guide for Interview and Intervention Planning (continued)

Social Support:	inadequate/absent	inadequate/present	adequate/present
Number Available			
Willingness to Use			
Isolation			

Family History:	high risk	moderate risk	low risk
Connectedness			
Suicide History			

Psychiatric History:	high risk	moderate risk	low risk
# of Diagnoses			
Types of Diagnoses			
Severity			
Response to TRT			
Hospitalization			

Medical History:	high risk	moderate risk	low risk
Chronic Illness			
Painful Illness			

Psychological Factors:	high risk	moderate risk	low risk
Trauma Presence			
Trauma Severity			
Loss			
Recency			
Prior Attempts			

Risk Assessment	high risk	moderate risk	low risk
Cognitive Distortion:			
Triggering Factors:			
Inhibiting Factors:			
Balance Assessment:			

Sample Suicide Progress Note

Clinician: Chris Brems

Client Name: Frederick X.

Date of Session: 24 October 1994; 2:00 P.M.

Session Number: 6

Payment Made: $60 (receipt provided)

Next Session: 26 October 1994

Client presented disheveled and poorly groomed today; he was sullen and noncommunicative, indicating he felt terrible and almost did not come to his session. Upon inquiry he revealed that he slept poorly for the five preceding nights, has missed work twice, once without even calling in. He has missed all of his classes and is talking about not going back. Sandra has been away this week for conference out of state and he has talked with her daily for brief periods over the telephone. Client expressed hopelessness about the likelihood of his depression lifting and began to discuss his brother's suicide, a topic that had not emerged in the past three sessions. Given his greatly dysphoric mood, hopelessness, his concern with his brother's suicide, his partner's absence, and deterioration in functioning as compared to the three prior sessions, the therapist raised the issue of suicidal ideation. The client responded affirmatively stating, "I have given serious thought to ending my life. I can't stand being without Sandra and I think she's going to leave me soon altogether. She's getting sick of being around my complaining and moping." Once this suicidal ideation was expressed, the therapist refocused the session on a suicide assessment that yielded the following information:

Assessment

Immediate Risk Factors

Suicide Ideation: present to a ruminative degree and strong

Suicide Plan: present and specific

Details of Plan:

   Method: client plans to take an overdose of antidepressants (which Sandra was prescribed a month ago by a general practitioner for depressive symptoms, and which she purchased but then never used) and plans to combine the pills with alcohol

   Availability: client reports being in possession of 42 pills of Tofranil (a form of imipramine, a heterocyclic antidepressant), 25 mg. each (a potentially lethal dose being 1000 to 1200 mg.); also is in possession of 2 fifth bottles of alcohol (one fifth of Tanquaray Gin; one fifth of bourbon)

   Place: client plans to take the pills in his home

   Time: client has not set a date but indicated he would carry out the plan at night when he was unable to fall asleep and was worrying about his relationship with Sandra

   Lethality of Method: moderately lethal

   Preparations: none made

Other Risk Factors

Coping:

   Internal Resources: limited resources at present time; adequate resources in past

   Daily Functioning: impaired at present; adequate to good in recent past

Sample Suicide Progress Note (continued)

Social Support:

Number Available: no close resources present in town at this time (Sandra out of state; due to return tomorrow night); there is a neighbor in the apt. next door with whom client is friendly but not close—has potential for temporary support; there is also a classmate who has approached client recently to form a study group—client has his phone number and address and likes him; there is a colleague at work with whom client has worked for several years and whom he has helped out with rides a couple of times when the colleague's car was broken down

Willingness to Use: hesitant but not unwilling to access

Isolation: isolated and withdrawn (by choice), but with some acquaintances

Family History:

Connectedness: one significant other; no contact with family-of-origin members

Suicide History: suicide attempt by mother; completed suicide by brother 2 months ago

Psychiatric History:

# of Diagnoses: diagnosis of major depression and dysthymia

Severity: moderate to severe acute depression; mild to moderate chronic depression

Response to TRT: good

Hospitalization: none

Medical History:

Chronic Illness: none

Painful Illness: none

Psychological Factors:

Trauma Presence: one trauma (brother's suicide) two months ago

Trauma Severity: severe trauma because it rekindled old grief over the accidental death of his best friend

Loss: multiple perceived immediate losses: brother, Sandra (emotional loss and current physical absence)

Recency: two months ago for the suicide; right now for the perceived loss of Sandra

Hopelessness: severe to moderate

Prior Attempts: none

Impulsivity: low

Other Risk Assessment

Triggering Factors: hopelessness

loss of brother to suicide

Sandra's physical absence

client's perception of a deteriorating relationship with Sandra

social isolation

hesitation to seek support and assistance

depression

poor coping

family history of suicide

presence and moderate lethality of means

Continued

Sample Suicide Progress Note (continued)

Inhibiting Factors: strong relationship and emotional connection with Sandra

                        no preparations

                        no history of attempts

                        verbalized plans for future (wants to improve social relationships, wants to bring Sandra to couples therapy to work on their relationship; wants to pass exam in his class next week; worries about his boss's reaction to his job absence and hopes his job is secure)

                        ambivalence about death

                        willing to sign a suicide contract

                        hesitant but willing to solicit help from neighbor and coworker till Sandra returns

                        willing to call Sandra to let her know about his ideation

                        willing to have means removed with assistance

Balance Assessment: Inhibitors are strong and may suffice to outweigh triggers; however, given Sandra's absence risk is moderate to high without intervention; once Sandra returns, risk will decrease significantly and interventions can become less restrictive

Intervention

Based on the risk assessment conducted and recorded above, the following interventions were made:

1. Client called and talked with Sandra at 2:20 P.M. and allowed therapist to talk to her as well (ROI is attached). Sandra agreed to reschedule her flight the next day to an earlier departure time and expressed great concern and caring; she reassured the client that although he was right in thinking that she had been somewhat dissatisfied with their relationship, she had no plans of leaving him (Sandra called back at 2:45 P.M. to confirm that she successfully rescheduled her flight and would arrive at the Anchorage International Airport at noon tomorrow).

2. Client agreed to and called his neighbor and his colleague in the therapist's presence at 2:50 P.M. and 2:55 P.M. respectively. Both expressed great concern and support (both to client and therapist who also talked to them with the client's permission; ROI is attached). Neighbor will stay with client tonight until 9 A.M. tomorrow morning at which time he has to leave to go to work. Coworker will come over at 9 A.M. to remain with client after that and will take client to airport to pick up Sandra at noon.

3. Client agreed to sign a suicide contract spelling out that he will not kill himself before his next appointment with the therapist and clarifying all contingency plans (one copy of the contract is attached; the original copy was given to the client).

4. Client agreed to turn all pills and alcohol over to the neighbor. Neighbor has agreed to destroy the pills by flushing them down the toilet and to drain the alcohol.

5. Client has agreed to schedule another appointment with the therapist in two days, on 26 October 1994 at 1 P.M.; the possibility of a referral for additional couples counseling for client and Sandra will be discussed at that time.

6. Client has agreed to allow the therapist to call him a taxi and neighbor has agreed to receive the client in the taxi at their apartment house and to pay for the taxi until Sandra can reimburse him when she returns. Client has given therapist permission to speak to the taxi driver to impress upon her or him that the client may not be dropped off anywhere but the specified address and to the care of the neighbor (ROI attached).

Session was ended at 3:05 P.M.; names and phone numbers for all involved support people are attached on a separate sheet.

Sample Suicide Progress Note (continued)

Follow-Up Notes

24 October

Taxi arrived at 3:10 P.M.; driver agreed to make sure client was only taken to the specified address and released to the care of the neighbor.

Therapist informed receptionist at 3:15 P.M. that the client would be making contact to leave messages for the therapist the next day to confirm his adherence to the suicide contract, giving her a copy of the suicide contract for dates and times (client's permission had been obtained; ROI is attached). Receptionist was instructed to notify the therapist on her cell phone immediately if the expected calls did not arrive on time. Client called (as agreed) at 3:35 P.M. to confirm that he had arrived safely at his apartment; neighbor spoke to therapist as well to confirm that he had taken possession of and destroyed the sleeping pills and alcohol, and would remain with the client for the agreed-upon time period.

25 October

Client called receptionist (as agreed) at 9:05 A.M. to let her know that the coworker had arrived (coworker confirmed this over the phone) and the neighbor had left.

Client called receptionist (as agreed) at 12:35 P.M. to let her know that Sandra had arrived at the airport and was willing to stay with the client until his next session (unless other arrangements were made in collaboration with the neighbor and coworker); Sandra confirmed this as well by speaking with the receptionist herself.

Notes about Possible Treatment Implications of the Intervention

- continue to monitor for suicidal ideation and general mental status
- explore impact of intervention on therapy
- explore the possibility of having reinforced client's dependence through the intervention (esp. through the fact that Sandra returned home on an earlier flight)
- find a referral for ancillary couple's therapy
- reevaluate therapy goals as needed

7

The Challenge of
Threats of Violence:
The Duty to Warn and Protect

A hurtful act is the transference to others of the degradation
which we bear in ourselves.

SIMONE WEIL

The threat of violence, such as homicide or aggression, in general mental health practice is a relatively rare, but upsetting, event for which all mental health care practitioners need to be prepared. The prediction of homicide is even more difficult than that of suicide, in part because of its less frequent occurrence. Mental health professionals, both clinicians and researchers, have not had as many opportunities to work with potentially homicidal clients as they have had with suicidal clients, both from a predictive and a postdictive perspective (Megargee, 1995; Meyer & Deitsch, 1996). Thus, in terms of risk assessment and intervention guidelines, much less reliable information is available for the clinician who is faced with a homicidal client. Prediction of homicide and aggression is imprecise and often assessments remain inconclusive despite the best efforts of the clinician or evaluator (Heilbrun, 1996; Slaikeu, 1990). The information that will be shared in this chapter is a compilation of data from various sources and yet it still no doubt leaves a large margin for error. At best, it presents the clinician with a repertoire of strategies for assessment and decision making that may enhance predictive power of the clinician as compared to a mental health care practitioner who does not have this information.

The prediction of homicide and other severe aggression toward others is not only more difficult because of its lesser occurrence, but also because it represents a human behavior that by definition is perpetrated in an interpersonal context and, hence, is an act that is highly influenced by situational variables and contextual factors (Heilbrun, 1996; Megargee, 1995). The action not only relies on the person making the threat (the person available to the clinician at the time of the assessment) but also on the person against whom the threat is made. This potential victim is generally not someone the clinician has access to or is familiar with and, hence, is not easily figured into the assessment equation. Additionally, situational circumstances are generally explored from the perspective of the perpetrator and may be inaccurate. Even if the client is an accurate historian and gives a precise description of the contextual variables, these factors may change, depending on the activity and behavior of the potential victim. Many homicidal or aggressive situations involve people whom the client knows. Specifically, it has been reported that 39% of homicides involved arguments between the victim and the perpetrator and 41% of victims knew their killer (Slaikeu, 1990).

The consequences of inaccurate prediction in a case of threatened homicide or extreme interpersonal aggression can be quite severe, perhaps even more devastating than incorrect prediction surrounding suicidal threats. Specifically, in the case of suicide, the gravest danger faced by client and counselor is that of a false negative, that is, the inaccurate judgment on the part of the clinician that it is safe to let the suicidal patient go. If risk for suicidal action is judged low when the client really presented high risk, the worst negative outcome is most obviously the client's death. The risks or consequences of a false positive are minimal compared to the risks presented by a false negative. If clinicians overestimate the risk presented by clients, they may place therapeutic rapport in jeopardy by breaching confidentiality, but usually neither a life-threatening situation follows, nor significant danger of negative consequences for the client from a social or economic perspective. In the case of a homicide threat, both false negatives and false positives represent potentially significant problems. The false negative can lead to the death of an innocent victim. The false positive can lead to severe social and economic side effects for the person who has been labeled dangerous by the clinician. Persons labeled as dangerous to others have experienced a variety of undesirable consequences such as loss of custody battles, jobs, and relationships; denial of bail, parole, or probation; and preventive criminal or psychiatric detention (see Megargee, 1995).

Prediction of homicide and severe aggression has to balance the threat of risk to the potential victim with the threat of negative consequences to the potential perpetrator and must be done with great sensitivity to both. Several legal and ethical factors must be considered as well, mak-

ing the issue of homicide threat assessment and intervention a sticky professional issue indeed. Thorough familiarity with legal and ethical issues, risk factors for aggression and homicide, procedures for assessment of risk, and possibilities for intervention will help the clinician weather this crisis with professionalism and relative freedom from panic.

LEGAL AND ETHICAL ISSUES FACING THE CLINICIAN OF THE VIOLENT CLIENT

As in suicide assessment, one central legal issue or decision the clinician has to make is whether to violate confidentiality. The same exclusionary clause quoted in the suicide chapter applies to threats of homicide or aggression against others: according to the American Psychological Association (APA), American Counseling Association (ACA), National Association of Social Workers (NASW), and many, if not all, other professional associations of mental health care providers, mental health care providers have a duty to warn and protect the potential victim from imminent and foreseeable harm that overrides the duty to maintain a client's confidentiality. This duty to warn and protect was added to professional codes of conduct in the aftermath of the *Tarasoff* case in California in the early 1970s. In the *Tarasoff* case, Prosenjit Poddar, a student at the University of California, disclosed to his psychologist that he intended to kill Tatiana Tarasoff. The psychologist consequently notified campus police verbally and in writing, and the officers took Poddar into custody. Because Poddar appeared rational to the police officers, they released him. The psychologist then consulted with a psychiatrist supervisor who directed that no further action was to be taken on the part of the psychologist. In fact, the supervisor also requested the return of the written statement the psychologist had sent to the campus police to protect Poddar's confidentiality. Tatiana was out of the country at the time of this occurrence (Beck, 1988; Boylan, Malley, & Scott, 1995). When she returned from Brazil two months later, Poddar killed her. Tarasoff's family then sued all university personnel involved in the case, alleging negligence on their part by failing both to warn Tarasoff and to detain Poddar. Initially, the case was dismissed in the lower court, but, when it was heard by the California Supreme Court in 1974, that court ruled in favor of the family. This decision by the court created a duty for mental health care providers to warn potential victims of crime, which resulted in great controversy among the mental health professions. Several reactions (including a brief by the American Psychiatric Association) were submitted to the California Supreme Court and triggered a rehearing of the case in 1976. This rehearing resulted in a further broadening of the duty to warn, directing clinicians not only to warn potential victims but also to take actions to protect them (Anderson, 1996; Beck, 1988; Swenson, 1997).

The court at that time also acknowledged the concerns expressed by the mental health community regarding the need for confidentiality in the therapeutic relationship. Thus, the court ruled that the duty to warn and protect a potential victim is only invoked if such disclosure is absolutely necessary to avert danger to others. Even in such a case, the court indicated, the disclosure is to be made discretely and in a manner that preserves maximal privacy for the client (Ahia & Martin, 1993; Swenson, 1997). Thus, the casual threat made by a client would not invoke *Tarasoff* (Anderson, 1996). Certainly, the *Tarasoff* ruling is anything but clear. On the one hand, counselors must warn and protect; on the other hand, they must preserve privacy as much as possible. Should clinicians fail to disclose because they misjudged the level of risk involved in the client's threat, and the client kills the potential victim, the clinician can be sued under *Tarasoff* for failing to warn and protect. Should the clinician disclose the threat to the potential victim (to warn) and to the police (to protect), and the client does not attempt to hurt the potential

victim, the clinician can be sued by the client for breach of confidentiality or negligence due to misdiagnosis (Ahia & Martin, 1993; Boylan, Malley, & Scott, 1995). However, in some states (e.g., Virginia) legislation has been enacted that holds clinicians harmless in the latter case, as long as they can demonstrate that their decision was based upon thorough assessment and clinically sound decision making (i.e., as long as they can prove that no negligence was involved [Anderson, 1996; Freemouw, Perczel, & Ellis, 1990]).

Although *Tarasoff* technically only has jurisdiction in the state of California, courts in at least 12 other states have made rulings that suggest that they have adopted a similar clinician duty to warn and protect. However, there have been subtle differences with regard to the judgment of danger, as well as with the specificity of the identity of the victim. Hence, the whole issue of duty to warn and protect remains in flux and the *Tarasoff* ruling cannot be taken as the final word on the issue (Freemouw, Perczel, & Ellis, 1990). In fact, some states (e.g., Ohio) have actually enacted legislation that absolves clinicians from the duty to warn and protect and gives them immunity from potential damages. Other states have enacted laws that limit *Tarasoff*. "These recent laws may mark a shift in legal duties toward further circumscribing the duties of therapists away from the extreme responsibility of having to foresee potential danger to any or all victims of each of their clients" (Freemouw, Perczel, & Ellis, 1990, p. 7).

With regard to the issue of judgment of danger, some states (namely, California, Colorado, Kentucky, Louisiana, Minnesota, Montana, and New Hampshire) have passed "communicated threat" laws that require the client to make a specific threat of violence before the duty to report and protect can be invoked. In other states (namely, Massachusetts, New Jersey, Oklahoma, and Rhode Island), however, the duty to warn and protect may need to be invoked even in the absence of a specific threat if the clinician has a reason to believe that the client represents a clear danger. Clearly, the legal conundrums faced by the therapist are even more difficult in these latter states where no recognition exists of a need for an objective, direct threat by the client (Ahia & Martin, 1993).

The issue of an identifiable victim, which has greatly expanded upon the original duties of the counselor, is a reaction to case law that developed after *Tarasoff*. *Tarasoff* specified that the mental health care provider must render a warning and protection if there is both serious danger and an *identifiable* victim (Boylan, Malley, & Scott, 1995; Freemouw, Perczel, & Ellis, 1990). All states in the United States require a report if a victim has been specified. However, some courts have held that the clinician has a duty to warn and protect even if the victim was only foreseeable, that is, not yet identifiable (e.g., in the states of California, New Hampshire, Kentucky, Montana, Massachusetts, and Oklahoma [Ahia & Martin, 1993]). These rulings have implied and state directly that clinicians can be held responsible for client's actions to an identified potential group of victims. For example, in one case, a ruling was made against a clinician who did not warn although the client had made threats of violence only against women in general (Boylan, Malley, & Scott, 1995). Similarly, some courts have decided that the duty to warn and protect may need to be invoked even in the case of threats against the general public, with no group or individual victim identified. Perhaps one of the most extreme cases in this regard was *Peterson v. the State* (1983) which involved a client with known historical and current drug and alcohol abuse who was released by a therapist from inpatient treatment. The client subsequently drove under the influence, collided with, and injured, a woman. The courts ruled liability on the part of the counselor who released the client, knowing his propensity for drug use. Arthur and Swanson (1993) therefore suggest that the duty to warn also applies in cases of a client who continues to threaten to drive while cognitively impaired (e.g., due to psychosis, seizure disorder, substance use) even after the clinician has attempted to get an agreement from this client to desist from the operation of a vehicle.

On the other hand, as previously noted, rejection of *Tarasoff* has also been documented through various case rulings (Swenson, 1997). A Florida court, ruling on a case that involved murder, concluded that "it would be fundamentally unfair to impose a duty to warn upon psychotherapists because psychiatry is an inexact science and a patient's dangerousness cannot be predicted with any degree of accuracy, . . . [and] added that the relationship of trust and confidentiality that is necessary for the therapeutic process would be undermined if a psychotherapist would be required to warn a potential victim" (Boylan, Malley, & Scott, 1995, p. 294).

Given the controversy and evolution since the second *Tarasoff* ruling in 1976, most legal resources for clinicians (e.g., Anderson, 1996; Canter, Bennett, Jones, & Nagy, 1994) suggest that when in doubt, clinicians best disclose their concern to the potential victim (to warn) and the police (to protect), although this may leave them vulnerable to prosecution for breach of confidentiality (Anderson, 1996; Slaikeu, 1990). These professionals conclude that because counselors or therapists discussed the limits of confidentiality with their clients at the outset, breach of confidentiality cannot be charged against them because they clearly defined its limits from the beginning of the clinician-client relationship. This argument once again demonstrates the great importance of appropriately phrased informed consents and their discussion at the moment of first contact with a client. Given a ruling in a court in Michigan, Anderson (1996) concludes that therapists are protected from legal liability if they can demonstrate having followed appropriate assessment procedures before making the clinical judgment to warn or not to warn. The procedures suggested in *Davis v. Lhim* (1988) include assessment and/or rendering of an opinion about clinical diagnosis, opportunity to follow through with the threatened act, history of violence, factors provoking the threat, likelihood of continuation of the threats, response to treatment, and relationship with the potential victim. The current consensus appears to be that the clinician in doubt should err on the side of caution and should disclose (i.e., should warn and protect); however, the clinician must also be able to document thorough assessment and informed decision making and should be familiar with state law and with whether its tendency is toward enforcement (and even expansion) versus rejection of *Tarasoff*.

A difficult issue related to threat of homicide or aggression is the threat of transmission of the Human Immunodeficiency Virus (HIV). The infection status of a client in and of itself is definitely protected information, not only because of confidentiality laws regarding psychotherapy, but also because of AIDS disclosure laws. However, a client who threatens to spread the virus through unprotected sex or through sharing "dirty" needles is, in effect, making a threat of harm against others. Anderson (1996) concludes that given the current state of legal case law and ethical guidelines, the clinician who identifies a clear intended victim for the planned transmission of the virus (either through sex or needles) has "an ethical obligation to take some steps to curb the behavior and to warn the potential victim(s)" (p. 33). If this action is necessary, clinicians can take steps to protect themselves from liability for violation of confidentiality. The first and most critical step is, once again, the discussion of limits of confidentiality and an explanation of how the planned spread of HIV falls into the category of danger/harm to others. Additional steps for therapist protection have been outlined by Ahia and Martin (1993) and include the following:

- being knowledgeable of the most recent medical information about HIV/AIDS and its transmission, and sharing this information with the client

- being knowledgeable about HIV/AIDS disclosure laws in the state where the clinician practices, and informing the client thereof

- educating HIV-positive clients about safe sex practices, safe needle sharing, information about needle exchanges, and other HIV/AIDS transmission prevention practices

- discussing all disclosures of the client's HIV/AIDS status to others with the client first, and obtaining releases of information, if at all possible, even when invoking *Tarasoff*

- keeping disclosures made under the duty to warn and protect to a minimum to preserve the client's rights to privacy as much as possible

The spirit of these guidelines about how to deal with a duty to warn or protect not only applies to situations involving HIV/AIDS, but is also applicable to most situations involving threat of homicide or aggression. In other words, counselors are always well informed and prepared, share all of their intentions about actions to be taken with their clients, always try to get releases of information from clients before making *Tarasoff* disclosures, always involve clients in every step of the disclosure process, and always keep disclosures brief and to the point. Further, the process leading to a disclosure always needs to meet at least three obligations to reduce the risk of negligence on the part of the clinician (see Appelbaum, 1985):

- assessment, consisting of data gathering about risk factors and judgment regarding dangerousness

- decision making as reflected in the selection of a course of action to warn and protect the potential victim

- implementation of the interventions to protect the potential victims and to continue to monitor the actions of the potential perpetrator

The duty to warn only applies to threatened future action, not to the disclosure of past aggression or criminal behavior. The confidentiality exception granted to the mental health care provider under the duty to warn and protect does not allow the provider to disclose any information beyond the circumstances of the future threatened act (Arthur & Swanson, 1993). The entire process, as well as all interactions with client, third party (potential victim and potential protectors), and others (e.g., supervisors and consultants) needs to be thoroughly and carefully documented and placed into the client's chart as soon as possible (Boylan, Malley, & Scott, 1995; Meyer & Deitsch, 1996). Trainees need to involve a supervisor throughout the process and clinicians with minimal experience should do the same with a consultant. Even seasoned clinicians should seek consultation, if not during, then immediately after, the disposition of the case, and all consultation needs to be carefully documented. In the cases of practitioners who are still required to work under supervision, all documentation should be cosigned by the supervisor who ultimately bears the legal responsibility of the decisions that were made in the management of the client. Throughout all documentation, decisions need to be discussed in terms of why a certain action plan was chosen over another option. It is not enough to document what was done, but the clinician must also document what was *not* done and why. If clinicians are able, they cite ethical standards or local laws that support their decisions and that reflect the thorough consideration given to each step in the assessment, decision-making, and intervention process.

One final interesting note with regard to the clinician's legal responsibilities in the case of a threat of violence will be quoted from Megargee (1995). This expert on interpersonal aggression points out that "few psychologists are aware of the fact that threatening the life of the President of the United States is itself a crime (18 USC 871). Therefore, psychologists who fail to report such threats could themselves be accused of misprision of a felony" (p. 406). A caution to keep in mind.

HELPFUL TRAITS DURING VIOLENCE ASSESSMENT AND INTERVENTION

A few mental health care provider characteristics that are helpful during a violence assessment will be mentioned briefly. Many of the assertions

made about the therapist characteristics that are helpful to a suicide assessment and intervention also apply to homicide assessment or intervention. Hence, rereading that section in the prior chapter may be helpful to the reader.

There are those who believe that homicide threats should never be dealt with by the counselor in training (Faiver, Eisengart, & Colonna, 1995). Although this claim has definite legitimacy, it does not excuse the trainee from understanding the issue. Even trainees must have basic knowledge about the issues involved so that in a crisis, they can begin an assessment before a supervisor arrives or can conduct the entire assessment should no other resource be available at the time of the crisis. Thus, a willingness to learn and to prepare for even only remotely possible events in a therapy setting is a primary trademark of an effective clinician. Having familiarity with the specific personality traits, familial background, social and socioeconomic factors, psychiatric history and medical predictors, and psychological traits, of potential perpetrators is very important among mental health care providers. Knowing which types of interventions are possible and indicated is similarly important. Above all, however, all mental health care providers should have a thorough understanding of the original *Tarasoff* ruling, its extensions and its rejections, as well as the conclusions that have been drawn about professional and ethical behavior of clinicians by professional associations such as the APA (1992), the NASW, and the ACA (1995) (Monahan, 1993). Unfortunately, although most clinicians are knowledgeable about the issues surrounding suicide, many do not consider homicide an issue with which they will ever need to contend. Hence, the threat of homicide or aggression may catch many a practitioner off guard. It is important to prevent this through thorough study and practice.

Besides being knowledgeable about data and information relevant to threats of homicide and aggression, clinicians also need to be mentally and emotionally prepared for meeting with a potentially violent client. They must be able to remain calm and collected even in a potentially frightening situation. They must be able to set limits and be able to deal with threats of immediate violence as well as future violence. How clinicians can deal with a client who becomes violent or aggressive in a therapy session is dealt with in Chapter 5 and the reader can refer back to this discussion. Generally speaking, mental health care providers encountering homicidal or aggressive clients need to be careful neither to overreact nor to minimize the threat. They must be willing to deal with the topic and to ask questions about it thoroughly and unthreateningly. They need to be aware of danger signs that signal the possibility for in-session violence, diffusing it if possible, setting limits as needed, and ending the session if their personal safety is threatened. Clinicians need to be knowledgeable about physical symptoms that signal increasingly angry affect, such as muscle twitching or restlessness, staring or lack of eye contact, shallow breathing, quivering or loud voice, clenched fists, and angry words (Morrison, 1995).

As with suicide, counselors best refrain from euphemisms and from moralizing. They must be able to retain control of the session, even if they decide to solicit help from a supervisor. In other words, if clinicians decide to involve another professional, this is presented to the client calmly and directly without apology or sign of wavering confidence. The issue is presented as one of seeking input and assistance from another professional, not one of being overwhelmed and frightened and having to find someone else to deal with the client. Another important trait is empathy. Empathy is easy to come by with clients who are suicidal; after all, a protective response usually emerges in the clinician that stimulates the desire to help the client overcome the self-destructive impulse. Empathy is often harder to come by if the client is angry and threatening. However, if clinicians can keep in mind that clients get angry and feel the urge for violence because of other underlying emotions that may be hurtful, frightening, or threatening to the clients' self-esteem, they may be able to empathize even with overtly hostile clients and to respond with caring and concern. Finding the

hurtful or shamed aspect of the client's self that is fighting for approval and raises the hostile urge will help both client and clinician toward a better relationship and a more likely positive outcome of the violence intervention.

Willingness to seek consultation and supervision is also an extremely important trait among therapists who are likely to encounter homicidal or aggressive clients in their practice (Baird, 1999). This is true for three reasons. First, consultation and supervision are critical during the process of assessment, decision making, and intervention, because input from another professional may increase the predictive power of the counselor. Second, with long-term clients the clinician may lose objectivity and may under- or overreact to risk level depending on prior interactions with and preconceived notions about the client. Third, homicidal action on the part of a therapy client is likely to have a profound effect on the therapist. As was noted in the suicide chapter, no amount of knowledge can prevent mistakes from occurring. Thus, this chapter does not guarantee that the clinician will never encounter a client who, after the therapist's best clinical judgment to the contrary, goes out and kills another human being. Such risk is inherent in the helping profession, and counselors must prepare themselves mentally for such trauma to prevent burnout. Hence, should a homicide occur, therapists need to be willing to seek consultation or help in processing the event to keep their personal and professional self-esteem intact, as well as to attempt to learn something from the experience in terms of future prediction of client behavior.

RISK FACTORS FOR THE PERPETRATION OF VIOLENCE

"Clients lacking impulse control may be the most difficult [clients] of all" (Kottler, 1992, p. 28). To make adequate decisions with regard to homicide or aggression threats, the clinician must obviously be knowledgeable about the predictors or risk factors that have been identified in the literature to date. Further, to document properly, the clinician must know which factors to assess and how to record in progress notes, not only that they were assessed, but what the outcome of this assessment was. A discussion of currently identified risk factors follows. This discussion is based not only on the homicide and aggression literature but also draws information from research on the precursors and predictors of specific DSM-IV diagnoses that are highly correlated with such behaviors, namely, conduct disorders, antisocial personality disorders, and intermittent explosive disorder (American Psychiatric Association, 1994), all of which will also be discussed in some detail below.

Demographic Characteristics

Significant consistency emerges for demographic variables. Thus, as in the suicide section, references are provided only for those demographics for which controversy remains or for which only one reference was found expressing a certain pattern. For all other demographic variables, information was drawn from a consistent set of sources (e.g., American Psychiatric Association, 1994; Borum, Swartz, & Swanson, 1996; Gilliland & James, 1993; Kamphaus & Frick, 1996; Meyer & Deitsch, 1996; Morrison, 1995). The most consistent finding about risk factors is identified with regard to gender, with males being significantly more likely than females to engage in homicidal or aggressive behavior. Criminal behavior occurs five times as often in males as in females, and with regard to aggressive crimes, the male to female ratio is as high as 50 to 1 (Meyer & Deitsch, 1996). A negative correlation is reported with age between early adolescence and age 35; in other words, for young people up to the age of 35, risk increases the younger the client's age. The highest risk in terms of age is identified for the age range between 15 and 30 (Gilliland & James, 1993). Ethnicity appears to be a predictor with greater risk reported for non-White clients, especially clients from disadvantaged minority groups. Probably highly correlated with the latter factor, living in inner cities, more than any other living situation, increases

risk. Similarly, as socioeconomic status decreases, risk increases. The same pattern is true for educational level and/or level of intelligence. Given these characteristics, not surprisingly, the most violent group of clients has been identified as disadvantaged inner city young males (Gilliland & James, 1993).

Personal Traits and History of Violence

Another strong set of predictors has to do with personal variables, such as increased hostility and anger, poor impulse control, and history of violence. Greater reported present and past anger (either self-reported or through observation by clinician or others) and hostility, as well as increased potential for rage and its related affects, significantly increase the risk for the acting out of aggressive impulses (Bellak & Siegel, 1983; Borum, Swartz, & Swanson, 1996; Gilliland & James, 1993; Megargee, 1995). This is true even if the anger and hostility are not displayed in session, that is, if they are not experienced or perceived by the clinician directly (Megargee, 1995). Poor impulse control, along with angry affect (either by history or currently), further appears to increase risk (Borum, Swartz, & Swanson, 1996; Faiver, Eisengart, & Colonna, 1995; Meyer & Deitsch, 1996), as does a history of stimulation seeking (Loeber, 1990). Poor frustration tolerance, irritability, temper outbursts, and recklessness are also features commonly associated with aggressive behavior exhibited as part of conduct disorders or antisocial personality disorders (American Psychiatric Association, 1994; Borum, Swartz, & Swanson, 1996). Violence, however, is not inevitably linked to anger or hostility, but rather may also arise from circumstances of fear or frustration (Baird, 1999). Regardless of accompanying affect, individuals who are likely to carry out a violent act tend to believe that their action is justifiable and that there was no alternative to violence (DeBecker, 1997).

According to some writers (e.g., Gilliland & James, 1993; Heilbrun, 1996), a history of violence (including juvenile violence) is one of the best predictors of future violence and always warrants more restrictive interventions if present. This is increasingly true with the severity of past offenses, which may range from threat of assault with a deadly weapon, to assault, to sexual attack, to attempted homicide, and homicide (Faiver, Eisengart, & Colonna, 1995; Gilliland & James, 1993). A history of violence is still considered predictive of risk even if it has not involved actual criminal convictions. For example, cruelty to animals, frequent fighting, truancy, and similar violent and oppositional behaviors also are believed to increase risk (American Psychiatric Association, 1994; Borum, Swartz, & Swanson, 1996; Meyer & Deitsch, 1996). Unstable school and job histories, though not necessarily synonymous with violence in the past, appear to be somewhat predictive of future violence as well (Meyer & Deitsch, 1996).

Psychological and Familial Factors

Meyer and Deitsch (1996) have indicated that fascination with weapons, as well as treasured ownership of weapons, may be predictive of violent acts. Acutely lowered self-esteem in the aftermath of a loss (literal, as in the loss of a job, or perceived, as in a rejection by another person) may also be related to more acute risk for aggressive behavior (Bellak & Siegel, 1983). Recent stressors may be related to increased risk, perhaps through biological changes or reactions induced by severe acute or prolonged experience of stress (Volavka, 1995).

The most powerful set of predictors in the psychological and familial arena, however, relates to family functioning. Specifically, great consensus appears to exist that harsh (authoritarian) or inconsistent parenting histories increase risk for aggression (American Psychiatric Association, 1994; Gilliland & James, 1993; Kamphaus & Frick, 1996; Meyer & Deitsch, 1996). Similarly, families with violent histories (e.g., domestic violence, child abuse) appear to be highly correlated with conduct-disordered offspring, as do families with high levels of emotional rejection (Meyer & Deitsch, 1996). Instability in the parental home, either through frequent parental fighting or through multiple divorces and/or separations, also

present a risk factor. Psychiatric history within a family system and social isolation of the family may also contribute to increased risk (American Psychiatric Association, 1994; Gilliland & James, 1993; Kamphaus & Frick, 1996).

Social and Environmental Factors

Modeling of violence may not only take place in the family context, as noted above, but may also occur in larger social contexts. Specifically, violence can be modeled through participation in violent peer groups or through observation of violence in the media, especially on television (Gilliland & James, 1993; Megargee, 1995). Risk also may be increased in the presence of others who condone violence or who encourage aggression by others. A community or peer group of violence can be a powerful motivator for aggression (Borum, Swartz, & Swanson, 1996; Gilliland & James, 1993), especially in the context of politics (see below) or organized crime (Faiver, Eisengart, & Colonna, 1995).

Quite to the contrary of the modeling factor, social isolation and peer rejection (or poor peer relationships in general) during the teenage and earlier childhood years may also increase risk for violence (Gilliland & James, 1993; Kamphaus & Frick, 1996; Loeber, 1990). Interpersonal conflicts in general, as well as poor communication skills, both of which suggest poor social relations, also have been noted as risk factors (American Psychiatric Association, 1994; Faiver, Eisengart, & Colonna, 1995; Gilliland & James, 1993; Loeber, 1990). Academic problems and poor social skills have been noted with regularity among aggressive antisocial clients (Meyer & Deitsch, 1996).

A variety of environmental contributors to violence and aggression have been identified fairly consistently by a number of writers (American Psychiatric Association, 1994; Gilliland & James, 1993; Megargee, 1995; Meyer & Deitsch, 1996). These include high ambient temperatures, crowding, noise, architectural design, and conflict. The availability of drugs and alcohol, weapons, and potential victims are additional risk factors. The drug and alcohol issue will be addressed later in more detail. Current political climate or strongly held political beliefs or agendas can contribute to the likelihood of aggressive behavior. The latter has certainly found expression in recent history through terrorism and militia membership.

Physical and Medical Variables

Purely motoric clues of impending violence that were briefly addressed earlier warrant repeating. They include signs such as tense or twitching muscles, darting eye movements, staring or avoiding eye contact, closed defensive postures, disheveled appearance, clenched fists, pacing, and so forth (Gilliland & James, 1993; Morrison, 1995). Therapists noting these behaviors in a client receive a clear signal that the client may become potentially violent in the immediate future, not only with regard to the threatened action, but also with regard to the therapist, and hence they need to take appropriate precautions for their own safety. These motoric clues also underscore the client's readiness for aggression in general and must be figured into the risk equation for the future as well.

Risk factors in the physical arena have also been identified. Neurological disease (according to some, including attention deficit disorder [Bellak & Siegel, 1983]), endocrine diseases or hormonal imbalances, organic brain disorder, and exposure to toxins, chemicals, or drugs have all been identified as potential risk factors for violence (American Psychiatric Association, 1994; Gilliland & James, 1993; Kamphaus & Frick, 1995; Morrison, 1995). Heilbrun (1996) notes that cognitive deficits (including lower intelligence, less effective thinking, poor empathic capacity, and limited psychological mindedness) are the second most accurate set of predictors of violence after the diagnosis of antisocial personality disorder. Specifically, he indicates that "a variety of cognitive flaws . . . not only make it more likely that crimes will be committed but also increase the likelihood that poor planning or faulty execution of the crime will be involved. The bungling criminal places the victim at

greater risk of serious harm by poorly conceived criminal activity and the failure to anticipate exigencies" (p. 192). Higher rates of prenatal, genetic, and birth disorders have been noted among aggressive clients with antisocial personality disorders (Loeber, 1990; Meyer & Deitsch, 1996). Unfortunately, the exact nature and mediating role of these various diseases and disease processes remains largely unclear.

Psychiatric History

A number of psychiatric disorders may contribute to a pattern of violence (e.g., bipolar disorder, schizophrenia, substance use disorders) and several have violence as their essence (i.e., conduct disorder, antisocial personality disorder, and intermittent explosive disorder). With regard to bipolar disorder, risk is increased if the client is in a manic phase of the disorder and has a history of aggressive or violent behavior during these episodes. Similarly, a schizophrenic client presents increased risk only if actively hallucinating and if hallucinations are of a paranoid or violent nature (Meyer & Deitsch, 1996). Common comorbidity between conduct disorders and attention–deficit hyperactivity disorder has also been reported, suggesting the possibility of increased risk for potential acting out with attention problems, impulsivity, and hyperactivity (Loeber, 1990; Kamphaus & Frick, 1996).

Substance use disorders have a slightly more complex relationship with violence or aggressive acting out. Violence may be mediated by substance use in a number of ways. First, it may be related to intoxication symptoms of substances such as amphetamines (which may produce grandiosity, excitability, and overreaction to stimuli), barbiturates (which, although they depress the CNS, have excitatory effects for some), cocaine, PCP, and alcohol. Toxic reactions to some drugs are another factor in the relationship of substance use with violence, especially for PCP, LSD, and barbiturates. Finally, withdrawal symptoms from alcohol, heroin, sedatives, and depressants can include excitability and overreactions that may lead to violence (American Psychiatric

Association, 1994; Baird, 1999; Borum, Swartz, & Swanson, 1996; Gilliland & James, 1993; Meyer & Deitsch, 1996; Morrison, 1995). Risk increases if substance use disorders occur in combination with psychotic disorders and/or personality disorders (especially antisocial personality disorder [Brems & Johnson, 1997]).

Finally, conduct disorder, antisocial personality disorder, and intermittent explosive disorder all have violence and/or aggression as the essence of their symptomatic picture (Heilbrun, 1996). If clients present with these disorders and make a threat of violence, they will likely be at a significantly higher risk for follow-through than clients without such a diagnosis. In fact, given that these diagnoses (especially conduct disorder and antisocial personality disorder) imply a history of violence, their presence suggests the possible need for more restrictive interventions (see Gilliland & James, 1993). If these diagnoses also coincide with other risk factors (which happens with great regularity), the potential for aggression increases substantially. Given their strong relationship to violence, these three disorders will be discussed in some detail and the reader is encouraged to seek more information about them through additional reading. The information provided below is primarily based on the DSM-IV (American Psychiatric Association, 1994) and secondarily on Meyer and Deitsch (1996), and should be used to guide assessment questions that rule out these diagnoses in the client who has made a threat of aggression against others.

Conduct Disorder. Although classified as a disorder of childhood, the conduct disorder label can actually be maintained into adulthood if a switch to antisocial personality disorder is not warranted (Meyer & Deitsch, 1996). Prevalence rates are not yet very clear and range widely from 6–16% for males and 2–9% for females in the general population. Onset is usually late in childhood or early in adolescence, though it may occur as early as age 5 and as late as age 16. Since earlier onset has a worse prognosis, two subtypes of conduct disorder are diagnosed: Childhood-Onset Type (onset before age 10) and Adolescent-

Onset Type (onset after age 10). Severity of the disorder is also rated and can range from mild (just meets criteria and causes minimal harm to others) to moderate (between mild and severe) to severe (has many more conduct problems than required to make the diagnosis and causes considerable harm to others).

The disorder is marked by a repetitive and persistent behavior pattern that violates the basic rights of others, age appropriate norms for behavior, and/or societal rules. This pattern of behavior must have manifested itself in the past year, with at least one symptom having been present in the past six months. A repetitive and persistent pattern is established if a client has exhibited at least three symptoms that reflect

- aggression to people and animals (e.g., bullying, physical fighting, cruelty, mugging)
- destruction of property (e.g., fire setting, willful destruction of others' belongings)
- deceitfulness or theft (e.g., breaking into someone's home, lying to obtain favors, shoplifting)
- serious violation of rules (e.g., breaking curfew, truancy, running away)

The behavior must have resulted in significant impairment of social, academic, or occupational functioning; and, if the person is older than age 18, does not meet criteria for antisocial personality disorder (American Psychiatric Association, 1994). Care must be taken not to confuse conduct disorder with oppositional defiant disorder. Although the latter disorder manifests itself as a behavior pattern that reflects negativity and hostility and includes symptoms such as anger, spite, vindictiveness, argumentativeness, and defiance, it usually does not involve physical aggression that is expressed through harm to others or willful destruction of property. Hence, it is not a disorder that would present a significant risk factor in the evaluation of a homicide or aggression threat.

Antisocial Personality Disorder. This disorder essentially represents the chronic manifestation of amoral and impulsive behavior that can lead to either an aggressive or nonaggressive path of criminal and other offenses (Borum, Swartz, & Swanson, 1996; Heilbrun, 1996; Meyer & Deitsch, 1996) and that manifests itself in a variety of contexts. Age of onset is in early adolescence or even childhood, where there has been a conduct disorder diagnosis in the past that is revised to the antisocial personality disorder diagnosis after the individual reaches age 18. Most common age of onset is approximately age 15, by which time the antisocial behavior pattern has been firmly established. The disorder tends to be more commonly diagnosed in men than women, with prevalence figures in the general population of 3% and 1% respectively. Prevalence rates are significantly higher among substance using populations and in forensic settings. Although symptoms tend to decrease over the life span (with regard to number and frequency of occurrence), aggressive acts do not appear to diminish (Meyer & Deitsch, 1996).

To diagnose antisocial personality disorder, a pervasive pattern of disregard for and violation of the rights of others has to be identified that extends back at least to age 15 and that resulted in or would have warranted a diagnosis of conduct disorder. This pattern is established through the identification of at least three antisocial behaviors included in the following list:

- failure to conform to social norms that lead to unlawful behavior that warrants arrest
- deceitfulness (e.g., lying, using aliases, conning others)
- impulsivity and failure to plan ahead
- irritability or aggressiveness (e.g., physical fights, assaults)
- recklessness that puts personal and other's safety at risk
- irresponsibility (e.g., poor work history, not repaying loans, ignoring financial responsibilities)
- lack of remorse

Antisocial personality disorder cannot be diagnosed before age 18 and differential diagnosis must be made with mania and schizophrenia. If either bipolar disorder or schizophrenia can account for all of the symptoms used to qualify the individual for antisocial personality disorder, the diagnosis of the latter would not be made. If the person's behavior does not warrant an antisocial personality disorder diagnosis, never resulted in or warranted a conduct disorder diagnosis, and cannot be accounted for by mania, schizophrenia, or intermittent explosive disorder, a diagnosis of adult antisocial behavior may be given. This diagnosis is used "when the focus of clinical attention is adult antisocial behavior that is not due to a mental disorder (e.g., conduct disorder, antisocial personality disorder, or an impulse control disorder). Examples include the behavior of some professional thieves, racketeers, or dealers in illegal substances" (American Psychiatric Association, 1994, p. 683–684). This diagnosis does not generally involve the level of physical aggression explored in this chapter.

Intermittent Explosive Disorder. This disorder is relatively rare and, hence, no reliable prevalence data exist. It is hypothesized that the disorder is more prevalent among males because episodic violence, the trademark of this disorder, is more common among men than women. Age of onset appears to be from late adolescence to the third decade of life, though data supporting this estimate are limited. Onset appears to be abrupt, with little predictability. It is important to differentiate intermittent explosive disorder from other disorders that may contain discrete episodes of acting out, such as is possible in some forms of dementia, substance use disorders, or personality disorders. Further, if the aggression is secondary to a physiological process or medical condition, the medical condition will be diagnosed instead.

The essence of intermittent explosive disorder is several discrete episodes of assaultive acts or destruction of property that manifest themselves due to an individual's failure to resist aggressive impulses even though that individual is normally capable of inhibiting impulses (or has not had them). The degree of aggression that is displayed is excessive given any possible psychosocial stressors that may have served as precipitants. Finally, as mentioned above, the behavior is not accounted for by any other medical or psychiatric condition. In this disorder, unlike antisocial personality disorder, regret and guilt are common features and the person perceives the behavior as a period of loss of control or "temporary insanity" (Meyer & Deitsch, 1996, p. 248). Enhancing awareness of anger impulses and teaching anger management strategies may be successful means of treating the disorder and preventing future acts of aggression.

Immediate Violence Predictors or Factors

As in suicide assessment, several factors need to be considered particularly valuable to predicting immediate risk of acting out. These risk factors are threats of aggression or homicide, a history of aggression, and the specificity of the plan that has been made with regard to the aggressive act. A very high correlation is present between threats of violence and the carrying through of the threatened plan (Slaikeu, 1990). This correlation has led to laws and legal decisions like *Tarasoff*. For this reason, once clinicians are aware of the threat, they must take some action to assess the severity of the threat and the likelihood that the threat will be carried out. The more specific the threat, the more likely that it will be carried out. A specific threat, in combination with personality factors and other immediate homicide predictors that are highly related to aggression, represents a severe risk that suggests restrictive intervention.

One of the best predictors of future violence against others is the presence of interpersonal aggression in the past (Borum, Swartz, & Swanson, 1996; Megargee, 1995). In addition, as the level of aggression involved in past violence increases, the risk for new violence also increases. Such a hierarchy of aggression could start with a threat of assault with a deadly weapon, and move on to assault, to sexual attack, and finally to homicide. However, some have argued that *any* background of involvement with the criminal justice system increases risk for follow-through on a threat (Gilliland & James, 1993). Similarly, it has also

been reported that any history of interpersonal aggression increases risk, even if it has never come to the attention of the criminal justice system (Megargee, 1995). In fact, exposure to interpersonal aggression alone might be sufficient to qualify in this category of history of aggression, in that risk may be increased also if the client who is making the threat has been the repeated victim of violence. Further, risk increases with the length and success of the history of aggression. Potential perpetrators who have a long history of interpersonal violence and who have perceived many advantages to this behavior (i.e., have rarely been caught and have often coerced their way through aggression) will present greater risks than clients who report only one prior act and were caught and punished for it (Megargee, 1995). This concept of habit strength will be revisited later in the risk assessment section.

As is the case with immediate suicide risk, the immediate risk for homicide or interpersonal violence increases with the specificity of the plan involved. There are five plan components that need to be explored: person, method, means, time and place (opportunity), and lethality of the chosen method. The more each component has been identified in detail, the greater the risk that the client is planning to follow through on the threat (Borum, Swartz, & Swanson, 1996; Gilliland & James, 1993). The first question, if it has not already been answered by the client in a spontaneous threat, will be about the victim. It is important to assess whether aggressors have a particular victim in mind or whether they are making a threat of aggression against a specific group of people. Some *Tarasoff* expansions may also suggest that a generalized threat against the public needs to be considered. Generally, however, risk increases the more the client has narrowed down the identity of the potential victim (Megargee, 1995). In this context, the mental health care provider must also assess the perpetrator's access to the victim. Recalling the original *Tarasoff* case, when Poddar first made his threat against Tatiana, she was in Brazil. Thus, although risk was clearly high, his limited access to her at the moment of the threat was very low.

However, warning Tatiana would have been essential anyhow, to let her know that, should she become accessible to Poddar in the future, her life might be in danger (as indeed it was). Additionally, it is important to assess clients' potential motives, by asking about whether they are angry with the identified victim or with anyone else (Beck, 1987).

With regard to the method, the clinician must clarify whether the client has thought about how to harm the identified victim. The more thought that has been given to exact procedures, the greater the risk. If the client has also given thought to how to secure the means to carry through the method, risk increases; if the client is already in possession of the means, risk is even higher. In the *Tarasoff* case, Poddar threatened the use of a firearm. He was in possession of a gun and planned to use it. Risk, hence, was considerably higher than if he had threatened Tarasoff's life without having either a chosen method or the means to carry it through. If, in addition to having method and means, the client also reports having chosen a time and place for the action, risk rises significantly. Time and place often relate to opportunity for carrying out the threat, an essential predictor of interpersonal violence (Megargee, 1995). Had Poddar decided to kill Tarasoff on the university campus on the day he made the threat, risk would have decreased, as the opportunity for contact with Tatiana would not have been there, given her absence from the place at the time.

Finally, the lethality of the method needs to be assessed. It must be noted that although lethality of method does not increase risk for follow-through, it is significantly related to risk of actual harm to the identified victim. For example, if the client reports that he plans to harm the potential victim by shooting her with a machine gun during a planned company picnic that will occur on a specified date, and the client has a weapon and ammunition in possession, risk is extreme because it is not only very likely that the client will follow through, but it is also likely that the attempt will be successful due to the lethality of the method. If, on the other hand, a client indicates that she plans to harm a potential victim by infecting him with a flu virus, has the flu herself right now, and knows when she will be in close enough physical contact to follow through on the threat, risk may be high that the client will follow through, but risk for harm to the identified victim is minimal. Risk for follow-through and risk for harm will both need to be assessed based on the specifics of the plan of action discussed by the potential perpetrator, and both have to be considered when planning intervention. Clearly, interventions will be most restrictive if both risk for follow-through and risk for harm are high (American Psychiatric Association, 1994; Gilliland & James, 1993; Megargee, 1995; Meyer & Deitsch, 1996).

ASSESSMENT OF VIOLENCE RISK

Once clinicians are familiar with risk factors for aggression as outlined in the previous section, they are ready to conduct homicide or aggression assessments, though always with some caution if they have little clinical experience. It must be noted again that there is some controversy in the literature about whether trainees should ever conduct homicide assessments (see Faiver, Eisengart, & Colonna, 1995). However, the reality is that there may not be a choice. Although in a perfect world a trainee would never be called upon for such an assessment without the immediate assistance of a supervisor, circumstances may arise when the trainee does not have a choice but to do the assessment alone. It is best for trainees to be prepared for this eventuality (see Boylan, Malley, & Scott, 1995), though they should call in their supervisor to assist with the assessment if this is at all possible.

Throughout the assessment, clinicians keep in mind that they are now functioning from a crisis perspective, not a therapeutic perspective. Thus, the particular interventions and expressed personality traits on the part of the clinician may differ from the provider's usual interactions with

therapy clients. The session will be more directive, more problem-solving oriented, more cautiously confrontational, more circumspect, and
less empathic than therapeutic interactions. Mental health care providers need to be able to shift
their behavior and attitude as needed for this
type of emergency assessment and intervention.
The focus of the assessment part of the session
becomes essentially one of evaluating whether
there is a verbalized or foreseeable threat of violence, imminent and foreseeable danger, and an
identifiable or foreseeable victim.

Being familiar with the risk factors for aggression and violence will help structure the assessment process because the clinician will know
which questions to ask and which areas of functioning to assess (Monahan, 1993). As was noted
in Chapter 6, the therapist or counselor who is
already familiar with the client through prior
work together will have an easier time of this
than the clinician who is working with a client
for the first time when the threat is made. If clinicians are already familiar with the client, they
may have much of the needed background information (such as family history, social environment, history of violence and criminal justice involvement, psychiatric diagnoses, substance use,
and so forth). The therapist, on the other hand,
who sees a person for the first time and is confronted with a threat of aggression will still have
to collect all of this information. In both cases,
providers will have to do a thorough assessment
of immediate risk factors and will have to conduct the other components of the risk assessment
outlined below.

In planning and conducting a homicide or
aggression assessment, the clinician must keep in
mind that for interpersonal violence to take
place, three conditions must be met: there must
be an aggressor, a victim or target, and the opportunity to attack (Megargee, 1995; Monahan,
1993). Thus, attention must be paid to all three of
these factors. Although this may seem to make
the assessment more complex, it actually creates
more options for intervention. Specifically, given
that all three components have to be in place, intervention can focus on any one of the three fac-

tors. The clinician can reduce potential harm not
just by intervening with the potential aggressor,
but also by removing opportunity, or by intervening with the intended victim. This provides
more flexibility in approach than in suicide, in
which the same person is the aggressor, the victim, and the creator of opportunity.

The assessment of risk for aggression and
homicide is similar to that of assessment of suicide risk. It concerns itself with the establishment
of inhibiting versus facilitating factors as well as
their relative balance. Megargee (1995) has proposed a model for assessment that will be presented and expanded upon here and that has
found corroboration through the writings and
recommendations of others. Specifically, in this
model, five factors need to be assessed: instigation,
habit strength, inhibitions, situational factors, and
reaction potential. Also added into the final risk
assessment decision-making process will be the
evaluation of threat presented by factors not covered in Megargee's model, such as the nature of
the plan and the opportunity for action.

Instigation and Triggers

Instigation refers to the motivating factors that
may drive the aggressive behavior. These factors
can be internal (e.g., anger, rage, annoyance) or
external (e.g., economic gain, revenge, political
conviction [Beck, 1987; Megargee, 1995]). Both
sets of factors should be assessed to get as complete a picture of the person and the potential
triggers of violence as possible. With regard to
internal factors, the clinician must attempt to get
a handle on the level of affective arousal present
in the client that may lead to aggression. Most
commonly, this implies an assessment of present,
past, and future anger, rage, and similar emotions.

Assessment of anger is best conducted
through relatively direct as well as more innocuous questioning (Gilliland & James, 1993;
Megargee, 1995). For example, it is better to
phrase questions about anger in terms of what
makes a person angry, rather than in terms of
whether a person is angry; better to talk in terms
of how the person responds, rather than how the

person plans to perpetrate. Questions are phrased to reduce the perceived level of blame placed on the perpetrator. This is done to maximize the information that can be obtained from the person, not to help the person shift blame from self to others. It must be remembered that what is done when a homicide assessment is conducted is crisis intervention, not therapy work. Thus, it is not the intent of the questions to increase insight or empathy on the part of the client, but to maximize information for the clinician to make the most reliable judgment about the level of risk presented by the potential perpetrator. Thus, the types of questions that are best for a homicide assessment are probably not good examples of the types of questions that are best for therapeutic intervention. To demonstrate this type of questioning, several paired sets of questions are presented below. First, a question is presented that is phrased appropriately for a therapy context but not optimally for a crisis assessment context; then an alternative is presented that is superior for a crisis situation. A final note about anger assessment: although there are ways to test anger objectively or projectively with a variety of psychological instruments (see Chapter 4), this is not practical during homicide or aggression assessment. Testing requires good rapport and calmness on the part of the client (and the clinician, for that matter), both factors that cannot be taken for granted during a homicide assessment. Also, it appears somewhat ridiculous to ask a client who is making a threat of homicide to sit down and take a self-report questionnaire about his or her level of anger!

- Therapy: When do you get angry?
- Crisis: What specifically makes you aggravated (or annoyed)?

- Therapy: In what kinds of situations do you tend to feel yourself getting angry?
- Crisis: What (or who) provokes you the most?

- Therapy: What do you do when you feel angry?
- Crisis: When someone provokes you, what do you want to do to them?

- Therapy: How do you react when you are angry?
- Crisis: What do you do when something makes you mad?

- Therapy: How do you relate to people when you are annoyed?
- Crisis: What do you want to do to someone who aggravates you?

- Therapy: Do you ever get enraged?
- Crisis: Have you ever lost your temper? When? With whom?

- Therapy: How do you act out your anger?
- Crisis: What happens when someone makes you lose your temper?

In addition to assessing anger per se, the sources of, or triggers for, the person's anger need to be evaluated. Both psychological factors and physical factors can be underlying the anger. A person may experience anger after having been frustrated or because of being required to delay gratification. They may respond angrily to actual or perceived physical or verbal threat or attack. They may respond with rage at intrusions of their territory (literally and symbolically). Physical factors that may lead to anger and its expression include drug use, endocrine illness or central nervous system disease, as well as genetic or other predisposition for irritability. Some of these issues are covered already through the assessment of risk factors. They may merely need to be entered into the anger assessment equation at this time, without additional questioning being necessary.

External factors or instigators refer to the anticipated rewards or reinforcements that result in the aftermath of an aggressive act. People may be motivated for aggression because of what they expect to happen afterward or because of the aggressive act itself. As such, some aggression may be motivated because people expect financial or economic gain (e.g., assault of a victim as part of a robbery), because perpetrators expect psychological release of some sort (e.g., feeling of having been vindicated after assaulting a person who has slighted them in the past), because they expect to win the victim or spectators over to their point of view (e.g., as in acts of terrorism or politically motivated crime), or because they seek to punish the victim for a perceived slight (e.g., an assault of a significant other who is suspected of infidelity). Thus, questions about what clients expect or hope will happen after the action is complete are important.

Habit Strength

Habit strength refers to the level of spontaneity with which the person tends to resort to violence or to respond in anger. It is largely dependent upon reinforcement history and follows the maxim that the best predictor of future behavior is past behavior. The more success the person has had as a result of past aggressive acts, the higher this person's habit strength is likely to be. If clients relate that they have often gotten their way by threatening or attacking others, this is a good indicator that their reinforcement history has rewarded violence, suggesting that this has become a powerful response choice for them (Megargee, 1995). The more aggression, anger, and interpersonal violence clients report in their history, the more likely they are to have a strong habit of responding aggressively (Gilliland & James, 1993). The therapist "can infer strong habit strength from a long history of aggressive behavior, especially if it has been successful. Habit strength is also the variable that most strongly predicts aggression; the longer and stronger the history of aggression, the more likely an individual will behave similarly in the future" (Megargee, 1995, p. 403).

Habit strength is formed not only through successful experience, but also through observation or modeling of successful violent acts. Specifically, habit strength can be developed through watching violent and successful role models of aggression in one's family, community, TV, or movies. Given the current level of aggression in many communities and peer groups, as well as on TV and in the movies, it is not surprising that many potential perpetrators have developed their habit strength less through personal perpetration of violence and more through its observation. This is particularly true given the many reports in the media of violence that is successful from the perpetrator's perspective, if nothing else, successful in terms of providing exposure and attention.

Habit strength is also generally not decreased by the occasional delivery of punishment as is imposed by the criminal justice system. Instead, just as intermittent reinforcement serves to increase or reinforce a behavior more thoroughly than constant reinforcement, punishment is only effective if applied after each aggressive act. Obviously, this does not usually occur. Instead, it is much more likely that aggression will be intermittently reinforced and only occasionally punished. Further, occasional punishment tends to serve to increase variables that are instigators or motivators of aggression (Megargee, 1995).

In summary, a thorough assessment of the client's history of violence is indicated, not only because it is a risk factor, but also because of its direct relationship to the issue of habit strength. The more aggression there is in clients' histories, the more likely that they perceive strong rewards for the action. If they can verbalize the perceived payoffs of violence in the past, habit strength can be assumed to be even higher. The more violence there is in a person's past with relatively low involvement with the criminal justice system, the stronger the likelihood that the person's reinforcement history supports a strong habit strength. Conversely, with a long history of criminal justice involvement and relatively little reported aggression or anger, habit strength will be considerably lower.

Inhibitions

Inhibitions refer to those factors that may counteract instigators and triggers and which may decrease habit strength. A number of possible inhibitors have been identified, including moral and practical factors (Megargee, 1995) and impulse control factors (Meyer & Deitsch, 1996). Moral factors refer to issues such as internalized values and morals (development of a superego) that are endorsed by a family, community, or culture. The internalization of morality and ethical behavior appears to be facilitated by a childhood history that includes proper role models. People from homes that instill values generally have better internalized controls than people coming from families that model violence. Homes that lead to the type of internalized morality that inhibits aggression tend to be nurturing and authoritative (not authoritarian or permissive), and tend to model and instill values, limits, empathy, and compassion (Baumrind, 1983; Megargee, 1995; Meyer & Deitsch, 1996).

Practical factors that inhibit behavior may stem from the fear of being caught. These factors have little to do with internalized controls or values and are much more situational than moral factors. They tend to reflect the opportunistic nature of violence and are therefore less reliable as inhibitors. If a client's sole inhibitors are of a practical nature, they should not be counted on as successful inhibitors of aggression in the face of a specific aggressive plan.

Internalized control in the form of the ability to delay gratification and to delay impulsive action is another important potential inhibitor to explore (Meyer & Deitsch, 1996). The ability to delay gratification and wait for more positive or preferable outcomes is a powerful inhibitor that tends to be absent in persons with antisocial personality disorders or conduct disorders (American Psychiatric Association, 1994). As with internalized morality, internalized self-control is fostered primarily by a childhood history of role models with proper inhibitors of their aggressive impulses. That is, clients with histories of child abuse or domestic violence more than likely had role models who did not encourage the internalization of impulse control, but quite to the contrary modeled aggression as a more powerful means of achieving desired outcomes than delayed gratification. Impulse control can be assessed through, not only the exploration of past aggressive acts, but also through other behaviors that show whether the person can delay gratification. This is the reason why it is more likely to have aggressive clients with histories of poor academic or job performance. Academic achievement and job security generally require delay of gratification and delay of impulses. Similarly, long-term success in maintaining intimate relationships generally suggests some ability to control impulses.

Finally, information about potential inhibitors may also be gleaned from an exploration of past situations in which clients reported that they were able to refrain from aggression. It is useful to explore why clients were able to suppress the aggressive impulse in those circumstances as compared to situations in which they acted out. Such inhibitors often are external rather than internal, but this information will nevertheless be useful when planning interventions. Potential external inhibitors will be discussed in more detail in the following section on situational factors.

Situational Factors

A variety of situational, or contextual, factors need to be explored, both as potential triggers and inhibitors (Borum, Swartz, & Swanson, 1996). The environment in which the aggression is likely to take place needs to be considered. Places such as bars or football stadiums may be more conducive to violence than places such as churches or office buildings. Thus, if the person who is being threatened is usually encountered by the potential perpetrator in an environment that is conducive to violence, this factor may increase risk; if it is an environment that is generally opposed to violence, risk may be decreased. Similarly, the setting in which the aggressive act is to take place is important to consider. A classical concert in a park may be less conducive than a rock concert in the same environment. Similarly, a peace demonstration in a public square may be less likely to result in violent action than an NRA rally in the same place. If the client has suggested a setting where his act will take place, it can then be evaluated in those terms. Next, the situation in which potential perpetrator and victim are likely to meet needs to be considered. If the situation involves the presence of security guards or police officers, risk may be lower than if it involves persons who are triggers for violence, such as drinking buddies, former coperpetrators, or estranged spouses. Finally, the stimuli involved in the situation need to be evaluated. If the potential victim is likely to present a stimulus that represents a trigger for the potential perpetrator, risk is higher than if the victim is more likely to present a stimulus that is inhibitory. For example, if intended victims are frightened of the perpetrator and try to leave the setting as soon as the perpetrator appears, they may be a low stimulus for violence. On the other hand, if the intended victims are aggressive themselves and make aggressive gestures toward the perpetrator, this may serve as a stimulus for violence.

A thorough analysis of the anticipated situation in which the victim is most likely to be encountered will help the clinician assess whether the situation will present more inhibitors or more triggers to the potential perpetrator. If more triggers are present, risk increases; if more inhibitors are present, risk decreases. Similarly, if the client is not clear about the anticipated situation, the mental health care provider can explore situational circumstances in past aggressive situations. If clear triggers emerge (e.g., if violence only occurs in bars without security guards), the counselor can use this information to build an intervention plan (e.g., make a contract with the client not to go to certain bars and potentially involve the owner of these establishments to deny access to the client if risk is otherwise high).

Reaction Potential

The reaction potential of the client is assessed based on the balance of triggers versus inhibitors and refers to the likelihood that the client will react aggressively. A response will be aggressive if the client's perceived reasons for the aggressive act outweigh inhibitors that would override the perceived reasons. Conversely, the client is less likely to react aggressively if the reasons, or triggers, for the aggression are outweighed by inhibiting factors. Assessing reaction potential is not as simple as adding the number of triggers and adding the number of inhibitors and seeing which is greater. Instead, the relative weight of each trigger and each inhibitor also needs to be considered. In the end, the therapist must decide which response will be perceived by the client as

meeting a greater need or purpose: the aggressive act or the restraint of the aggressive act. It is clear that the assessment of reaction potential is somewhat subjective. But, it too can be judged more reliably if assessed in a historical context. In other words, if clients can be asked how they responded in a similar situation in the past, when both similar or identical triggers and inhibitors were present, clinicians can make a sounder judgment about future potential.

Decision Making Based on Risk Assessment

The final decision about the level of risk for follow-through and the risk of harm to the potential victim will need to take into consideration not only reaction potential (i.e., the balance of triggers and inhibitors), but also habit strength, lethality and specificity of the plan, and opportunity for action (see Borum, Swartz, & Swanson, 1996; Meyer & Deitsch, 1996). It is the combination of all of these issues that will ultimately best predict the client's behavior. Given that each of these factors was determined based upon a set of both objective and subjective impressions, it becomes clear that the margin for error is considerable. However, the main goal of the risk assessment is to make the best possible clinical judgment with the largest possible number of pieces of data or information. This is important to protect client, potential victim, and therapist. The protection of the former two is self-explanatory: clients need to be protected from engaging in an action that can have a multitude of negative effects on their lives and victims need to be protected from harm. Clinicians need to be protected from vulnerability to lawsuits of negligence or breach of confidentiality. Following the proper procedures for assessment and carefully documenting the decision-making process about all actions will protect clinicians from such liability.

In making the decision about risk of follow-through, the counselor or therapist must revisit the issues of clarity and specificity of the plan and the identified victim, the balance of triggers and inhibitors, and the history of violence (Mon-

balancing this with the client's rights to privacy. Also figured into the equation is the safety of the mental health care provider throughout assessment and intervention, as well as after the client leaves the session. Interventions will be most restrictive if both reaction potential and lethality to the potential victim are high.

INTERVENTION IN CASE OF VIOLENCE RISK

As noted earlier, because an aggressive act requires the presence of an aggressor, a victim, and the opportunity for action, three different methods for intervening can be used separately or can be combined. This provides some flexibility, which is not always possible in other intervention situations (for instance, preventing a suicide). The first route tends to be the one that involves the aggressor, because this is the client. Interventions involving the potential victim and opportunity generally require either releases of information or breach of confidentiality and, hence, are used only if less restrictive options of intervention are not possible or available.

It should be noted here that trainees should probably never plan interventions without supervision. Although they may have conducted the assessment without the direct input from a supervisor, once the intervention stage is reached, a trainee probably does not have sufficient experience to make the necessary clinical judgments independently. In terms of a learning experience for the trainee, it would be best if a supervisor could join the trainee in session with the client to model appropriate negotiation of intervention with the client. This may not always be possible if supervisors are not physically present. In such a situation, however, the supervisors should be present in the room with the client via telephone. Even seasoned clinicians may want to give serious consideration to consultation with a colleague during or before intervention planning. The threats of a homicide situation are more severe than those of a suicide situation because the

person at greatest risk is not even involved in the decision-making process. Thus, the clinician has greater protection responsibilities than in a suicide situation. All supervision and consultation needs to be carefully documented in the client's chart. If the need for supervision or consultation arises, the client is merely informed thereof, not asked for permission. In the situation of a trainee, it is generally clear to the client that a supervisor will review all work of the clinician anyhow.

In the case of a seasoned clinician who wishes to seek consultation, the client is informed that the consultation is sought in the best interest of the client and does not require the release of personal data that would identify the client. The consultation then may be conducted over the telephone in the presence of the client. If no consultation is sought about risk assessment and intervention while the client is with the clinician, it is best to debrief with a colleague after the crisis situation has been resolved. Such debriefing serves to verify that proper procedures were followed and allows therapists to document that another professional agreed with their professional judgment, leaving them less vulnerable to lawsuits for negligence or breach of confidentiality. The consultation form provided in the intern handbook by Boylan, Malley, and Scott (1995) can be used for supervision and consultation purposes, and it provides an excellent means for charting risk information about the client.

If the risk assessment suggested high reaction potential and high level of potential harm, intervention is clearly necessary. Such intervention is generally conducted in the context of a crisis intervention, not a therapy session. As such, if intervention is necessary, counselors will maintain the directive stance already developed during the assessment process. They will conduct the session with much structure and guidance, outlining options and supporting problem-solving approaches with the client.

Such crisis intervention relationships are described in greater detail in Chapter 5. Suffice it to say that in a homicide or aggression intervention situation the clinician is clearly in charge of the agenda, directs the session, and makes all the important decisions about disposition of the client. However, despite this directive and relatively controlling approach to the session, clinicians will attempt to involve their clients as much as possible in the decisions that are being made, leaving them some sense of control over their destiny (Boylan, Malley, & Scott, 1995). Asking clients to participate in decision making, even if it involves disclosure to others (such as family members, victim, or police) will maximize the chances to preserve whatever therapeutic rapport had been established (Beck, 1987, 1988). Also, being directive and in charge does not equate with taking all choices and options away from the client. It merely means narrowing choices down to acceptable options. Being directive also does not mean being rude or confrontational; in fact, confrontation with an aggressive patient may not be a wise choice. Respect and caring need to be communicated throughout the process, much in the same way a parent communicates caring while setting firm safety limits for a child.

Intervention options in the case of threatened aggression or homicide range from intensified outpatient treatment to warning the potential victim (also referred to as "target hardening" [see Borum, Swartz, & Swanson, 1996, p. 213]) to involuntary commitment. The therapist will move through interventions from least to most restrictive, stopping at the level that imposes enough restrictions to guarantee some level of safety for the potential victim, while imposing the least number of restrictions on the client to guarantee the right for the least restrictive treatment alternative.

The least restrictive alternative for intervention is increased outpatient intervention that will decrease the client's angry affect and inhibit impulses sufficiently to prevent action. Such increased outpatient intervention may consist of increased number of appointments, scheduled telephone contacts between sessions, availability of a crisis line phone number, and other means available to a mental health professional. Outpatient intervention is best supplemented with a contract not to commit a lethal act, much in the

same way as clients and therapists make suicide contracts (Slaikeu, 1990). This contract outlines the client's willingness not to engage in the harmful action without prior face-to-face consultation with the counselor. It also lists scheduled times when therapist and client will meet, any other agreed-upon means of intervention, as well as consequences of the client's failure to meet these conditions. A sample is provided in Figure 7-1. Generally, outpatient intervention contracts include clauses such as the removal of weapons or other threatened means (Slaikeu, 1990), the avoidance or removal of provocative stimuli (e.g., agreement to avoid the bar where the client sees the potential victim [Bellak & Siegel, 1983]), abstention from drugs or alcohol which otherwise may lower resistance to impulses (Meyer & Deitsch, 1996), possible medical or psychiatric referral (e.g., to rule out physiological causes for the aggressive impulse or to obtain medication [Bellak & Siegel, 1983; Meyer & Deitsch, 1996]), and the enlistment of help from others (see below).

If risk appears too high to allow the client to function without supervision between scheduled outpatient contacts, a support person may need to be involved who can provide more consistent supervision or monitoring of the client. If this is the case, the contract between client and clinician can be expanded to include the name of this person and the specific terms of supervision to which all parties have agreed. Enlisting the help of others is a good option if the client has some impulse control but the chosen method is so lethal that the clinician is uncomfortable relying on the ability of the client to refrain from the threatened act between contacts with the clinician or other crisis intervention personnel. If such assistance is sought, it is best to attempt to get the client to agree to this arrangement and to provide a signed release of information that permits communication between the therapist and the assisting person selected by the client (Boylan, Malley, & Scott, 1995). If clients refuse such a release of information, it is most likely that a more restrictive level of intervention is necessary and this can be presented to them as an alternative. If clients agree to involve

FIGURE 7-1   Sample Violence Contract

I, _____ *(insert client's name)* agree not to kill, harm, or attempt to kill _____ *(insert name or identity of potential victim)* for the time period beginning now, _____ *(insert current date and time)* until _____ *(insert agreed upon next meeting or contact time).*

I agree to call the crisis numbers on the crisis card given to me today by my clinician, _____ *(insert clinician's name),* should I have the urge to kill or harm the above-named person at any moment during this time period.

I agree to come to my next appointment on _____ *(insert day and date)* at _____ a.m./p.m. *(insert exact appointment time)* with _____ *(insert clinician's name).*

I agree to get rid of all things that I have thought about using to kill or harm the above-named person. Specifically, I will discard _____ *(insert means identified as necessary to the client's method).* I will contact my clinician at _____ *(insert phone number where clinician can be reached)* as soon as I have discarded the means to confirm that I did this. If I don't call by _____ *(insert a time later today),* my clinician will alert my family and the police.

I agree that I will make the following changes that will reduce my likelihood to become aggressive toward the above-named person: I agree not to/to _____
_____
_____
_____
(insert all agreed-upon environmental modifcations; e.g., abstention from drugs)

I agree to seek the following addtional services to ensure that I will not kill or harm the above-named person: I will _____
_____
_____
_____
(insert all agreed-upon additional terms of treatment; e.g., supervision by a support person; referral for medication)

I realize that this contract is part of my therapy contract with my clinician at _____ *(insert name of clinic/hospital/etc.).* I am aware that my clinician can break the agreement of confidentiality if I do not comply with this contract to refrain from aggression against the above-named person.

_____   _____
Signature of Client   Signature of Clinician

_____
Date and Time

another person to monitor their behavior, the contact with this person is made while client and mental health care provider are still together in the session. The clinician will not rely on the client's word that he or she will seek out the person and ask for assistance after leaving the session. Clinicians initiate the contact with the support person in the presence of the client and negotiate this person's involvement in detail. They need to assure themselves that chosen resource persons are not potential perpetrators and that they are capable and willing to restrain the client in the event of an attempted aggressive act. Further, the counselor must ensure that the support person is sufficiently available to monitor the client during times of potential provocation. It is possible to elicit the help from more than one support person and to get several supporters to share the supervision of the client. Clearly, to go the route of intensified outpatient treatment either with or without a support person indicates that the therapist assesses risk for follow-through as relatively low. Not involving supports also suggests that lethality is low so that should a misjudgment have been made about follow-through, the harm that will come to the potential victim will be relatively minor.

Intensified outpatient treatment can also be accompanied by a referral for psychotropic medication. Some evidence has accumulated that medication may be helpful in some cases of violence and aggression. Most importantly, it appears that clients with chronic difficulties in modulating affect can greatly benefit from a medication referral. Specifically, lithium has been found helpful in moderating affective expression for these clients; similarly, carbamazepine has been found helpful in controlling aggression related to organic brain disease and psychosis (Beck, 1987). Involvement of a psychiatric consult or a physician thus appears greatly supportive when the client is to be managed on an outpatient basis. Other referrals may be helpful as well, such as anger management groups, and support groups (perhaps as related to concurrent substance use issues).

In cases of greater threats that either do not appear to warrant, or in which the client refuses, voluntary hospitalization, the clinician will take steps to warn and protect the victim. This warning process needs to occur immediately and preferably in the presence of the client. As was true for the elicitation of the support person, it is best to warn the potential victim after having obtained a signed release of information from the client. However, there is no option here for clients should they refuse. If they refuse, the clinician must still proceed with the warning if the threat is taken seriously (Ahia & Martin, 1993). The client is informed of this; in fact, this is merely a reminder about the duty of the clinician because the issue was in the informed consent and was discussed during the initial contact with the client. Often when faced with the alternatives, clients will consent to the release of information to retain some sense of control over the situation (Boylan, Malley, & Scott, 1995). The mental health care providers must also consider taking steps not only to warn potential victims, but also to protect them. This protection can take on various forms. Therapists may feel satisfied that the victim is properly protected as long as the client is willing to make a contract and to be released to the supervision by a support person. However, if counselors have any doubt about the success of such protective procedures, they must call the police and inform them of the client's threat (Ahia & Martin, 1993). This is again best accomplished in the presence and with the agreement of the client. However, given the likely frame of mind of the client with whom the clinician needs to go to such intense levels of intervention, such cooperation may be asking too much. In all disclosures, clinicians must take care to protect the privacy of the client as much as possible (Ahia & Martin, 1993), while sharing enough information to give support persons, potential victims, and protectors the necessary data to take whatever actions they deem appropriate (Anderson, 1996).

Another possible intervention at this level of risk is to obtain the client's consent for voluntary hospitalization until such time as the client can once again feel in control over impulses. As with suicidal clients, voluntary hospitalization can be presented as an alternative to involuntary commitment, a presentation that generally increases client compliance with the idea. Commitment to a hospital is critical if both risk for follow-through and lethality are high and the client refuses other means of intervention, including the refusal to seek inpatient services voluntarily. Commitment procedures have been covered in depth in Chapter 6 and are identical for this type of hospitalization.

As was noted earlier, for a homicidal or aggressive act to occur, there needs to be a perpetrator, victim, and opportunity. This can be factored into the intervention plan for the aggressive client in that the removal of opportunity may be possible without restrictive intervention, even in a case of high risk. For example, the warning of victims may lead to the revelation that they are no longer living where the client can locate them, that they have made arrangements for self-protection because of prior knowledge about the aggressive threat by the client, or that a situation has changed and thus the client's anger is diffused. Thus, environmental modifications, that are not an option in suicide intervention, may be possible to diffuse the threat. Such possibilities need to be considered but certainly cannot be relied upon. Ultimately, the therapist must make the same judgment as with suicide prevention: breach confidentiality and warn and protect, or maintain the client's privacy and deal with the threat in the least restrictive manner, taking a risk that harm may occur. Whichever course of action is chosen, clinicians best ensure that they can back up their decisions with sufficient assessment data and knowledge about aggression and homicide risks as well as interventions that they can feel comfortable defending their actions in a court of law. The best rule to live by can be summarized as follows: Assess intensely, make a data-based decision, attempt to intervene at the least restrictive level, and, when in doubt, warn and protect.

DOCUMENTATION AND RECORD KEEPING

As pointed out in the previous chapter, keeping adequate records is essential in a case in which a breach of confidentiality is indicated or in which a misjudgment on the clinician's part can result in harm to another human being. It is best to include verbatim quotes from the client in the progress note that will be written for a session dealing with the duty to warn and protect (Beck, 1987, 1988). Such notes also need to reflect the clinician's understanding of the standards of care for this emergency situation and show an assessment of the basic process outlined in this chapter.

At the end of this chapter is a sample of a behavioral contract that would be prepared during a crisis involving a threat against another person by a client. This sample will outline for the reader how such a contract would look once the specific information has been included (a generic sample of such a contract was presented above in Figure 7-1). A copy of this contract would be handed to the client to take home; another copy would be filed in the client's chart, along with a progress note that would follow all the rules for appropriate duty to warn and protect charting (see sample of the suicide progress note in the documentation section of the previous chapter), covering all of the cautions mentioned earlier. Also attached to the progress note and the behavioral contract would be a listing of all of the crisis and contact numbers that were presented to or provided by the client. Any other written documents prepared with the client during the session should also be copied and attached. Any phone calls made in addition to the ones included in the sample should be recorded in the same way. Any message slips received about the case (e.g., messages from security or other health and mental health professionals returning calls about the client) also need to be attached (with time and date of receipt).

SUMMARY AND CONCLUDING THOUGHTS

The threat of homicide or aggression is a difficult one for many mental health care providers to deal with. Fortunately, it is a rare event because its consequences are so severe. Even if risk is low, it is likely that the clinician will leave the session

---

FIGURE 7-2   Violence Assessment Guide for Interview and Intervention Planning

Immediate Risk Factors

Violent Ideation: 7 ---------- 6 ---------- 5 ---------- 4 ---------- 3 ---------- 2 ---------- 1
                         extremely strong   extremely weak

Violent History: 7 ---------- 6 ---------- 5 ---------- 4 ---------- 3 ---------- 2 ---------- 1
                         extremely prevalent   virtually absent

Motive for Violence: 7 ---------- 6 ---------- 5 ---------- 4 ---------- 3 ---------- 2 ---------- 1
                         extremely specific   extremely vague/none

Violence Plan: 7 ---------- 6 ---------- 5 ---------- 4 ---------- 3 ---------- 2 ---------- 1
                         extremely specific   extremely vague/none

   Details of Plan
   Identity of Victim   _____
   Access to Victim   _____
   Method   _____
   Availability   _____
   Place   _____
   Time   _____
   Lethality of Method   _____
   Preparations   _____

Level of Anger: 7 ---------- 6 ---------- 5 ---------- 4 ---------- 3 ---------- 2 ---------- 1
                         extremely hopeless   hopeful

Impulsivity: 7 ---------- 6 ---------- 5 ---------- 4 ---------- 3 ---------- 2 ---------- 1
                         extremely impulsive   good self control

Other Risk Factors

Demographics: high risk   moderate risk   low risk
   Age   _____
   Gender   _____
   Ethnicity   _____
   Education/SES   _____

FIGURE 7-2   Violence Assessment Guide for Interview and Intervention Planning (continued)

Social Factors:	encourage violence	tolerate violence	discourage violence
Role models/Peers			
Isolation			
Environment			

Family History:	high risk	moderate risk	low risk
Parenting			
Violent History			

Psychiatric History:	high risk	moderate risk	low risk
# of Diagnoses			
Types of Diagnoses			
Severity			
Chronicity			
Substance Use			

Medical History:	high risk	moderate risk	low risk
Relevant Illness			
Motoric Clues			

Psychological Factors:	high risk	moderate risk	low risk
Sensation Seeking			
Recent Stressors			
Fascinated with Weapons			

Risk Assessment	high risk	moderate risk	low risk
Instigation/Triggers:			

Inhibiting Factors:

Habit Strength:

Situational Factors:

Reaction Potential:

Sample Behavioral Contract with Janet Doe

I, <u>Janet Doe</u> *(insert client's name)* agree not to kill, harm, or attempt to kill <u>Jonathan Doe</u> *(insert name or identity of potential victim)* for the time period beginning now, <u>23 October 1997</u> *(insert current date and time)* until <u>26 October 1997</u> *(insert agreed upon next meeting or contact time).*

I agree to call the crisis numbers on the crisis card given to me today by my therapist, <u>Chris Brems</u> *(insert therapist's name),* should I have the urge to kill or harm the above-named person at any moment during this time period.

I agree to come to my next appointment on <u>26 October 1997</u> *(insert day and date)* at <u>2:00 P.M.</u> *(insert exact appointment time)* with <u>Chris Brems</u> *(insert therapist's name).*

I agree to get rid of all things that I have thought about using to kill or harm the above-named person. Specifically, I will discard *(insert means identified as necessary to the client's method):*
- my handgun
- and all ammunition for said gun

I will do so by handing the gun and ammunition to my mother after this session, when she picks me up from the Psychological Services Center to take me home and take my children (see below).

I also vouch for the fact that I have no other violent (e.g., butcher knives) or nonviolent (e.g., poisons) means in my possession that I could use to kill Jonathan Doe.

I will contact my therapist at <u>(123) 555-4567</u> *(insert phone number where therapist can be reached)* as soon as I have discarded the means to confirm that I did this. If I don't call by <u>3:30 P.M. today, 23 October 1997</u> *(insert a time later today),* my therapist will alert the above-named person, my family, and/or the police.

I agree that I will make the following changes that will reduce my likelihood to become aggressive toward the above-named person.

I agree *(insert all agreed-upon environmental modifications; e.g., abstention from drugs)*
- not to use alcohol between today and my next appointment as noted above
- to leave my two children with my mother, who was contacted by me and my therapist today and who agreed to take my son, Derek, and daughter, Renee, for a week to relieve my stress
- not to seek out the presence of Jonathan Doe, who moved out of my home one week ago and now has an apartment in town
- not to go to the bar where Jonathan and I usually hang out at night for at least the next 8 days
- not to go to Jonathan Doe's parents' house where I may run into him for at least the next 8 days
- to seek a restraining order against Jonathan Doe with the assistance of Kate Doe, who has agreed to help me with this process because she is a legal aid and knows the procedure
- to call the police if Jonathan Doe shows up at my home or my place of employment

I agree to seek the following additional services to ensure that I will not kill or harm the above-named person: *(insert all agreed-upon additional terms of treatment; e.g., supervision by a support person; referral for medication)*
- I will call the crisis line or Chris Brems if I have thoughts of harming Jonathan Doe before my next appointment
- I understand that Chris Brems has a Release of Information which I signed today to call Jonathan Doe should I call her with the information that I once again have the urge to harm him or should I fail to contact her to inform her that I have discarded the gun and ammunition
- I will move into the abused women's shelter if I begin to feel severely depressed again or if I once again have a strong urge to harm Jonathan Doe
- I will be in the presence of Kate Doe, my sister-in-law and best friend, who was contacted by me and my therapist today, to stay with me for the next three days

---

Sample Behavioral Contract with Janet Doe (continued)

I realize that this contract is part of my therapy contract with my therapist at <u>the Psychological Service Center</u>
(insert name of clinic/hospital/etc.). I am aware that my therapist can break the agreement of confidentiality if I do not comply with this contract to refrain from aggression against the above-named person.

_____   _____
Signature of Client   Signature of Therapist

<u>23 October 1997, 3:00 P.M.</u>
Date and Time

---

with the client in emotional turmoil, wondering about whether the best course of action was chosen. Being well aware of risk factors, knowing how to conduct a violence assessment, and being able to plan an appropriate intervention can be useful in reducing a clinician's anxiety when faced with an aggressive client. To assist with guiding the therapist through a violence assessment, Figure 7-2 provides a guide for assessment interviewing and intervention planning with violent clients.

Given the high likelihood that homicide and aggression interventions leave the clinician in some state of upheaval, one cannot overemphasize the importance of consultation and debriefing after such a situation. This will not only help clinicians reevaluate whether they have taken the best possible action, but it will also discharge some of the emotional distress they may feel and will help them regain psychic equilibrium and confidence. If an aggressive act should occur after clinicians have intervened in a manner they considered prudent, they should go back over their data even if no lawsuit is initiated. Such reevaluation of data may give them a better understanding of possible future prediction of the behavior of similar clients and may also serve to assuage nagging doubts about the appropriateness of their chosen actions. Debriefing with a colleague would also be helpful.

8

The Challenge of Child Abuse and Neglect: The Duty to Report

There was an old woman who lived in a shoe. She had so many children she
did not know what to do. She gave them some broth without any bread,
and whipped them all soundly and put them to bed.

TRADITIONAL NURSERY RHYME "OLD WOMAN IN THE SHOE"

The issue of child abuse is perhaps as old as hu-mankind, though this has not been acknowl-edged until recently. Accounts of child abuse can be found in historical documents as well as in fairy tales and myths. However, it was not until the late 1800s that child abuse was recognized, and it was not until the mid-to-late 1900s that child protection laws were enacted. Since that time, child abuse and neglect of any form has be-come a criminal offense in all states of the United States and in most, if not all, European countries. Yet reported cases have skyrocketed: current statistics suggest that at least 12% of chil-dren in the United States are physically abused by a parent and 6–10% are sexually abused by a parent (Haugaard, 1992). These statistics have been criticized as sorely underreporting the true incidence of abuse and neglect in American so-ciety; further, these statistics do not reflect the proportion of clients that will present with abuse or neglect issues in a general psychotherapy prac-tice. Because a larger proportion of therapy clients than the general population has histories of childhood abuse or neglect, clinicians tend to

encounter a large number of victimized clients in their general practice. Also, because abuse and neglect tend to involve cross-generational trans-mission, a large proportion of clinicians will con-front adult clients who themselves have become perpetrators of abuse or neglect of children in their care.

In the United States, reporting laws require professionals who are likely to become witness to or gain knowledge of incidents of child abuse to turn the offending family or individual over to child protection service agencies. All states re-quire professionals who have regular contact with children to report suspicion of child abuse. These professionals always include physicians, nurses, dentists, emergency room personnel, coroners, law enforcement personnel, medical examiners, mental health professionals, social workers, teachers, and day care workers. Addi-tionally, in some states reporting laws also apply to clergy, camp counselors, film processors, and foster parents (Bulkley, Feller, Stern, & Roe, 1996). Clearly, mental health care providers can-not do their jobs without having at least some familiarity with reporting laws and procedures and without a basic knowledge about when to refer clients or their children for additional assessment and treatment as relevant to abuse issues that may be perpetrated in the home (Ammerman & Hersen, 1992). It is the focus of this chapter to provide clinicians with information about child abuse that will help them make informed decisions about when the duty to report a family to child protection should be invoked. The chapter is written from the perspective of the perpetrator, not the victim. It is assumed that the readers will become aware of the abuse through working with the perpetrator (or another adult in the home), not the victim, given that the focus of this book is on therapy with adults. Abuse and neglect issues may be dealt with differently if the client is a child (cf., Brems, 1993) or if the mental health professional is not a therapist or counselor but rather an evaluator or investigator of child abuse (cf., Oates, 1996). In fact, the role played by the mental health professional is critical in terms of the proper, ethical handling of a child abuse case. It differs vastly depending on whether the professional became involved as an advocate for the child, as a court-appointed evaluator, or as an advocate for the perpetrator. Role clarification will be addressed in some detail later in the section on clinician traits.

The issue of reporting child abuse is a controversial one for many mental health professionals, and as many as 50% of mental health practitioners fail to report incidents of child abuse that they identify through their practice (Sedlak, 1990). The reasons for such nonreport (which is in clear violation of state laws) are multifold. Some practitioners indicate that they do not report in cases when they do not have sufficient evidence to substantiate the report; others suggest ambivalence about reporting if the given case is not clearly or only ambiguously covered in state statutes; some practitioners are unclear about definitions of abuse and neglect and fail to report because of these inadequacies in their own understanding of the issues (Barnett, Miller-Perrin, & Perrin, 1997). In addition to nonreport

by clinicians, many cases of child abuse probably remain unreported because of nondisclosure about the abuse by the victim, and because of legal definitions of who can be called a child abuse perpetrator (e.g., if the perpetrator is an adolescent, the case may not be reported as a case of child abuse [Barnett, Miller-Perrin, & Perrin, 1997]). Some cases of child abuse are not considered as such because child protection agencies may classify some reports as unsubstantiated because they are too minor to justify resource allocation to them (Barnett, Miller-Perrin, & Perrin, 1997). Finally, "insufficient staffing of child welfare services in many communities has caused some reports that might have been accepted to be screened out" (Zellman & Faller, 1996, p. 371). Failure to report has a number of negative consequences and implications. First, it may result in denial of services to needy children; second, it exposes professionals to liability and anxiety; third, it may punish professionals who do report by directing clients to professionals who are known not to report; and fourth, it distorts the national database on child maltreatment (Zellman & Faller, 1996). In other words, the factors resulting in nonreport or nonclassification of certain abusive child-adult interactions suggest that current estimates of the number of child abuse cases are underestimates that fail to reveal the depth and breadth of this problem in contemporary society. The figures provided in the following section must be interpreted with this recognition in mind.

DEFINITIONS AND INCIDENCE OF VARIOUS FORMS OF CHILD ABUSE AND NEGLECT

No chapter on child abuse reporting laws can be complete without first defining the behaviors in question, though this may not be as easy as it would appear. Efforts at defining the various forms of child abuse have been made repeatedly in the literature and legal codes, and good consensus appears to have emerged with regard to physical and sexual abuse definitions. Emotional abuse and neglect, however, are somewhat more difficult to describe. The main forms of child abuse that have been addressed in the literature and the law (and that will be referenced here) are physical abuse, sexual abuse, emotional abuse, and neglect (with all of its various forms), in that order. This order was chosen because it represents the chronological succession of when these various forms of child misuse became of societal and legal concern. Finally, it must be noted that the issues of etiology and causation of child abuse and neglect are not addressed in this chapter, given their complexity and controversial nature. Many theories have been developed and offered to explain the occurrence of abuse and neglect. Barnett, Miller-Perrin, and Perrin (1997) offer an excellent overview of the various theories that have been proposed (ranging from social learning theories, interpersonal interaction theories, socialization, social exchange theories, to attachment theories and others) and the interested reader is referred to this resource for more information. Excellent information is also presented in the *American Professional Society on the Abuse of Children (APSAC) Handbook on Child Maltreatment* that was edited by Briere, Berliner, Bulkley, Jenny, and Reid (1996).

Physical Abuse

Walker, Bonner, and Kaufman (1988) define physical abuse "as inflicting injury such as bruises, burns, head injuries, fractures, internal injuries, lacerations, or any other form of physical harm lasting at least 48 hours. . . . [This] may also include excessive corporal punishment and close confinement" (p. 8). This definition is consistent with the one provided by Oates (1996), which emphasizes that the receipt of nonaccidental injury that was caused either by an act of commission or omission on the part of the parent qualifies as child abuse. Oates (1996) points out that, despite the relative clarity of definitions of physical abuse, the identification can be ren-

dered difficult once one considers situational relativity (e.g., an accidental versus a deliberate act of commission), differences in cultural values that may condone some but not other forms of interpersonal violence (e.g., permissiveness with regard to corporal punishment versus prohibition of physical punishment [as in Scandinavia for example]) and across incidents (e.g., if an incident results in injury and comes to the attention of care providers, it is recognized as abuse; if the same incident results in minor injury and does not come to the attention of a care provider, is it any less abusive?).

According to figures complied and summarized in Barnett, Miller-Perrin, and Perrin (1997), in 1994, 3,140,000 reports of child abuse or neglect were made to social service agencies across the United States (affecting 47 out of 1,000 children). Of these reports, one million were substantiated as constituting physical abuse. Of reports made to child protection agencies, 27% are for child physical abuse (Kolko, 1996). In 1993, 715 children died due to physical abuse and an additional 520 died due to physical neglect. Of identified physical abuse victims, 51% are younger than 6 years of age; 26% are aged 6 to 11, and 23% are aged 12 to 17. If the victim is under the age of 12, he is more likely to be male than female; if the victim is over the age of 12, she is more likely to be female than male. There is a reported higher risk for children from lower socioeconomic strata and some figures suggest that minority status (especially being of African American descent) increases risk of physical abuse. However, this finding must be interpreted with caution. Abuse and neglect occur across all socioeconomic strata. The overrepresentation of poorer families may merely reflect that these individuals tend to be more vulnerable to being reported and are often unable to seek legal counsel in a way that prevents prosecution and distorts statistics. Children with special needs due to premature birth, disability, or mental retardation are also more likely to be victimized by their caretakers. Perpetrators of physical abuse tend to give birth to children at a younger age than the average parent and often have their first child in their teens. Although data appear to contain some contradictions, a cautious conclusion has been drawn that more women than men become perpetrators of physical abuse toward their children, a reality that may merely reflect the greater number of women in child-care roles. Single parents appear to be overrepresented as perpetrators, though there is some evidence to suggest that this results from poverty more so than from poorer parenting skills (Barnett, Miller-Perrin, & Perrin, 1997). Further, the size and quality of a single parent's (especially a mother's) social and emotional support network appear to be important mediating variables. There is a correlation between the experience and perpetration of physical abuse on children: approximately 30% of children who were physically abused become abusive parents themselves (Kolko, 1996).

Sexual Abuse

Sexual abuse has been defined as "the involvement of dependent, developmentally immature children and adolescents in sexual activities they do not fully comprehend, are unable to give informed consent to and that violate the social taboos of family roles" (Schechter & Roberge, 1976, p. 129). Such definitions stress that there cannot be informed consent on the part of the child for developmental and knowledge-based reasons, even if the child is claimed by the perpetrator to have consented to the sexual act. Further, sexual abuse involves the violation of a power relationship between a child and an adult that is exploitative and coercive in nature, given the child's vulnerable position vis-à-vis the adult. This definition excludes sexual play and exploration among children from the category of child abuse because such behavior involves neither noninformed consent nor power differentials (unless of course the ages are highly disparate, in which case sexual abuse may need to be reported). Definitions of sexual abuse have been further expanded to account for levels of severity of the abuse. Lawson (1993) suggests that there

are at least five levels of sexual abuse ranging from subtle to sadistic forms. Subtle abuse is defined as noncoercive interactions that may not involve the genitalia but inadvertently lead to the sexual gratification of the adult at the expense of the child's emotional or developmental needs. This may also involve verbal harassment, such as sexual name calling, especially of adolescent girls who are beginning to date. Seductive abuse is defined as nonphysical sexual stimulation of the child (e.g., through exposure to sexual displays or verbal sexual arousal) that is inappropriate for the child's age and is engaged in for the gratification of the adult. Perverse sexual abuse is defined as behavior that humiliates the child sexually through making fun of sexual development during puberty, questioning the child's sexual orientation, or forcing the child to dress in cross-gender clothing, all for the gratification or amusement of the perpetrating adult. Overt sexual abuse involves the actual sexual interaction between child and adult, as in intercourse, fondling, sexual play, and so forth. The final and most severe form of sexual abuse is sadistic abuse, which is defined as any sexual behavior that is intended to harm the child.

Of the reports of child abuse or neglect in 1994 (which exceeded 3 million), Barnett, Miller-Perrin, and Perrin (1997) conclude based on their review of sexual abuse data that 11% of cases involved sexual abuse, for a total of 330,000 children in 1994 alone. However, rates of child sexual abuse are extremely unreliable, because there is no national reporting system of crimes against children (Barnett, Miller-Perrin, and Perrin, 1997). Additionally, sexual abuse of female siblings by older male siblings is chronically underreported and not generally included in sexual abuse estimates. Children aged 7 to 12 appear to be most vulnerable to becoming victims of sexual abuse, though this statistic may be confounded by the fact that younger children may merely not be able to report such abuse (which is not easily visibly identified by adults in their environment). More females are identified as victims, although Barnett, Miller-Perrin, and Perrin

(1997) suggest that male underreporting of sexual abuse may confound this proportional distribution. All in all, less than half of all sexual abuse victims reveal their abuse at the time of its occurrence (Berliner & Elliott, 1996). The likelihood of a child becoming a victim of abuse increases with the presence of a stepfather in the home; with the presence of an unemployed, ill, or disabled mother; or with the presence of parents who are diagnosed with a mental illness or with substance use disorders. Sexual abuse appears to occur more commonly in rural than urban areas and in families where the children have poor relationships with their parents. Data regarding socioeconomic status and related variables remain inconclusive (Barnett, Miller-Perrin, & Perrin, 1997) because of the greater likelihood of abuse in poorer families to be reported and the poorer family's inability to retain effective legal counsel. The average age of the sexual abuse perpetrator is 32.5 years, though this mean age may be too high due to the underreporting of juvenile offenders. Men tend to be more likely to become sexual perpetrators than women, with 4–17% of the male population in the United States admitting to having molested a child in self-report surveys. Women as perpetrators may be underreported due to more subtle behaviors that can be disguised in the course of routine child care (e.g., bathing of young children). Of children for whom a report of sexual abuse has been substantiated, 11% were abused by a father or stepfather, 45% by a friend or acquaintance of the family, 20% by another family member (e.g., brothers, uncles, grandparents), and 11% by strangers. Thus, the majority of sexual abuse cases involve perpetration by a member or a friend of the family of the victim.

Emotional Abuse

Emotional abuse, also referred to as psychological maltreatment (Hart, Brassard, & Karslon, 1996), has been defined as "the use of excessive verbal threats, ridicule, personally demeaning comments, derogatory statements, and threats . . .

to the extent that the child's emotional and mental well-being is jeopardized" (Walker, Bonner, & Kaufman, 1988, p. 8). As straightforward as this definition may seem, Oates (1996) has pointed out that emotional abuse tends to be defined more through the consequences to the child than through the behavior of the perpetrator. This form of abuse is most difficult to prove from a legal perspective and is less likely to arouse the interest of child protection agencies that are already overloaded with reports of physical and sexual abuse. Consensus does exist among health care and mental health care providers that emotional abuse has occurred if the child evidences parentally-induced social or medical problems, behavioral disturbances, impaired emotional development, low social competence, and/or poor self-esteem (Barnet, Manly, & Cicchetti, 1991; Iwaniec, 1995). Parental behaviors that may be involved in creating such consequences have been described by Garbarino, Guttmann, and Seeley (1986) as fitting into five categories. Namely, rejecting behaviors communicate abandonment or rejection of the child through refusal of touch and affection. Terrorizing behaviors threaten the child with terrible consequences for the child's behavior, making the world hostile and unpredictable for the child. Ignoring behaviors deprive the child of interaction, stimulation, and adequate opportunities for learning and growth. Isolating behaviors remove the child from normal social interactions and opportunities to experience being part of a larger human community. Corrupting behaviors reinforce behaviors in the child that are deviant or antisocial, as well as potentially harmful for the child (e.g., forcing the child to drink alcohol, involving the child in criminal acts). Brassard, Germaine, and Hart (1987) added three additional behavioral patterns to Garbarino, Guttman, and Seeley's (1986) list by indicting that missocializing, exploitation, and deliberate deprivation of the child of emotional responsiveness also can be considered abusive. Further, verbal harassment through disparagement, criticism, threat, ridicule, rejection, or withdrawal constitutes emotional abuse

(Iwaniec, 1995), especially if it is of a repetitive or continuous nature (O'Hagan, 1993).

Barnett, Miller-Perrin, and Perrin (1997) report based on their thorough review of the literature that 3–28% of reported abuse and neglect cases take the form of psychological maltreatment or emotional abuse. The wide discrepancies in reported rates of emotional abuse may be related to the great definitional difficulties reported in this aspect of child abuse. It is in the area of psychological maltreatment that underreporting of cases may be most prevalent due to the less visible nature of this type of victimization. Self-report surveys of parents have resulted in figures that suggest that between 15% and 63% of parents admit to at least one interaction with their children that may be classified as emotionally abusive (Daro & Gelles, 1992; Vissing, Strauss, Gelles, & Harrop, 1991). The average age of the emotionally abused child is 8–8.5 and more than half are female (47% male, 53% female). White children are at greater risk for emotional abuse than children of other cultural backgrounds and low income increases the risk of psychological maltreatment in a home. Risk increases in homes with single mothers or other unemployed single caretakers; in other words, single parenting status may be confounded with poverty variables.

Neglect

Finally, neglect is defined through "acts of omission in which the child is not properly cared for physically (nutrition, safety, education, medical care, etc.) or emotionally (failure to bond, lack of affection, love, support, nurturing, or concern" (Walker, Bonner, & Kaufman, 1988, p. 8). It refers to the "passive ignoring of a child's emotional needs; to lack of attention and stimulation; and to parental unavailability to care, to supervise, to guide, to teach, and to protect" (Iwaniec, 1995, p. 5). Through neglect of a child, the parent fails to provide the age-appropriate and necessary amount of nurturance, stimulation, protection, and encouragement that would help the child grow into an emotionally and behaviorally mature adult. Neglect should be defined, not only through parental failure to provide for a child's needs that are due to parental limitations of a psychological or emotional nature, but also due to limitation imposed by social or socioeconomic factors, such as poverty. For example, neglect may occur in a home in which the parents cannot afford to provide their children with needed medical care due to financial difficulties and/or the lack of proper medical insurance (Barnett, Miller-Perrin, & Perrin, 1997). Another issue in the assessment of neglect is that of intentionality which is often presented as a dichotomy (intentional versus unintentional perpetration of neglect). However, the simplicity of this idea has been challenged. For example, Erickson and Egeland (1996) pointed out that a Hispanic mother may have a cultural value of maintaining physical contact with her infant at all times. Such a mother may not use a car seat and instead hold her child on her lap while riding in a car. Erickson and Egeland question whether this constitutes intentional neglect or a cultural behavior that needs to be condoned and understood as such. Another recent example emerged in the popular media when a Scandinavian couple dined in a New York City restaurant, leaving their infant in a baby carriage outside the restaurant below the window where they were seated. They had visual contact with the child at all times. However, other patrons of the restaurant were appalled and alerted child protective services who arrived promptly, took the child in custody and apprehended the parents, charging them with neglect. The parents were released once it was clarified that they were engaging in a behavior that was not only condoned, but encouraged, in their home country where infants are not taken into restaurants but rather are left outside under the vigilant eyes of their parents.

Clearly, neglect is difficult to define. Its definition is further complicated by the fact that it can take many forms and, depending on the form, may come to the attention of different care providers. If the lack of proper care is perpetrated in the physical realm, there may be physical and medical symptoms; neglect may also be emotional or psychological, resulting in symptoms coming to the attention of mental health care providers; neglect may occur with regard to the proper monitoring of a child's safety, resulting in accidents or injuries; and neglect may be expressed through indifference to or interference with the child's educational needs (Oates, 1996).

According to Barnett, Miller-Perrin, and Perrin's (1997) summary of the literature, up to 55% of reported cases of child abuse and neglect are substantiated as cases of child neglect. Figures range widely because, although neglect is the most common form of child abuse reported in the United States, the wide range of manifestations it can take makes it difficult to define. For example, in the late 1980s the National Incidence Studies (as cited in Barnett, Miller-Perrin, & Perrin, 1997) revealed 103,600 cases of physical neglect and 174,000 of educational neglect during one studied time period and 507,700 cases of physical neglect and 285,900 cases of educational neglect during a later phase of the study. Such discrepancies result in changing definitions of neglect that have evolved from including only children who have been harmed to including children who are at risk for harm. The average age of neglected children is 6 years of age. Of these children, 51% are under the age of 5, and of these children 34% are under the age of 1, indicating a decreasing risk for neglect with aging. Younger children are also more likely than older children to receive serious injuries or fatalities due to neglect. Gender differences in the victims of neglect remain largely unsubstantiated, suggesting that boys and girls are equally at risk. Similarly, cultural and ethnic differences are inconclusive, pointing to no single ethnic or racial group that is at greater or lesser risk. Neglect rates do increase, however, with unemployment, decreasing income, and increasing dependence on social assistance. Socioeconomic status is perhaps the strongest predictor of child neglect of all forms. Children in single parent households may be at higher risk, though the issue of poverty may again confound this pattern.

LEGAL AND ETHICAL ISSUES FACING THE CLINICIAN OF THE ABUSIVE CLIENT

Awareness of the definitions provided so far for the various forms of abuse and neglect starts therapists on their way to being able to recognize when the legal and ethical issue of child abuse or neglect must be raised in a session with a client. These definitions may not always coincide exactly with the language used in a practitioner's state's child reporting statutes, underscoring once again the need for mental health care providers to be familiar with state law relevant to their practice.

Child abuse has been in existence since ancient times, with infanticide being the most common crime in Europe until the 1800s. Infanticide, although rarely occurring in the United States and Europe, remains common in some countries today (e.g., China [Barnett, Miller-Perrin, & Perrin, 1997]). The first formal case of physical abuse of a child was reported and prosecuted in 1874 in England. It involved a child, Mary Ellen Wilson, who had been abandoned by her mother and was abused by her foster family. She had come to the attention of a concerned citizen who did not know how to protect this child, given that there were no child protection laws at the time. She turned to the Society for the Prevention of Cruelty to Animals (SPCA), arguing that the child was indeed a member of the animal kingdom and, as such, deserved the same protection as other animals. This argument was accepted and the case was successfully tried through this venue. Mary Ellen's stepmother was convicted of assault and battery in a trial that lasted less than an hour, a conviction that later led to the founding of the Society for the Prevention of Cruelty to Children (SPPC). The SPPC was an advocacy group for abused children that ultimately led to laws criminalizing child abuse and neglect (Barnett, Miller-Perrin, & Perrin, 1997). The first legislation in the United States appeared around 1900 in the state of New York (Boylan, Malley, & Scott, 1995). Despite this early effort in England and in the United States, child protection did not become a nationwide or global concern until the mid-1900s. At that time, physicians began to notice and talk about injuries and illnesses in children that could not be explained through plausible physical causes. Understanding of the phenomenon of child physical abuse began to grow throughout the 1960s due to the efforts of Dr. Kempe of the American Academy of Pediatrics. In fact, a paper coauthored by Kempe in 1962 (titled "The Battered Child Syndrome") finally set in motion the process that would lead to consistent child protection laws throughout the United States by 1964 (Azar, 1992). The efforts to criminalize child abuse culminated in 1974 when the United States Congress enacted the Child Abuse Prevention and Treatment Act that made federal funds available to those states in the Union that enacted reporting laws for child abuse and neglect. Once reporting laws were federally encouraged, they were passed in all states of the United States within five years. "No legislation in the history of this country had been so widely adopted in so little time" (Zellman & Faller, 1996, p. 361). The first reporting laws were written primarily for medical doctors; however, by 1986, most states had expanded the mandate to include other professionals with regular contact with children as well.

Although most Western countries have defined child sexual abuse and exploitation as illegal, a few cultures have existed or continue to exist that have practices that involve sexual relationships between adults and children (e.g., the Sambia of New Guinea [Barnett, Miller-Perrin, & Perrin, 1997]). However, these practices generally involve complex religious ceremonies or rites of passage and do not represent general standards of day-to-day conduct. Even cultures in which such sexual practices have been or are used for various reasons, adult-child sexual relationships outside of the prescribed ritual or religious context are generally not condoned. In the United States, child sexual abuse was not taken

seriously until the late 1970s, through the efforts of the feminist movement (Oates, 1996). It was not until 1978 that the protection of Children against Sexual Exploitation Act was passed, followed in 1986 by the Child Sexual Abuse and Pornography Act. These acts of Congress made it a federal crime to exploit children sexually or to use children for purposes of pornography. They did not address the issue of incest, which falls under the purview of the Child Abuse Prevention and Treatment Act. As with physical abuse, child sexual abuse has a long history, dating back to classical Greece. In the United States, an early account of prosecution in a case of incest is available in Connecticut, in which a father was sentenced to death for having violated the incest taboo with his daughter. Reflecting the bias of the times, which also held the victim accountable, the daughter was also severely punished. Despite taboos and legal actions against the practice of child sexual abuse and incest, the actual practice continued through the centuries. Certainly, child sexual abuse and exploitation was a common topic presented by women to their physicians in Freud's times in the early 1900s. Although Freud originally took these claims seriously and literally and decried these actions by adults against children, he later bowed to the pressure of his physician colleagues. In fact, he changed his entire early theorizing to accommodate the taboos of the time, erroneously revising his theories to suggest that the recollections of childhood sexual abuse were in essence false memories reflecting a neurotic personality adjustment of his female clients. It is possible that through this revision of his thinking, Freud unwittingly contributed to decades of continued sexual perpetration against children.

Fortunately for today's children, state-level child abuse reporting laws, covering both sexual and physical abuse, have now been in place for almost three decades. During this time there have been tremendous increases in the number of reports of victimization of children, bearing witness to the reality that children have been abused and neglected severely and prolongedly (Haugaard,

1992). All states in the United States have child protection laws that require professionals who come into contact with abused children to report their abusers to the relevant child protection agency in the state. Despite this uniformity of the requirements, state laws differ in details with regard to how, where, and to whom incidences of child abuse are reported. Hence, it is critical that all individual practitioners become thoroughly familiar with the child protection laws of the state in which they practice (Azar, 1992).

However, a few consistent procedures do appear to emerge that apply in all states. Virtually all states require that care providers not only report incidents of child abuse that are proven, but also cases which merely harbor the suspicion of abuse or neglect (Azar, 1992). Only a judge can determine if abuse or neglect is taking or has taken place and decide who will have custody or control over a child (Bulkley, Feller, Stern, & Roe, 1996). Reports must also be made immediately upon forming the suspicion or gaining knowledge of an incident. It is best to make the report before the family leaves the premises, not only to protect the children, but also to protect practitioners from liability for injuries to the child victims should further abuse occur after the family leaves and before care providers report their suspicion or knowledge (Boylan, Malley, & Scott, 1995). Similarly, virtually all states require that if the child is on the premises of care providers' offices and they believe the child to be in imminent danger of violence or abuse because the family may flee or engage in other behavior to avoid prosecution, care providers have the responsibility to protect the child until the child protection agency can take over. In other words, care providers are not allowed to release the child to a severely abusive parent once they have determined that significant suspicion exists that the child will become a victim of violence or that the family would attempt to flee prosecution (Walker, Bonner, & Kaufman, 1988).

Virtually all states have a stipulation in their reporting law that holds the practitioner who re-

ports an incident or suspicion of abuse in good faith harmless from prosecution for breach of confidentiality; that is, they have a statutory exemption to client-therapist privilege, as well as immunity from any civil or criminal liability that might arise from the report (Ahia & Martin, 1993; Anderson, 1996; Bulkley, Feller, Stern, & Roe, 1996). This immunity from prosecution and exception to confidentiality must be covered in the informed consent that is presented to new clients before they begin therapy. For clients who resist this aspect of informed consent or who attempt to challenge the care provider not to conform to the reporting laws, it may be pointed out that all but one of the United States (the exception being Oregon [Ahia & Martin, 1993]) make child abuse reporting mandatory. In fact, state laws impose penalties and legal liability on practitioners who fail to report an incident or suspicion of abuse that has come to their attention (Anderson, 1996; Corey, Corey, & Callanan, 1988). These penalties can include fines ranging from $100 to $1,000 and imprisonment for five days to one year (Azar, 1992). The mandatory nature of reporting laws in the United States has been criticized not only by clients, but also by some practitioners who believe that the reporting laws are a double-edged sword (e.g., Meyer & Deitsch, 1996). Some evidence has accumulated that reporting laws appear to keep abusive families from entering treatment for fear of being reported or to induce those who have started treatment to drop out because of the strain on the therapeutic relationship (Berlin, Malin, & Dean, 1991). This is a particular tragedy when therapeutic intervention had hopes of creating change in a family by reducing the abusive behavior on the part of the parents. Regardless of clinicians' opinion on this matter, at the current time there is no room for judgment: suspicions and incidents of abuse must be reported to the appropriate child protection agency in the state, lest providers are willing to bear the legal consequences of their noncompliance with state law.

It is often helpful to remember that first reports, and especially first reports of minor (if there is such a thing) incidents of abuse or ne-glect do not result in much action by the child protection agency. These offices tend to be so overworked that they respond with intervention only in cases of severe first abuse incidents or multiple incidents. Thus, not every report will lead to intervention; most reports will not lead to removal of the child. Keeping this in mind and informing the perpetrator thereof may leave the clinician some leeway with regard to salvaging a workable therapeutic relationship. Further, it is important to keep in mind that Federal Law PL 96-272 requires that all reasonable efforts be made to maintain the family unit or to reunify any family that has been separated due to child abuse reporting (Bulkley, Feller, Stern, & Roe, 1996). In other words, it is the mandate of child protection services to keep families intact and to provide assistance to families-at-risk, that is, to families who have been reported for child abuse or neglect. Such assistance (at least in theory) consists of access to crisis nurseries, respite care, family resource centers, counseling or therapy, parent education, and home visits, as well as of referrals to Big Brothers/Big Sisters, Parents United, and other self-help groups (Barnett, Miller Perrin, & Perrin, 1997). Thus, families may become eligible for support and therapeutic services and resources exactly because of a report that resulted in child protective and/or legal action. Consequently, despite potential negative outcomes of child reporting laws (e.g., risk of unnecessary intervention, removal of the child instead of the perpetrator, long and/or multiple out-of-home placements [Bulkley, Feller, Stern, & Roe, 1996]), positive outcomes are possible that may protect a vulnerable child from further abuse or neglect in the future (Azar, 1992). How to make a report and how to do so while main-taining rapport will be thoroughly discussed in the intervention section of this chapter.

Reoffending is an important legal and hu-manitarian issue to consider as well. A counselor who becomes aware or suspicious of a client who is once again reverting to behaviors that may qualify as abusive or neglectful must report

the client all over again (i.e., even if a report was made for the original offense). Nonreport by the clinician because the reoffending client shows re-morse (real or feigned) is illegal. If the client re-offends after a clinician worked with a family in good faith, saw significant improvement, and therefore did not report, the clinician becomes legally liable and faces the humanitarian issue of having chosen an action that led to the revictim-ization of the child. Child abuse cases are gener-ally prosecuted in civil court if the perpetrator is a family member. They can be prosecuted in criminal court if the perpetrator is not a family member. Some child sexual abuse cases may be prosecuted twice; once in civil court and once in criminal court, depending on the circumstances of the case (see Brems, 1993).

HELPFUL TRAITS DURING CHILD ABUSE AND NEGLECT ASSESSMENT AND INTERVENTION

Any mental health professional working with adults, adolescents, or children must be aware of the legal, therapeutic, and assessment issues in-volved in working with perpetrators or victims of abuse (Brems, 1993; Fantuzzo & McDermott, 1992). However, these professionals also must be aware that the assessment or evaluation of child abuse and its effects on the victim and legal im-plications are strictly in the realm of experts who have delved into this topic through continued education and applied, supervised learning and practice. In other words, there is a clear demarca-tion between mental health professionals who deal with child abuse on a regular basis and make this an important aspect of their practice and practitioners who only deal with the issue as it emerges in a more generalist setting. Given the focus of this book, the issue will be dealt with from the latter perspective by helping clinicians recognize the difference between the incidental assessment and reporting of child abuse and the systematic assessment of child abuse for forensic or child/victim treatment planning purposes. Awareness of this difference will prevent clinicians from working outside their bounds of expertise or in roles that are inappropriate to the relationship that has been or needs to be established with a given client.

Specifically, at least three distinct roles can be held by a mental health professional in relationship to a perpetrator or victim of abuse. The professional can be a therapist of the perpetrator before having had any knowledge of the presence of abuse or neglect; or the professional can be a counselor of the victim before knowing about the abuse or neglect; or the therapist can be a formal evaluator who was called by the courts or through other circumstances to evaluate or investigate the presence and effects of abuse that has been reported previously. These three roles are very distinct. If care providers already have a preestablished role with the perpetrator, they are most likely to be an advocate for that person once the abuse has been uncovered and reported. That is, professionals should then not be called to court to testify in behalf of the child and should not be involved in the investigation of the action or the evaluation of the child. If care providers had a preexisting relationship with the victim at the time of recognition and reporting of the abuse, they then become advocates for the child, who should neither speak out against perpetrators, nor evaluate them formally for legal reasons. If, on the other hand, the professional is called in by the courts or the child protection services to evaluate either victim or perpetrator, the role is not that of an advocate for either child or adult, but rather that of an objective, impartial evaluator who establishes facts and findings that are then relayed to a decision-making body (such as a judge or court-assigned child advocate). It is important to differentiate these roles so that mental health care providers who have incidental contact with perpetrators or victims do not allow themselves to be placed in roles of impartial or objective evaluators. Differentiating the roles of therapist versus forensic evaluator is crit-

ical because there are essential differences in several arenas of the client-care provider relationship (Greenberg & Shuman, 1997). For example, Greenberg and Shuman (1997) eloquently point out that there are important differentiations that must be made regarding whose client the litigant is (the clinician's versus the attorney's) with respect to the mental health care provider (supportive and therapeutic versus objective and evaluative), the area of expertise (therapy versus forensic evaluation and assessment), the type of relationship established with the client (helping versus adversarial), and the goal of the joint work (to benefit the client versus to benefit the court).

Related to role clarification, the counselor or therapist who recognized abuse through treatment of the victim or perpetrator is not responsible for authenticating the report; this role is left to child protection workers, the court system, or another mental health professional who is called in for that specific reason. Optimally, clinicians will retain their original roles as mental health care providers for the initial client. If they can maintain this role with the client, therapeutic rapport can be salvaged while the child is being protected through the intervention of other care providers who are mobilized in response to the report made by the original clinician. Thus, the therapist's role is essentially to make the report while maximizing the likelihood of maintaining a therapeutic alliance with the client and while assuring that other care providers engage in authentication of the incident or suspicion (Boylan, Malley, & Scott, 1995).

It is important for counselors to have some idea about how they feel about or react to perpetrators before beginning to see clients. The possibility of being confronted by a client who has abused or neglected a child is significant, and all mental health professionals must be aware of their personal and professional biases and attitudes about perpetrators. If these attitudes are highly negative and strongly emotionally charged, countertransference issues need to be explored and the possibility of transfer of the client to another clinician seriously considered. Throughout working with a perpetrator (i.e., not just while assessing whether a report is necessary but also during subsequent intervention and therapeutic work), mental health care providers are clear that it is not their role to judge but to understand; that is, it is not their role to authenticate but to gather enough information to make an informed decision about whether a report is necessary. It is important to realize that the role of care provider differs significantly from that of judge or court-appointed child advocate, and that, as therapists or counselors, providers have to be on the side of the clients who present for treatment to attempt to help them deal with the presenting concern as well as with the unexpectedly reported problem of being a perpetrator. Mental health professionals who work with children know that they have to come to terms with having to deal with parents who are abusive or neglectful. However, this is an issue that is often ignored among clinicians who deal with adults. Nevertheless, they need to learn to develop empathy for the abuser (or, if they cannot, to be up front with clients that a transfer may need to be made). Such empathy for the abuser is best developed by exploring the abuser's own personal history, which is often filled with problematic and/or abusive relationships as well. As the risk factors section will show, the majority of abusive or neglectful parents have experienced significant abuse or neglect in their own childhoods and are perpetuating generational patterns. If clinicians understand and appreciate this, they will become more capable of empathizing and working with the perpetrator, because, in essence, most perpetrators are also victims. Finally, empathy for abusing parents is also often critical for the welfare of victims who themselves may have a conflicted and ambivalent relationship with the offending parent. A child still loves the abusive parent in most cases and may be greatly upset if involved care providers are not sensitive to this fact by demonstrating caring and concern not just for the victim but also for the perpetrator.

In other words, it is important for clinicians to be aware of their own biases and not to allow these biases to interfere with treatment or assessment (Meyer & Deitsch, 1996). If nothing else, clinicians need to know when to transfer clients because of their own blind spots or attitudes. The maintenance of respect and caring for clients must be assured regardless of the actions reported by them (Brems, 1994). Finally, it is of course essential that counselors be able and willing to broach the topic of child maltreatment. Any mental health professional who is unable to raise the topic with clients is likely to miss the duty to report many times over. Therapists must grow comfortable with asking specific questions about how parents treat their children in a variety of settings and circumstances. Such inquiry can often be combined with an assessment of parenting skills and can be done easily without inappropriate confrontation, hostility, or rudeness.

Well-prepared clinicians are also familiar with the child abuse literature, at least to the extent of being able to recognize abuse or neglect when it occurs. They have the legal and ethical awareness to know how much information needs to be gathered to make a reasonably sound judgment about the necessity of a report to a child protection agency. It is the goal of the remainder of this chapter to provide information that will help clinicians become knowledgeable about just these issues. First, it will provide information about risk factors and signs or symptoms that can trigger the suspicion in the therapist that direct questions regarding child abuse or neglect may be necessary. Then, it provides some guidelines about how to ask the relevant questions and how to intervene if the counselor either learns of a proven incident of abuse or neglect or has a strong suspicion that it is occurring despite a client's denial.

RISK FACTORS FOR AND WARNING SIGNS OF CHILD ABUSE AND NEGLECT

The clinician must be familiar with two sets of information to be ready to assess and report incidents or suspicions of child abuse. First, any mental health (and health) professional needs to know the risk factors that are correlated with the occurrence of child abuse. Such possible precipitating or contributing factors to the occurrence of child abuse can help trigger suspicion and more in-depth questioning on the part of the clinician so that the topic of child abuse and neglect is covered thoroughly and appropriately with clients who present a particularly high risk. These risk factors can be divided into three distinct sets or categories, consisting of traits of the perpetrator, traits of the family in which perpetrator and child currently reside, and larger societal traits that contribute to violence in families, especially violence toward children. Second, clinicians must be familiar with warning signs in children so that, if clients report such characteristics, behaviors, or events in their children, clinicians know to follow up with questions about parenting and parent-child relationships that can lead to assessment of the risk factors and ultimately to a discussion with clients about abuse and neglect in the home. It should be noted that the following elaboration of risk and warning signs can also be used to help adult clients identify whether they were victims of abuse or neglect themselves during their own childhood. If this is the case, the counselor or therapist then has the responsibility to help the client deal with this childhood experience appropriately in the context of the adult's current therapy by incorporating this issue into the existing treatment plan.

Risk Factors and Warning Signs of Physical Abuse

Several traits have been identified that are common among perpetrators, families, and communities that report the presence of physical abuse of children. These traits appear to point toward the disrupted childhood experience of the perpetrator, lack of knowledge in many arenas concerning the child victim, and the emotional disturbance that is often part and parcel of the adult's current life experience. At the family level, the traits that emerge suggest a certain level of discord and disharmony that has resulted in other forms of vi-

olence and/or nonconcern for the younger members of the family. With regard to communities, societies that generally condone violence are conducive to the perpetration of violence in the home. Perhaps for this reason, corporal punishment has been rendered illegal in Scandinavia, and this law was accompanied by increased efforts at parent education. This initiative was successful in reducing abuse rates toward children (Oates, 1996), suggesting that societal intervention may indeed be necessary to improve the welfare of a country's children. The risk factors outlined below have been reported very consistently and hence are not referenced individually. The interested reader can refer to Barnett, Miller-Perrin, and Perrin (1997); Oates (1996); Helfer, Kempe, and Krugman (1996); Ludwig and Kornberg (1992); and Walker, Bonner, and Kaufman (1988) for more information and detail.

Traits of the Perpetrator

- personal experience of abuse or neglect in childhood
- poor ability to attach or bond
- poor ability to express physical affection
- emotional immaturity as evidenced by egocentrism, poor self-esteem and self-worth, and poor emotional regulation
- strong dependency needs
- inability to express negative emotions (especially anger) appropriately
- high impulsivity
- presence of other violent or criminal behavior
- experience of stressful life events
- inability to tolerate stress
- emotional and interpersonal isolation
- social immaturity
- guardedness with regard to talking about family relationships
- lack of an appropriate parenting role model in childhood
- poor parenting skills

- favored use of corporal punishment
- anger at or frustration with children
- lack of awareness of children's needs
- age-inappropriate expectations of children (either too high or too low)

Traits of the Community and/or Culture

- acceptance of violence (also through the condoning of violent media)
- acceptance of the use of corporal punishment
- emphasis on competition rather than cooperation
- low socioeconomic support for schools and low cultural valuation of education
- presence of a power differential between adults and children

Traits of the Family

- presence of socioeconomic stress
- lower socioeconomic status and/or lower educational level
- single-parent or merged-family home
- unwanted pregnancy
- marital/domestic violence
- presence of substance use
- presence of a child with a difficult temperament
- mismatch of parental and child's temperament
- presence of a medically ill or physically demanding child

In addition to needing to be knowledgeable about and aware of these risk factors, the clinician should also know what warning signs child victims of abuse may express. These warning signs can be used as the springboard for additional data gathering from the adult who is being seen and may also lead to a more thorough assessment of parenting skills. Not all of the warning signs provided below are unique to children who have been physically abused; many are also notable in sexually or emotionally abused children; some are to be found in children who have never experienced abuse.

Hence, the presence of any of these signs and symptoms is never conclusive, it is merely a red flag that the clinician has the responsibility to probe further. The commonly agreed upon signs and symptoms listed below are collated from various sources, such as Gil (1991), Meyer and Deitsch (1996), and Oates (1996) and can be grouped into physical, behavioral, and emotional symptoms.

Behavioral Signs in the Child Victim of Physical Abuse

- accident-proneness
- academic problems
- social problems or poor peer relationships
- social isolation and withdrawal
- shrinking from physical contact
- self-injurious or suicidal behavior
- destructiveness, disruptiveness, hostility and/or delinquency
- excessive clothing for weather or unusual layers of clothing
- resistance to normal dressing (not due to poor parenting)
- drug use
- avoidance of parents or one parent
- frequent absence from school

Emotional/Psychological Signs in the Child Victim of Physical Abuse

- neediness
- low self-esteem
- fearfulness
- hypervigilance
- oppositionality
- mistrustful

Physical Signs in the Child Victim of Physical Abuse

- unexplained injuries
- frequent accidents and/or frequent trips to the emergency room
- unexplained bruises, burns, bald spots, or bleeding (also known as the four Bs)
- substance abuse

Risk Factors and Warning Signs for Sexual Abuse

As with physical abuse risk factors, consistent traits of the perpetrator, the family, and the community have emerged that suggest sexual abuse of children. These traits have much in common with the traits listed for physical abuse, but some notable differences are present as well. For example, there is a stronger focus on inappropriate sexuality within the family, both in terms of lack of proper boundaries and poor communication about sexuality and sexual relationships. With regard to the perpetrator, the great disregard for the violation of the trust placed in the adult by the child is notable, as is the indifference to the child's emotional and physical needs in favor of over-concern for personal gratification and need-fulfillment. With regard to societal issues, it is remarkable that violence against women is condoned, as is a general power differential between women and men, and children and adults. Despite these special traits that are related specifically to sexual abuse, it is probably fair to conclude that there are more similarities than differences and that a common set of traits exists that cuts across all spectrums of abuse (and perhaps even neglect). The most commonly cited risk factors, or traits, for sexual abuse can be studied further in Barnett, Miller-Perrin, and Perrin (1997); Burkhardt and Rotatari (1995); Finkelhor (1986); Kempe, Helfer, and Klugman (1996); Oates (1996); and Walker, Bonner, and Kaufman (1988) and are summarized below.

Traits of the Perpetrator

- personal experience of sexual abuse in childhood
- poor ability to attach or bond
- poor ability to express physical affection
- emotional immaturity as evidenced by egocentrism, poor self-esteem and self-worth, and poor emotional regulation
- strong dependency needs
- strong needs for immediate gratification

- guardedness with regard to talking about family relationships
- lack of an appropriate parenting role model in childhood
- poor parenting skills
- poor awareness of children's psychological and emotional needs
- placement of personal needs above those of the children in the home

Traits of the Community and/or Culture

- acceptance of violence (also through the condoning of violent media), especially violence toward women
- poor valuation of women
- presence of a power differential between men and women
- sexualization of women in the media
- acceptance of sexual exploitation of women

Traits of the Family

- home without a father or with a stepfather or maternal boyfriend
- frequent maternal absence
- employed, ill, or disabled mother
- lack of appropriate communication about sexuality leading to lack of knowledge in the child
- poor relationship between child and at least one parent
- high acceptance of family nudity and lack of privacy
- marital/domestic violence and other parental conflict
- presence of substance abuse

A number of signs and symptoms are present in the child victim of sexual abuse that require further probing about the possibility of the existence of incest or similar sexually inappropriate behavior in a family. These signs and symptoms again have some overlap with those of physically abused children. But a number of signs are entirely unique to sexual abuse victims and are very indicative of a sexual abuse problem in the home. These will be obvious from perusing the following list, but they may be summarized by pointing out the emphasis on sexual acting out, sexualized behavior with other children or adults, and illnesses involving the genitalia. Mental health care providers need to take care to be aware of the importance of these symptoms and to conduct a full assessment when these signs emerge from the conversation with adult clients about their children. The following list of warning signals was collated from sources including Burkhardt and Rotatari (1995); Gil (1991); Kempe, Helfer, and Klugman (1996); and Oates (1996).

Behavioral Signs in the Child Victim of Sexual Abuse

- sexualized or sexually provocative behavior (such as promiscuousness, sexual precociousness, or sexualized play activity with other children)
- precocious sexual understanding
- sexual language
- sexual acting out with other children
- excessive bathing
- sudden unexplained change in behavior
- avoidance of parents or a parent
- runaway behavior
- suddenly possessing money or unaffordable items

Emotional/Psychological Signs in the Child Victim of Sexual Abuse

- neediness
- low self-esteem
- insecurity
- excessive secrecy

Physical and Medical Signs in the Child Victim of Sexual Abuse

- sexually transmitted diseases
- elimination disorders

- illnesses, diseases, and injuries involving the genitalia (e.g., pain, rashes, itching, urinary tract infections)
- psychosomatic concerns (such as headaches, stomachaches)
- eating disorders

Risk Factors and Warning Signs for Emotional Abuse or Neglect

Risk factors and warning signs of emotional abuse and all forms of neglect are sufficiently similar to be collapsed for the purposes of this discussion. The risk factors that emerge for emotional abuse and neglect of all forms have some overlap with those for physical and sexual abuse but point toward more severe attachment and developmental problems among the perpetrators. These families tend to be socially isolated with poor supports available within and outside of the family. Social isolation, however, is not always as easy to identify as noting the absence of contacts with others. Emotional isolation or lack of intimacy can also occur in contexts of what appears to be an active social life. Finally, although substance use disorders are associated with all forms of abuse, neglect and emotional abuse appear to be more strongly related to other psychiatric illness (especially depression) than other forms of child maltreatment. Emotional abuse and neglect also appear to be less likely to be socially or culturally bound, and more strongly predicted by the poor attachment and bonding history of the perpetrator and the nuclear and extended family. However, all have a strong systemic flavor and origin that often transcends generations. The risk factors have been noted with enough consistency in the literature that they are not individually referenced. The list below was collated from sources including Gil (1991); Iwaniec (1995); Kempe, Helfer, and Klugman (1996); and Oates (1996).

Traits of the Perpetrator

- personal experience of abuse or neglect in childhood

- poor ability to attach or bond
- poor knowledge about human development, especially with regard to children's developmental needs and milestones
- unrealistic expectations of children (either too high or too low)
- impulsive aggression
- poor ability to experience and express affection
- social isolation
- increased experience of stressful life events coupled with poor coping ability
- psychiatric illness, especially depression
- substance use
- emotional immaturity as evidenced by egocentrism, poor self-esteem and self-worth, and poor emotional regulation
- strong dependency needs
- guardedness when talking about family relationships
- lack of an appropriate parenting role model in childhood
- poor knowledge of parenting strategies

Traits of the Community and/or Culture

- acceptance of violence (also through the condoning of violent media)
- low respect for or valuation of children
- poorly developed support systems
- little emphasis on educational support and day care regulation

Traits of the Family

- marital/domestic violence and other parental conflict
- presence of substance abuse
- presence of psychiatric illness
- punitive or inconsistent discipline
- chaotic lifestyle
- poverty and unemployment

- lack of social or extended family supports
- fussy, difficult-to-manage infant

In addition to these risk factors that point to the need to assess parenting and relationship issues between adult and child, a number of traits in the child may suggest the presence of emotional abuse or various forms of neglect. These traits are often quite notable in interactions with these children, unlike the traits of the sexually abused child, which tend to be more subtle. For example, these children tend to draw attention due to their small size, poor hygiene, inappropriate dress, or clear physical deprivation. Nevertheless, emotional abuse and neglect remain much more difficult to document sufficiently to warrant significant intervention by child protection agencies. Thorough assessment is of utmost importance to have enough evidence for a suspicion to move child protection workers to an active level of intervention. Warning signs of emotionally abused or neglected children include a variety of symptoms that have been reported consistently in many sources, including Gil (1991); Iwaniec (1995); Kempe, Helfer, and Klugman (1996); and Oates (1996).

Behavioral Signs in the Child Victim of Emotional Abuse or Neglect

- academic problems
- social problems or poor peer relationships (including violence)
- social isolation and withdrawal
- shrinking from physical contact
- inappropriate clothing for weather (especially insufficient clothing when cold)
- frequent complaints of hunger and/or thirst
- frequent absence from or tardiness to school
- bed-wetting
- wetting and soiling of clothing

Emotional/Psychological Signs in the Child Victim of Emotional Abuse or Neglect

- neediness
- low self-esteem
- parentified child (e.g., taking care of siblings; concern over parents)
- attachment problems
- fearfulness and anxious relating
- repression of feelings and denial of family problems
- hypervigilance
- mistrust

Physical Signs in the Child Victim of Emotional Abuse or Neglect

- poor grooming and hygiene
- developmental delays
- poor weight gain
- sleep problems
- untended injuries or illnesses
- untreated medical needs
- recurring infections

Summary of Risk Factors and Warning Signs

A number of traits cut across perpetrators, their families, and their communities regardless of type of abuse or neglect that should make the therapist wary and concerned when they are disclosed. Many of these traits can be related to other (not abuse-related) psychological or family systems issues and, hence, none is conclusive in and of itself. Similarly, the behavioral, physical, and emotional traits one can expect among abused or neglected children also serve as warning signs that abuse or neglect may be present in a client's home, though again not conclusively so. It is the responsibility of the clinician to recognize the risk factors and warning signs as such and to probe further to assess whether it is nec-

essary to invoke the duty to report. Hopefully, if some therapeutic rapport has been established, denial on the part of the adult client will be minimal. However, often this is not the case and the counselor must proceed cautiously and with some circumspection to attempt to glean the most reliable and accurate information about the parent-child relationship.

ASSESSMENT OF THE NEED TO INVOKE THE DUTY TO REPORT

To rule out the duty to report, the clinician needs to explore whether an actual incident of abuse or neglect has occurred or whether there is a suspicion of past or current abuse or neglect that is being denied by the client. This assessment is quite different from suicide or homicide assessments, for which the primary focus of the assessment is on evaluating the risk factors, because the client presented with a threat. In the case of the duty to protect, the risk factors have largely been established, and it now becomes the counselor's responsibility to assess whether there is a threat or danger. In other words, in the case of the duty to report, the reason for the assessment comes about because, in the course of a regular session, enough warning signs emerged that the clinician now is concerned with uncovering the possibility of abuse. The assessment for the duty to protect can be relatively brief and to the point by merely asking direct questions about abuse or neglect. However, the process tends to be more successful if a slightly more circuitous route is chosen (if there is adequate time remaining in the session to do so; because the duty to report must be carried out with immediacy, the luxury of a slow assessment may not always exist).

As such, an assessment for the need to invoke the duty to report is best initiated through a suggestion to the client that the clinician would like to find out more about the parent-child relationship by conducting a formal parenting assessment. The parenting assessment then will move into an assessment of the parent's ability and willingness to meet the child's basic physical and psychological needs, and finally explores the presence and circumstances of the abusive situation directly. The latter issue may be easy or difficult depending on the level of denial by the parent about past abuse or neglect and on the amount of trust that has already been established in the therapeutic relationship.

Parenting Assessment

Most simply put, a parenting assessment consists of three components: parenting strategies, parenting philosophies, and parenting goals. With regard to assessing parenting strategies, counselors had best be aware of a number of points. First, most parents have never taken classes that teach parenting strategies and are merely operating on belief systems that stem from their own respective childhood (Dietzel, 1995). Thus, they may not be familiar with all different types of parenting strategies, their appropriate applications, and complex components. Second, given that parents often have not had any particular parenting education, they may either be inconsistent or rigid with regard to how they implement the parenting strategies with which they are familiar (Brooks, 1994, 1996). They may be overly rigid or overly inconsistent with regard to when, how, or with whom they apply them, in terms of who enforces or implements them, and with regard to having negotiated their use and appropriateness in certain situations (Darling & Steinberg, 1993). Third, disagreements between coparents about what constitutes adequate disciplinary strategies are not at all uncommon (Johnson, Fortman, & Brems, 1993). When this is the case, it is equally likely that children are aware of the conflict and use it to their advantage (though generally neither consciously, nor maliciously so).

It is generally necessary to ask very specific questions of clients to gain an understanding of how they (or their coparent) parent and discipline their children. Also, it is generally worthwhile to investigate whether the same strategies are employed with all children, by all coparents, in all situations, and at all times. Whenever clinicians ask about whether clients know of a certain strategy, they also may want to ascertain that the client knows the correct definition and application of that technique. Knowing about a strategy and using it correctly are often two completely different things. If a client is unfamiliar with a given strategy, the clinician defines it briefly, as it is often true that a parent may use a certain strategy without being aware of its label. Finally, it can be illuminating to ask the clients about their favorite and best parenting strategy, as well as about what they perceive to have been their worst interventions as a parent. A brief overview of the most commonly taught parenting strategies is provided in Table 8-1; for more information the reader is referred to parenting resources (e.g., Brems, 1993, 1999a; Johnson, Fortman, & Brems, 1993).

Another important aspect of parenting that needs to be assessed with the client who has potential abuse or neglect issues is that of parenting values or philosophy (Brooks, 1996; Elkind, 1995). Parenting philosophy can translate into at least three styles: authoritarian, permissive, and authoritative (Baumrind, 1971, 1973, 1983). Authoritarian parents tend to be restrictive, expecting their children to follow rules without questioning. They are generally strict disciplinarians who have little tolerance for negotiating a compromise or for letting children work toward their own solutions to problems. These parents have strict rules and expect their children to abide by them. Violation is punished without asking about purpose or reason for the transgression.

Permissive parents, on the other hand, have overly loose rules so that children are often not clear about what is expected. Any rules that do exist are not consistently enforced, and limits are notoriously unclear. This leaves little room for structure and often has children guessing about appropriateness of behavior and acceptability of affect. Because children are not born with an innate sense of rules, this permissive style can be quite unsettling and anxiety provoking for the

TABLE 8-1 Comparison of the Four Primary Parenting Education Programs

PROGRAM	ASSUMPTIONS	ENVIRONMENT	STRATEGIES	GOALS
Logical Consequences (Dinkmeyer and McKay, 1976)	■ Children have the inherent capacity to develop in healthy, effective ways ■ Children grow within the family system ■ Children want acceptance ■ Children want the common good	■ Encouragement that fosters self-respect ■ Democratic process ■ Joint work toward common goals	■ Natural consequences ■ Logical consequences ■ Encouragement	■ Foster competence to function in and contribute to the family ■ Foster competence in the area of ability to communicate effectively
Humanistic Parent Effectiveness Training (Gordon, 1970)	■ Children are a competent problem solver ■ Children need acceptance	■ Parental honesty, acceptance, and openness ■ Open communication	■ Active listening ■ I-messages ■ Environmental modification ■ Parental change	■ Effective problem solving ■ Mutual acceptance
Behavior Modification Training (Krumboltz and Krumboltz, 1972)	■ Children learn to adapt to the environment ■ All behavior is learned ■ Feelings are of secondary importance	■ Parents as authority ■ Maximization of learning opportunities ■ Parents control behavior through setting contingencies	■ Reinforcement ■ Shaping ■ Modeling ■ Extinction ■ Time-out ■ Satiation	■ Development, maintenance, and elimination behavior ■ Emotional responses are secondary to behavior
Self Psychological Parent Education (Brems, 1990a)	■ Children have needs for strength, nurturance, and companionship ■ Children use the interaction with the environment to develop a self	■ Nurturing, strong caretakers who are available for modeling ■ Shared responsibility for problem solving ■ Availability of support and empathy ■ Acceptance	■ Nurturing strategies ■ Guidance strategies ■ Companionship strategies	■ Cohesive, orderly, strong self ■ Awareness of affect ■ Empathy with and caring for others

child. Permissive parents often have a peer relationship with the child, rather than a parental relationship. Children need adults in their environment to figure out what is right and wrong in their society and neighborhood, so in such permissive situations they are often left guessing and unsure of themselves.

Authoritative parents have developed a compromise between having no rules, like permissive parents, and having too many rigid rules, like authoritarian parents. They generally have several sensible rules that are explained to the child and are often agreed upon and renegotiated as the child develops and matures. Although these parents are in charge of the family, they are tolerant of input from their children and are more likely to run a democratic household in which every member is perceived as having input about what is appropriate and acceptable and what is not. Authoritative parents are not personally threatened by children's misbehavior or stubbornness, but rather they see them in perspective and can deal with them constructively. Of the three parenting styles, the authoritative is the most psychologically sound approach to parenting. The therapist needs to assess which style the client favors and whether any coparent has the same style. Disagreements in styles are a major problem that often results in conflict and arguments for parents.

Finally, parent goals need to be explored briefly. Parents have various goals for their children and themselves as parents. These goals often play a significant role in how parents and children interact and may lead to problems in the parent-child relationship. Goals also have to be developmentally appropriate and have to reflect children's needs at various ages (Roberts, 1994; Stern, 1985). Tailoring parenting strategies to age and need is critical and reflects sensitivity and caring to the child. Parenting questionnaires that have been developed (largely for research purposes) often address the dimension of developmental appropriateness and how it relates to parental empathy and intervention (e.g., the Adult Adolescent Parenting Inventory by Bav-

olek [1984]; and the Child-Rearing Practices Report by Block [1965]). A very promising new scale that investigates the relationship between child and parent on several dimensions is the Child-Parent Relationship Inventory by Gerard (1994). Any or all of these scales can be administered in addition to conducting a parenting assessment interview, though they cannot replace it. If counselors use these scales, they can become the foundation of a discussion with the client about parenting skills and are used for feedback to the parent.

When only a parenting assessment is done, only the previously mentioned categories of parenting are addressed, and questions about abuse or neglect of children play a marginal or no role. In a situation of potential abuse or neglect, however, other warning signs have moved the clinician to a different plane of intervention; namely, one in which a potential legal or ethical issue must be explored that has direct implications for the therapeutic relationship due to the potential duty to protect. Hence, the assessment then must move on to address additional components of the child-parent relationship.

The Parent's Ability and Willingness to Meet the Child's Needs

As previously pointed out in the risk factor section, parents who are abusive or neglectful often have a poor understanding of children's developmental needs, emotional states, and physical concerns. They often lacked proper parenting role models and had histories of abusive or neglectful childhoods themselves. They are either unable or unwilling to meet their children's basic physical and psychological needs, be it because they themselves had never had them met or because they are preoccupied with other events in their life (e.g., stressful life events, psychiatric illness, substance use, self-care plans). It is also possible that these parents are unaware of their children's needs and hence do not meet them, not out of hostility, anger, or lack of concern, but purely out of lack of education and knowledge. For the clinician to be able to make a proper assessment, he or she must have some basic understanding of child development and children's needs. Details about child development clearly go beyond the focus of this book and the reader is referred to Brems (1993, 1998), Kloss (1996), Sigelman and Shaffer (1995), and Stern (1985, 1990) for more information. However, a very brief discussion of children's basic needs does appear to be warranted here to give the mental health care provider some idea about the types of questions that need to be asked.

Children have at least five basic needs: they depend on their parents for physical care and protection, for affection and approval, for stimulation and teaching, for discipline and control over behavior, and for the encouragement of gradual autonomy (Kloss, 1996; Kohut, 1984; Winnicott, 1958). Well-prepared parents are aware of these five needs of children and are therefore able to provide for them. Parents may not necessarily verbalize all five of these needs when asked about what their children need from them, but should show some recognition or awareness if questioned directly about any one of these topics. As such, mental health professionals who are probing into the parent–child relationship can formulate questions that relate to the five topic areas. They can inquire about what parents do to keep their child physically safe and how this may have changed over the years. Questions may be asked about how parents communicate caring and affection to the child, especially in difficult situations. Inquiries can be made about opportunities for role modeling and teaching, such as joint activities that may help the child become familiar with and learn about new situations. Discipline strategies are assessed and hopefully reflect the parents' understanding of the fact that children need healthy limits and boundaries to feel psychologically safe and cared for. The acquisition of autonomy is ascertained by asking questions about how the child has changed in the relationship with the parent over the years and how much freedom that child has gained over the developmental span.

In all five categories of need, an assessment must be made about the parent's recognition of the child's needs, as well as the parent's ability and willingness to meet these needs. It is entirely possible for a parent to recognize a need but to do nothing about it. Perhaps the worst prognosis is for the parent who recognizes the child's need, would be able to meet it, but chooses not to. This parent is less likely to respond to intervention than the parent who either fails to recognize the need independently or the parent who recognizes the need, would like to meet it, but does not know how.

Assessment of the ability and willingness of parents to meet their children's needs does not imply that parents have to meet their children's needs at all times. In fact, this is impossible. All parents make mistakes or miss opportunities for helping their children. The issue that needs to be evaluated is whether the parent is good enough (Winnicott, 1958) and fails the child only occasionally, with opportunities later to make up for the failures (optimal empathic failures [Brems, 1999a; Kohut, 1984]). Further, the issue is also not one of doting on the child. Far from that, overconcern and overinvolvement can be quite harmful for the child as well (though not a form of reportable abuse). What the clinician must ascertain is whether parents have a healthy recognition of their children's needs and do their best to attempt to meet these needs appropriately most of the time.

Closely related to the issue of recognizing and meeting a child's needs well enough, is the issue of appropriate bonding and attachment between child and caretaker (Stern, 1985). In some cases of neglect, the most strongly contributing factor is the lack of attachment or bonding; in other cases of abuse (such as in sexual abuse), the problem may be one of inappropriate attachment or bonding. It has been suggested that "because an unhealthy attachment relationship can be directly observed, even if the actual abuse cannot, one may conduct an assessment analysis that first identifies and appropriately eliminates healthy attachment relationships for sex abuse and then examines unhealthy attachments for specific indicators of abuse" (Haynes-Seman & Baumgarten, 1994, p. 6). Unhealthy attachments in that context are defined as parent-child relationships wherein the parent fails to maintain appropriate generational boundaries, fails to meet a parental role, and instead uses the child for the gratification of personal emotional needs.

If the assessment suggests that the parent fails to meet the child's needs and is either not attached or unhealthily attached to the child, it becomes necessary to initiate questions about abuse and neglect. It is safest to move into this topic area by exploring whether the parent personally has ever had concerns about the relationship with the child or about their own behaviors and skills as a parent. This can be followed up with questions about whether others (e.g., teachers, physicians) have ever expressed concerns. Finally, the counselor must express concern about the patterns that have been revealed and must explain that this concern is sufficiently great to warrant some direct questions about abuse and neglect. Although counselors certainly want to maintain rapport and do not want to be hostile, overly or inappropriately confrontational, or rude, they must nevertheless be straightforward about the assessment of the situation as abusive or neglectful. Once the transition has taken place, more direct inquiry may need to be made about specific incidents in the past if this information is not yet available. A sample transcript of the transition process is provided here.

Clinician: All right, let me see if I understand. You are telling me that your daughter is responsible for getting herself up in the morning to go to school and then has to let herself back in when she gets home in the afternoon. You are pretty much gone from 4 A.M. to go to work until 2 P.M., and then you have classes until 6 P.M. three times a week. On weekends you say you have a couple of study groups with colleagues from work who are also getting their master's.

Client: Yeah, that's right. I study a lot!

Clinician: So Mary is alone a lot!

Client: For an 8 year old, Mary is very capable of taking care of herself. As I said, there have only been a few times when we've had problems. . . . *(sounds defensive)*

Clinician: And those problems led Mary to the emergency room once because she burnt herself when she made her dinner, to the police once because a neighbor called when she noticed a stranger in your backyard, and to the same neighbor three times when Mary either forgot her key, got hurt and was bleeding severely because she had an accident on the way home from school, or was crying because she missed the school bus.

Client: That's it for the most part. . . .

Clinician: Do you see a problem with any of these events?

Client: What do you mean? Mary or someone else handled it just fine each time. . . . I never even got called at work or at school. . . .

Clinician: Because Mary does not have a phone number for you. . . .

Client: Well yeah—I don't want her bugging me at work.

Clinician: Do you have any worries about Mary's physical safety when you leave her without supervision for this much time? Especially given that there have been a few times when stuff has happened?

Client: Are you accusing me of something here?

Clinician: I suppose I am questioning how right it is for you to leave Mary at home alone at her age. I believe that you are neglecting her needs for physical safety and supervision. At 8 years old she isn't really able to handle crises yet—a child her age does not yet have the cognitive ability to think on her feet. . . . *(client interrupts)*

Client: Mary is smarter than most 8 year olds and I have taught her about dialing 911!

Clinician: That just isn't enough to keep her safe, though. What do you suppose might have happened if your neighbor hadn't noticed the stranger in your backyard and hadn't called the police? After all, they did take him in because he was someone they had been looking for in a criminal case.

Client: But see that's why it's okay to leave Mary at home. It's a very safe neighborhood and there is always some other adult around who can keep an eye out for her.

Clinician: But ultimately it is your responsibility to take care of Mary and to watch over her. You cannot really rely on neighbors, especially since you haven't formally talked to them to baby-sit Mary. . . . *(client interrupts)*

Client: I've told you I can't afford a baby-sitter!

Clinician: *(ignores client's angry outburst)* Aside from just keeping Mary safe, have you ever thought about her emotional needs? I believe that with your not being around to help her with homework or other difficult stuff in her life, she may be scared and unsure a lot of the time. Kids need to learn from their parents and they need company.

Client: She has plenty of company in school all day.

Clinician: *(clinician derails)* Yet you told me she is very isolated and has no friends.

Client: Well yeah, but that's her fault, not mine. I keep telling her she shouldn't be so shy.

Clinician: *(clinician gets back on track)* When you took Mary to the emergency room for her burns, did the physician say anything to you about how Mary got the burns?

Client: No, she just said to be more careful about allowing Mary to handle hot foods—I didn't really have the chance to tell her how it all happened; she was in a rush because there were a ton of other people in the waiting room.

Clinician: Have you ever had teachers express concern about Mary? Either about her being

so shy and isolated or about how she looks and feels when she gets to school?

Client: Yeah, her first grade teacher kept calling and complaining because Mary said she was hungry or forgot to bring her snow boots and stuff like that. I've had talks with Mary that she shouldn't complain so much or people may think we're too poor to feed her right. I also told her to be more careful to pick out the right clothes when it's cold. She's just got to learn to take care of herself. It's like the logical consequences I learned about in a parenting seminar at work. They said that if a kid screws up, let her feel the consequences. So I told Mary, you don't wear the right clothes, you freeze. It's just natural.

Clinician: There is a big difference between a natural consequence and neglect. I believe you are neglecting Mary. You are not keeping her safe, you are not taking care of her emotional needs, and you are not making sure that she learns and has social contacts that are so important for kids her age. . . . *(client interrupts)*

Client: What are you saying? *(client gets angry)*

Clinician: What I am saying is that you are not doing what you need to do to keep Mary safe and that I am very concerned about Mary's welfare. I believe she is in danger much of the time and that you are not supervising her as you need to as her parent. I realize that you are a single parent and that things are tough. But Mary needs more help and guidance than you are giving her. In fact, she probably could also do with more affection and time together. But right now I am most concerned just about her physical safety.

Client: Well that's really none of your business. I mean I was sent to you because I am having trouble with my boss who claims I'm drinking too much—I am here because *I* need help, not because of *Mary.* Why is it always about her? What about me? Who gives a shit about me?

Clinician: Please understand that I am very worried about *you* too. *And* I am glad that you came to see me, and I am looking forward to working with you. And you are right, your and my work together will be about you, and only sometimes about Mary. But I guess, right now is one of those "sometimes." Remember when we talked earlier about privacy and confidentiality and I talked about the only times when I have to violate your confidentiality is if you tell me about things that may harm you or others, like suicide or child abuse?

Client: You're saying I'm abusing Mary? . . . *(client is very angry)*

Clinician: Yes, I am saying that you are neglecting Mary and are putting her in dangerous situations. I am concerned about her, just as I am concerned about you. I can help *you* because we will be meeting here every week and will work on how we can help you improve your life. I can help *Mary* by making sure that you and others will begin to look after her a little bit more.

Client: What others?

Clinician: Well as you remember, I told you earlier that if I think that there is evidence of child abuse or neglect I need to notify the authorities. So, the people I'll involve are the case workers at the Department of Family and Youth Services.

Client: I can't believe you are doing this to me. . . . *(client is calming down; sad now)*

Clinician: I don't have a choice in the matter. There are laws that require me to do this to protect Mary. If I don't call DFYS, I am breaking the law. . . . *(client interrupts)*

Client: What's DFYS?

Clinician: Sorry; that's the Department of Family and Youth Services. . . . *(client interrupts)*

Client: Well, screw the law; if you call them I'll never come back!

Clinician: I hope you'll change your mind about that because I really have no choice. Has anyone else ever called DFYS about you and Mary?

Client: Of course not, I'm not an abuser. . . . Only you seem to think so. . . . *(client is wavering between a range of emotions)*

Clinician: I believe you are neglecting Mary; that's a little different from abuse but just as dangerous for your daughter. Again, I do want to help you and I think I can—if you give me a chance.

Client: What's gonna happen if you call these people? Are they gonna take Mary? . . . *(client appears scared now)*

Clinician: No, they won't. Here is what I think will happen. . . .

From here the conversation will move into the prediction of what will happen and then the discussion of how to go about making the report. Also, more specifics may need to be collected about the neglect situation in this home before the clinician feels ready to make the report along with the parent (or rather to get the parent to make the report along with the clinician; see below). This transcript merely serves to point out how the issue of neglect (or abuse) is broached with a parent who is somewhat resistant about recognizing a neglectful (or abusive) role in the relationship with a child.

Specifics about the Abuse or Neglect Situation

Abuse-related information needs to be collected that explains the history of the abusive or neglectful situation (Meyer & Deitsch, 1996). As such, the clinician must inquire about how often the abuse or neglect has occurred, when it tends to occur, and where it tends to occur. It is important to explore patterns that may emerge, such as abuse that occurs only after the parent has been drinking. Once the when, where, and how often have been answered, the specifics of any past events need to be explored. The clinician needs to gather information about what happened during past abusive incidents, getting details about the behaviors or actions that were involved, the players that were part of the event, the level of force that was involved in the action, and the representativeness of this interaction between parent and child of their usual relationship. Exact descriptions of antecedents and consequences of prior incidents are helpful, along with descriptions of witnesses and professionals who may have become involved. To summarize, the following details need to be explored:

- Who perpetrates the abuse or neglect?
- Who is the victim?
- Who is involved or present (in addition to the victim and perpetrator)?
- Is there a witness?
- Where does it take place?
- When does it occur (times and situations)?
- How often has it occurred?
- How severe is it?
- What behaviors and actions does it involve on the part of the perpetrator, victim, and witness?
- Is it situation-specific?
- Has it come to anyone's attention before?
- Has it been reported before?
- Has it resulted in the involvement of other professionals (e.g., emergency room physicians, teachers, neighbors)?
- Has it resulted in injuries, illnesses, or diseases?
- Have there been attempts to stop it?
- Is it predictable?
- What else can the client tell about the abuse or neglect situation?

It is important to clarify whether the client is the perpetrator of the abuse or merely a witness thereof. Both instances are reportable and result in the same basic intervention. However, clearly,

the child protection agency needs to know the name of the actual perpetrator, which is not necessarily the same as the name of the client. It is important for a therapist or counselor to understand that even if another adult in the household perpetrates the abuse, the duty to report has to be invoked because the essence of this duty is child protection. If the client is the witness, not the perpetrator, a whole additional set of circumstances needs to be explored. Namely, the clinician must gain appreciation of whether the client is a willing or a reluctant witness. In other words, is the client upset by the abuse that is being perpetrated but does not know how to stop it? Is the client experiencing abuse as well? There is ample evidence that child abuse and domestic violence go hand in hand (McKay, 1994). Thus, if clients are witnesses to abuse, they may also be victims. Other questions that are raised by clients who are witnesses, rather than perpetrators, relate to the safety of not just the child, but also the clients themselves after the report has been made. If the abusive adult also engages in domestic violence, it may be necessary not only to involve child protective services, but also to help clients seek temporary safe housing (such as in a women's shelter). It is possible that clients are quite fearful of the perpetrator in the aftermath of the report even if no prior incidents of victimization by the abuser of the client have occurred. These concerns need to be assessed and attended to.

INTERVENTION IN THE EVENT OF A DUTY TO REPORT

Table 8-2 reiterates the criteria that can be used to assess and prepare for a report of child abuse or neglect. Once a client has been assessed using these criteria, the clinician needs to decide whether the duty to protect must be invoked. If the counselor believes that a child in the client's care has been victimized in some form, either by the client or by another adult in the home (with the client as a witness), a report to the local child protection agency must be initiated. To be ready

TABLE 8-2   Assessment of the Need to Invoke the Duty to Report

CRITERION	EXAMPLES OF RISK FACTORS*
Parental Knowledge of Children's Needs	Unaware of children's needs for: ■ physical care and protection ■ affection and approval ■ stimulation and teaching ■ discipline and control over behavior ■ encouragement of gradually increasing autonomy and independence
Parental Meeting of Children's Needs	■ unwillingness to meet children's needs in any or all categories of need ■ inability to meet children's needs in any or all categories of need ■ expectations for the child to meet the adult's needs ■ preoccupation with personal needs ■ impaired ability for bonding and attachment ■ inappropriate attachment
Parental Knowledge of Child Development	■ lack of knowledge about developmental changes ■ lack of knowledge about developmental expectations ■ age-inappropriate expectations of children with regard to behavior, affect, independence, and so forth ■ equal household rules for all children regardless of their developmental ages
Parental Knowledge of Parenting Skills	■ lack of knowledge about discipline strategies ■ inconsistent implementation of strategies ■ inappropriate implementation of strategies ■ overly authoritarian or permissive (as opposed to authoritative) parenting style ■ inappropriate household rules and values ■ parent-focused or stereotypic (not child-focused) goals and expectations for children ■ lack of clarity regarding rules, schedules, routines, and expectations
Traits of Potential Perpetrators	■ personal childhood experience of abuse or neglect ■ poor emotional maturity and regulation ■ strong dependency needs ■ guardedness about talking about family ■ lack of appropriate parenting model in childhood ■ impaired expression of affection ■ stressful life events with inability to cope ■ impulsivity and need for immediate gratification ■ self-preoccupation due to factors such as substance use, psychiatric illness, stressors
Traits of an Abuse-Condoning Community	■ acceptance of violence ■ low respect for or valuation of children and/or women ■ emphasis on competition, not cooperation ■ low socioeconomic and political support for schools, education, daycare facilities, women, and/or family issues ■ power differential between men, women, and/or children

TABLE 8-2   Assessment of the Need to Invoke the Duty to Report (continued)

CRITERION	EXAMPLES OF RISK FACTORS*
Traits of a Potentially Abusive Family	■ marital or domestic violence ■ parental conflict ■ presence of substance abuse or psychiatric illness ■ socioeconomic stress ■ lower educational level ■ temperament difference among parents and children ■ difficult-to-manage child(ren) ■ chaotic family life ■ parental absence ■ divorce-affected family (e.g., single, blended) ■ lack of social support network (e.g., extended family, group memberships)
Emotional Signs of Victims	■ neediness ■ fearfulness ■ low self-esteem and/or insecurity ■ mistrustfulness and/or secretiveness ■ attachment problems
Physical Signs of Victims ■ *Physical Abuse* ■ *Sexual Abuse* ■ *Emotional Abuse/Neglect*	■ unexplained injuries ■ frequent accidents ■ 4 B's (bruises, bleeding, bald spots, burns) ■ substance abuse ■ sexually transmitted diseases ■ injury or illness involving the genitalia ■ psychosomatic concerns ■ eating disorders ■ poor grooming and hygiene ■ developmental delays ■ sleep problems ■ untreated medical needs
Behavioral Signs of Victims ■ *Physical Abuse* ■ *Sexual Abuse*	■ accident prone ■ shrinks from physical contact ■ self-injurious behavior ■ destructiveness and hostility ■ social problems ■ sexualized behavior and excessive bathing ■ sudden unexplained change in behavior ■ avoidance of a parent or parents ■ sudden possession of money or items

Continued

TABLE 8-2   Assessment of the Need to Invoke the Duty to Report (continued)

CRITERION	EXAMPLES OF RISK FACTORS*
■ Emotional Abuse/Neglect	■ academic problems
	■ social isolation
	■ bedwetting; soiling
	■ frequent complaints about hunger or thirst
	■ tardiness for and absence from school

*For additional specific risk factors with regard to perpetrator and community characteristics in the areas of physical, sexual, and emotional abuse, also refer to the listings in the relevant text pages.

for this process, counselors must be prepared and must have decided how they want to handle such calls. Both preparation and a plan of action are important to assuring the greatest likelihood of continued client compliance and rapport. Finally, follow-up care for the purposes of prevention needs to be attended to.

Preparation for the Report

The best preparation for having to make a duty-to-report call is for clinicians to have investigated the necessary procedures beforehand. Just as was discussed in relation to the duties to protect and warn, conscientious clinicians familiarize themselves with local agencies and procedures for reporting abuse and neglect. It is good practice to phone the local agency and make an appointment to discuss procedures and to glean a sense of how families are treated once a report has been made. Specifically, mental health care providers need to have immediate access to (or have memorized) the phone number of the local child protection agency. This means that they need to know the name of this agency. This may seem like a ridiculous point to make, but the labels of child protection services vary widely across the nation. In Alaska, the relevant office is called the Division of Family and Youth Services; in Oklahoma it is the Department of Health and Human Services. Thus, knowing the name of the agency is important to locating the proper phone number, and the unprepared clinician's fumbling may only make the client more nervous and less compliant!

Second, mental health care providers need to know what information will be needed during the phone contact so that no unnecessary questioning of clients occurs while the phone call is being made. Knowing what the child care workers will ask helps clinicians prepare by asking these question of the client beforehand so that the information is readily available during the phone conversation. Although readers are encouraged to call their local child protection service to determine what answers they require, some pieces of information appear to be agreed upon as necessary to a report. Depending on the state and community, these include the following information (Brems, 1993):

- biographical data of the alleged victim (e.g., name, date of birth, address)
- current location of the alleged victim
- current location and addresses of the family of the victim
- biographical data of the alleged perpetrator (e.g., name, date of birth, address)
- current location and addresses of the alleged perpetrator
- current location and address of the person making the report
- ease of access of the perpetrator to the victim
- nature, severity, and chronicity of the situation/incident
- current status of the victim (e.g., extent and description of bruises or injuries)

- dates of recent incidents
- names of witnesses
- immediate safety issues concerning the victim

Third, the clinician should have some sense of the level of intervention that is likely to occur in the type of circumstance that is being reported. Some cases will result merely in written documentation that a report was received; others result in interviewing the children and/or other family members; some result in the immediate removal of the child from the home. Being able to give clients some idea about what to expect is helpful in alleviating or reducing their anxiety. It is, of course, important to make these predictions with the caveat that clinicians are merely making an informed guess about the level of intervention by the child protection agency given their prior experience with them. It is entirely possible that a different action may be taken, because it is often difficult to predict how a child protection agency will respond. This can happen for many reasons, one of which may be that a client has not been completely truthful with a care provider and the counselor finds out that many prior reports have been made. Suddenly what appeared to be a minor report that should have resulted in written documentation at most becomes a major concern that leads to the removal of the children from the home. One helpful practice in preparing for giving the client an idea of what action may be taken is to call the child protection agency before making the report. During this call the clinician provides the agency with a hypothetical case scenario that resembles the one in question and receives information about the need to report and the most likely intervention that will be taken. However, even if this hypothetical scenario is presented, the clinician should not report the likely reaction provided by the child protection worker as a definitive course of action, since things may still change once the actual report has been made.

A mental health care provider who has detailed information about the process for invoking the duty to report is less likely to flounder and feel incompetent at the time of an actual report. Such a prepared therapist is better able to put the client at ease by being able to answer knowledgeably and reliably the inevitable questions about the reporting process itself and about what will happen. Being able to impart information to clients about what will happen next and what it all means will help them anticipate more exactly what will happen to them and their family and will leave them feeling more trusting of the clinician. As mentioned earlier, there is some controversy among clinicians as to whether reporting laws interfere with therapeutic rapport with clients, and for this reason significant noncompliance with the law has been reported (e.g., Bromley & Riolo, 1988; Kalichman, Craig, & Follingstad, 1990; Watson & Levine, 1989). Although the question about the impact of reporting on the therapeutic relationship is an important one, it remains somewhat academic, because the laws are completely clear: there is no leeway; if there is a suspicion or evidence of neglect or abuse, a report *must* be made. However, there are some things clinicians can do while invoking the duty to report that may serve to salvage the therapeutic relationship. One of these is being able to share information and to demonstrate some level of knowledge about the reporting process.

The Plan of Action for the Report

The plan of action must reflect respect and caring, not judgment and contempt, for clients if it is to maximize the likelihood of a continued positive treatment alliance (Boylan, Malley, & Scott, 1995). Even under the trying circumstances of having to report clients to child protection services, clinicians take care to work with their clients, not against them. Most authors and clinicians suggest that when a suspicion or evidence has been discovered, perpetrators (or witnesses to the action) are first reminded of clinicians' duty to report. Clients are then asked to make the report to the child protection agency themselves from the clinician's office in the clinician's presence. Generally clients can be reassured that this procedure will result in more leniency from the child protection agency toward the perpetrator. Further, it places the responsibility for the behavior squarely in the perpetrator's (or witness's) lap and often serves to preserve a relationship between clinician and client. There is consensus in the clinical literature that getting clients (whether perpetrators or witnesses) to make the report themselves in the clinician's presence is preferable to the therapist making the report because this practice tends to maintain the therapeutic relationship (see Boylan, Malley, & Scott, 1995; Dodds, 1985; Walker, Bonner, & Kaufman, 1988). Sometimes clients are actually relieved to make the report; this is particularly true when they have been witness to the abuse, rather than perpetrator.

If clients are willing to make the report, clinicians dial the telephone and inform the child protection worker of the situation, that is, of the fact that they are sitting with a client who needs to make a report. A speakerphone is ideal for this purpose because both client and clinician will be fully informed about the entire conversations with the child protection worker. Clinicians obtain the name of the child protection worker, provide their own name and name of the clinic to the worker, and then hand the phone over to the client. Clinicians remain with the client during the entire phone call for support as well as monitoring that the report is made in its entirety. Before hanging up the phone, clinicians ask to talk to the child protection worker one more time to inquire whether any additional information may be needed from them. Clinicians unfortunately cannot rely on the honesty of their clients to make the call later or to make the call from home. They need to be convinced through direct witnessing of the call that the report has been made as necessary. If clients refuse to make the call themselves, therapists will need to make the entire call in the client's presence at this time, informing the child protection worker that this is the case. Once the phone call has been initiated, clinicians best give clients one more chance to talk to the child protection worker themselves. Sometimes, once the inevitability of the report sinks in for clients, they are willing to take

over the phone, although they were reluctant to do so before.

It is best to have the client's chart available during the phone conversation and to have ascertained that sufficient detail is contained within it so that specific questions can be answered. Counselors must be prepared to identify the victim, the perpetrator, the client's family, potential witnesses, and themselves, as well as the exact circumstances of instances of abuses and neglect and the victim's current state of health and safety (see the earlier preparations section). Clinicians are sensitive to the fact that reporting laws provide an exception to client-clinician confidentiality only around the incident(s) of abuse and neglect that gave rise to the report. Any other information they have about the family and the client (e.g., the client's own child abuse history or employment situation) is not to be disclosed to child protection workers (except with a signed release of information obtained from the client), even if agencies request it (Azar, 1992). Finally, given that most state protection agencies are currently overwhelmed with work, it may be that a phone call initiated during a session cannot be brought to completion due to the lack of availability of a child protection worker who can respond to the call. If this occurs, mental health care providers impress upon their client the need to continue to attempt to make the report once they have left the session. Further, the clinicians themselves will need to do the same and let their client know that they will also continue to try to make contact with a child protection worker to make the report.

Once the phone call has been completed, clinicians take some time to debrief with their clients. Clients are given the opportunity to express how they feel about having made or witnessed the report, what their fears are about the report, and about their plans for the future with regard to the relationship with the victimized child and any other children in the same home. If the client is not the perpetrator, but rather a witness to the abuse (perhaps by a spouse or significant other who shares the home), that person needs to take some time to deal with fear and guilt issues related to letting the perpetrator know that a report was made. Although all of these issues are tied directly to the reporting, they are also therapeutic issues that will continue to be dealt with repeatedly in future therapy sessions. If clients remain committed to treatment, clinicians will have the opportunity to follow up with them with regard to changes in the home, consequences, and other issues that emerge in the aftermath of the report. If clients appear to have been extremely reluctant about the call and noncooperative with the whole process, chances are good that they will not return for future sessions. In this case, the debriefing is the only opportunity clinicians may have to attempt to salvage the relationship and to effect positive change in the client's and child's life. Hence, the importance of taking the time for a debriefing must not be underestimated.

Once the report has been made and the debriefing is completed, it may be advisable to schedule a quick follow-up appointment with the client to enhance the chances of keeping the individual in treatment. This follow-up appointment would serve to help the client deal with any investigations initiated by the child protection agency in response to the call, but more importantly it will also address issues of trust and rapport between the client and the care provider. The therapist must be very sensitive to hurt feelings and feelings of rejection and betrayal by the client and as well as the client's family members. There may be pressure on the client from family members (especially from the perpetrator if the client is the witness) not to return to treatment, and this needs to be anticipated and discussed. Dealing with clients in the wake of a report of abuse or neglect is not easy. However, not dealing with them is even more detrimental to everyone involved.

Follow-Up after the Report

As was true with regard to the duties to warn and protect, merely making it through the crisis of having had to breach confidentiality by involving various other people is not enough. The problems underlying the crisis then have to be-

come therapy issues that have to be incorporated into the regularly planned treatment of the client. If child abuse issues emerged and had to be reported, they now cannot be ignored by the counselor in future sessions just because the report has been made (Haynes-Seman & Baumgarten, 1994). In fact, more importantly, they need to become an integral aspect of the client's treatment to prevent future incidents of abuse and neglect in the client's child's life. With this in mind, the clinician needs to take some time to plan additional treatment interventions before the client's next session to be prepared to make all necessary and appropriate referrals. For example, more than likely, child victims need to be seen in treatment themselves and clinicians best have some names and phone numbers ready for their clients to initiate a referral. Often, in child abuse cases, given the far-reaching systemic issues implied, a number of professionals have to become involved due to the wide-reaching consequences of the action (Oates, 1996). Often school personnel become involved, as do health care providers, parent educators, support groups, and case managers. Making appropriate referrals for parental and child support is important to make inroads with regard to prevention of future occurrences of abuse (Finkelhor & Dziuba-Leatherman, 1994).

A team approach is most beneficial when other care providers have to become involved in a specific case. Such an approach prevents contradictory information from being given to the family from different care providers, works to avoid splitting or other means of manipulating the involved mental health care providers by the family, and in general results in a more effective and comprehensive treatment plan for all parties involved. Not coordinating treatment of an abusive or neglectful family across all providers tends to result in cases that rapidly spin out of control, often leading to the involvement of an unnecessarily large number of care providers and resulting in uncoordinated and ineffective care. In fact, even if the clinician decides that no report of abuse or neglect was necessary or if the child protection agency does not open a file after a report of suspected abuse or neglect, it is likely that given the risk factors and warning signs that became red flags for the clinician, coordinated intervention and referral is important to prevent potential future occurrence of abuse and neglect in families-at-risk who have not yet perpetrated abuse.

DOCUMENTATION AND RECORD KEEPING

An assessment for or intervention due to risk regarding the duty to protect has to be carefully documented in the client's chart for the reasons outlined earlier. At the end of this chapter is a sample of a progress note that was written after a report of child abuse was initiated with a client. This sample outlines for the reader how appropriate duty to report charting is done, covering all of the cautions mentioned above. Attached to the progress note would be copies of any written documents prepared with the client during the session. Any phone calls made in addition to the ones included in the sample should be recorded in the same way. Message slips received about the case of the client (e.g., messages from child protection workers or other health and mental health professionals returning calls about the client) also need to be attached (with time and date of receipt).

SUMMARY AND CONCLUDING THOUGHTS

The issue of child abuse and neglect results in a great number of responsibilities for the counselor or therapist. The duty to report is only the be-

ginning of these, though it is often the most crucial component. This is true because how the reporting is handled may have a great impact on continued treatment compliance on the part of the client. Handling the situation with caring, respect, concern, and understanding is much more likely to meet with success in the sense of continued opportunities for treatment than dealing with the client in a judgmental, contemptuous, or hostile manner. The clinician is, after all, the advocate for the client, even if the client has engaged in reprehensible action. Developing empathy for clients who have been perpetrators of abuse or neglect is an important growth process in every clinician's development.

The issue of abuse and neglect cannot be ignored once the report has been made. Abuse and neglect are merely symptoms of an underlying problem. It becomes the responsibility of client and therapist to modify the treatment plan (if necessary) to address these underlying issues in order to prevent future incidents. It is also important to keep in mind that the duty to report represents an opportunity as well, both for the client and the clinician. Very often a report can be the turning point in a family's life when the parent or parents finally recognize that a change is necessary. The report can be used to mobilize resources for the family, making sure that they have a case worker who helps them with applications for social services they may not have known about, but are eligible for. Keeping this in mind, both therapist and client may feel more comfortable about proceeding with the report. Finally, of course, it must be kept in mind that the duty to report is about protection: it may well help a child have a better future and freedom from abuse. That alone is worth the effort.

Sample "Duty to Report" Progress Note

Progress Note

Clinician:	Chris Brems
Client Name:	Sandra B.
Date of Session:	21 January 1997; 11:00 a.m.
Session Number:	1 (Intake Interview)
Payment Made:	$50 (receipt provided)
Next Session:	28 January 1997; 10:00 a.m.

Special Note: See Intake Report for Sandra B. for complete detail of the data collected during this session; this note is written to document the reason for and process of reporting a suspicion of child abuse.

In the course of the Intake Interview the client reported information about her 9 year-old daughter, about the family's dynamics, and the mother's relationship with her significant other that aroused suspicions of possible child abuse in the client's home.

Assessment

The client described her daughter as being socially isolated and lonely, prone to accidents and mishaps (e.g., frequent falls when bicycling, running across the street in front of cars, being burned by a hot stove at age 6, falling off her bed during the night, cutting her hand when sewing or working with scissors), expressing unwillingness to join the family for family meals or other joint activities, having low self-esteem and little self-confidence, being mistrustful and reticent with adults, occasionally bullying younger children, and having had multiple visits to the emergency room during her short life.

At the same time the client reported that her nuclear family is currently struggling economically because she was recently laid off from work and her significant other (who lives with the client and her daughter) is currently unemployed. Her significant other (of 2 years) has two children (a girl aged 4 and a boy aged 3) who were permanently removed from her care and custody two years ago because of physical abuse and neglect.

The client herself presented as a meek woman with poor assertiveness skills and low self-esteem. She had difficulty making eye contact with the counselor and sat in a guarded position. She had two prominent bruises on her right lower arm and a small, faded set of bruises on her neck. When asked about the bruises on her body, she indicated that she wasn't sure how they got there but conceded spontaneously that she and her significant other have a tendency to get into physical fights.

The presence of domestic violence was then used to open inquiry into the possibility of child abuse, starting with the relationship between the client and her daughter. The client denied ever being physically abusive with her child, denying the use of corporal punishment, shoving or pushing her daughter, or shaking her when she was an infant. She described adequate physical care behavior of the child, and although she showed some clear deficits with regard to knowledge about child development and parenting strategies and skills, no information emerged about abusive interactions between her and her daughter. The inquiry then shifted to an exploration of the daughter's relationship with the significant other. The client was hesitant to talk about their relationship, indicating merely that the two had some problems getting along and that her daughter never liked the significant other. When specific questions were asked about physical abuse (again asking questions about corporal punishment, hitting, shaking, pushing, shoving, etc.), the client began to cry. She revealed that she had never witnessed any inappropriate behavior between her daughter and significant other but that she had noticed that her daughter was avoiding the significant other, often had unexplained bruises, and seemed to be more afraid than she used to be. The client also revealed that her daughter had recently begun to withdraw from the client, being less open with her mother about her life, feelings, and thoughts than she used to be. The issue of possible abuse of the daughter by the significant other was then broached with the client, pointing out the patterns in the daughter's behavior that are typical of abused children and the family dynamics that are often present in abusive homes.

Continued

Sample "Duty to Report" Progress Note (continued)

Intervention

The client was reminded of the counselor's duty to report suspicions of child abuse and was informed that a report needed to be made. She was then presented with the option of making the report of her suspicion herself, right now with the counselor in her office. This procedure was described thoroughly and its advantages were discussed with the client. The client was also given information about the child protection role of the DFYS and about likely outcomes of the report. Once these pieces of information were shared with the client and she was confronted with the reality that a report would be made, whether she agreed to do it herself or not, she reluctantly consented to make the phone call herself. To get ready for the call, the following information was collected from the client:

- **Biographical Data of the Alleged Victim**: Melissa B., age 9 (DOB: 13 July 1987), living at 1234 Any Road Street, Small Town, Alaska, 99999
- **Current Location of the Alleged Victim**: at this moment the client is in school at Small Town Elementary
- **Current Location and Addresses of the Family of the Victim**: same address as above for the family consisting of Melissa, her mother Sandra, and the mother's significant other, Doris F.; address for the child's biological father is unknown; his name is Andrew Y., and he was last in contact with the victim's mother in 1993
- **Biographical Data of the Alleged Perpetrator**: Doris F., same address as victim, age 27 (DOB: 3 January 1970); biological mother of Tom S. and Tammy M., currently in foster care waiting for adoption (DFYS custody)
- **Current Location of the Alleged Perpetrator**: at home (same address as above)
- **Current Location and Address of the Person Making the Report**: mother of the alleged victim and her therapist, Chris B., both currently at the Psychological Services Center, (997) 123-1234
- **Ease of Access of the Perpetrator to the Victim**: very easy because they live in the same household
- **Nature, Severity, and Chronicity of the Situation/Incident**: not clear; alleged perpetrator has lived with the child and her mother for just over two years; evidence of bruises on the child is reported to have been present sporadically for at least two years; recently the child has become withdrawn from her mother in combination with other significant recent (last 4–6 months) behavioral changes and characteristic emotional problems
- **Current Status of the Victim**: generally healthy, but mother reports that the child had a new bruise on her left thigh which she only noticed this morning and that was both large and painful; the daughter refused to tell her mother how she obtained the bruise; significant behavioral and emotional disruptions have been noted by the mother; the child is also significantly socially isolated and withdrawn, and her grades have been deteriorating over the last academic year; the mother disclosed that she has been called three times by the daughter's teacher in the last two weeks but has not yet made an appointment to meet with him
- **Dates of Recent Incidents**: suspected incident last night (that may have caused the bruise)
- **Names of Witnesses**: none known
- **Immediate Safety Issues Concerning the Victim**: victim and alleged perpetrator live in the same home; victim's mother is nonassertive and is a victim of the same perpetrator; victim's mother has failed to confront the possibility of abuse of her daughter by her significant other despite having harbored suspicion herself; alleged perpetrator has had two children removed from her custody in the past for severe physical abuse and emotional and physical neglect

The phone call to DFYS was made at 10:45 A.M. by the therapist who introduced herself and asked for the name of the DFYS intake worker (Jamie N.). The phone was then handed to the client who made the report of her suspicions. The client then returned the phone to the therapist who finalized the report, assuring that all of the above information had been conveyed to the DFYS worker.

Sample "Duty to Report" Progress Note (continued)

The remainder of the session was utilized to process the report with the client and to deal with her feelings vis à vis the therapist and her significant other. It appeared that the therapeutic relationship remained intact and that the client's primary concern was over her significant other's reaction. The client voiced little concern over her daughter in this regard, but expressed more concern over the possibility of the significant other ending the relationship with the client. The therapist also assessed the client's perception of the likelihood of retaliatory violence by the significant other. The client was unsure about this possibility but was not considered to be an accurate predictor by the clinician. Thus, she was supplied with information about the local shelter for abused women and was encouraged to seek its services should the significant other react aggressively. It was also impressed upon the client that it was her responsibility to protect her daughter. The possibility of treatment for the daughter was broached with the client, but will have to be taken up again at a later date, because the client currently was nonresponsive to this suggestion. The client was scheduled for another appointment with the clinician. This appointment was scheduled for January 28, 1997 at 10 A.M.

Plan

1. Follow-up regarding events that may have occurred in the aftermath of the report.
2. Complete intake.
3. Discuss possibility of treatment for the daughter.
4. Explore the relationship between significant other and client in more detail, with focus on violence between the two women as well as violence toward the child in the family.
5. Discuss the possibility of ancillary treatment services, such as support groups.
6. Make final treatment plan.

PART III

Transcending the
Challenges

9

The Challenge of Preventing Burnout and Assuring Growth: Self-Care

He who knows men is clever;
He who knows himself has insight.
He who conquers men has force;
He who conquers himself is truly strong.

He who knows when he has got enough is rich,
And he who adheres assiduously to the path of Tao is a man of steady purpose.
He who stays where he has found his true home endures long,
And he who dies but perishes not enjoys real longevity.

LAO TZU, *TAO TEH CHING*

It has been my pleasure to write this book, and I hope that sharing my thoughts about challenges in counseling and psychotherapy has encouraged at least some readers to try to embrace the challenge of being a mental health care provider for particularly troubled clients. I hope all readers have learned something new and have been challenged in their thinking. In concluding this book, I would like to draw attention to the fact that no clinician working with difficult clients can remain healthy without a strong repertoire of professional and personal self-care skills. The final chapter of this book will be much more personal than its first eight chapters have been. This is due to the fact that, at least for me, self-care is inherently a very personal thing and is difficult to write about in an impersonal manner. Self-care reflects personal values and beliefs; thus, it is difficult to objectify and quantify. It is impossible to do the type of work that is described in this book if one does not recharge and take care of oneself. As Walker and Matthews (1997) point out, when therapists and counselors begin practice, "they enter an arena that is at once highly exciting, stimulating, and rewarding, but also intense, strenuous, and emotionally draining. Physical and emotional burnout are real dangers" (p. 11).

Burnout comes in two forms: Fatigue and exhaustion from chronic overwork; or the loss of a life dream that permeates all aspects of personal and professional life, including body, mind, and spirit. Some writers have hypothesized that therapists and counselors are particularly vulnerable to burnout because of a number of personality traits that have guided them into their chosen profession—traits that tend to be highly correlated with burnout. Specifically, Glickauf-Hughes and Mehlman (1995) express the belief that common traits among mental health care providers are parentification (willingness to take responsibility for others' feelings and action), perfectionism (excessively high expectations about one's own competence and performance), imposter feelings (self-doubt and the perception that others think too highly of the person), and audience sensitivity (heightened sensitivity to others' feelings and reactions). All of these traits render the individual who has them vulnerable to emotional depletion and self-doubt if the feedback from the environment is negative or excessively challenging. Other factors that relate to burnout among mental health care providers include client behaviors (e.g., client suicide, client verbal aggression against therapists), working conditions (e.g., paperwork, excessive caseloads), psychic isolation (i.e., too much listening and not enough emotional disclosure), therapeutic relationships (i.e., self doubt about efficacy in helping a client), and personal disruptions (e.g., upsetting life events [Bayne, 1997]).

Burnout of either form (exhaustion or the loss of a life dream), but especially its more profound existential manifestation, often has an insidious onset, with symptoms that develop so gradually that they may escape the attention of the person experiencing them. Clues about potential burnout include feeling *less* (Jevne & Williams, 1998): less enthusiastic, less idealistic, less valued, less able, less connected, less involved, less energetic, and less creative. The professional begins to experience disillusionment, detachment (including loneliness, isolation, and withdrawal), dread, depression, worry, despair, hopelessness, and a variety of possible physical symptoms, such as poor or decreased sleep, lowered immunity, increased aches and pains, digestive problems, or sexual dysfunction (Jevne & Williams, 1998). Early warning signs about possible impairment include changes in thoughts and emotions (such as difficulty concentrating, boredom, loneliness), changes in physical well-being (such as tics, tight throat, aches), and changes in behavior (such as irritability, accidents, eating too much or too little; Bayne, 1997). If they are ignored, these experiences and affects quickly translate into distress and impairment leading to problems with work and clients, as well as life and intimate others.

Impairment is best defined as "the interference in ability to practice therapy, which may be sparked by a variety of factors and results in a decline in therapeutic effectiveness;" distress is the "subjective experience of discontent" (Sherman & Thelen, 1998, p. 79). Impairment comes in many forms; the most common symptoms have been grouped into three primary categories: substance use, mental health problems, and sexual misconduct. Clearly, the consequences of impairment are potentially severe and measures need to be taken by practitioners not to imperil their clients through their own reaction to the stresses of life and work. Unfortunately, this does not always happen. In fact, depending on the survey, somewhere between 30–60% of psychological providers have reported seeing clients when they were really too impaired due to stressful life events (Sherman, 1996). It is not surprising that "malpractice suits are more common when the psychologist is experiencing stress" (Sherman, 1996, p. 300), and that there is a high positive correlation between the number of stressful life and work events and the level of distress and impairment of the care provider (Sherman & Thelen, 1998). The primary personal life factors identified as related to distress and impairment by Sherman and Thelen (1998) are problems in close personal relationships and major personal illness or injury; primary work events are malpractice claims by clients, changed work situations (e.g., managed care fall-out), and inadequate time to meet all obligations. Common problems that have been identified as related to distress have been reported among mental health care providers to a surprisingly large degree.

Specifically, 30–46% of providers have indicated that they have experienced irritability and emotional exhaustion, concerns about case loads, insufficient or poor sleep, self-doubt about therapeutic effectiveness, problems in intimate relationships, chronic fatigue, loneliness and isolation, anxiety, or depression (Mahoney, 1997).

Recognizing that many therapists and counselors practice while impaired, 22 U.S. states and Canadian provinces have developed programs for impaired mental health care providers (Coster & Schwebel, 1997). These programs, however, are geared toward intervention, not prevention. Focus may better be placed on making suggestions about self-care behaviors that can be incorporated into a career in mental health practice early on so as to prevent the development of burnout, distress, and impairment and to maintain well-functioning. To support well-functioning, therapists have been encouraged to seek interpersonal support from peers, friends, and family members. They have also been encouraged to take part in intrapersonal activities, such as self-awareness and self-monitoring, as well as self-regulation of the distribution and balance of various life activities (e.g., work versus play); professional and civic activities, such as professional activism and civic activism, that reflect interest in community and society; self-care, including attention to personal well-being and professional development; and the recognition of impending impairment along with the ability to cope with such recognition and the prevention of such impairment (Coster & Schwebel, 1997). Sherman (1996) identifies time management classes, knowledge about warning signs of burnout and impairment, coping skills training regarding work stress, relaxation techniques, personal therapy, periodic supervision and consultation, professional networking, and balanced caseloads as variables that are helpful in reducing distress and impairment. Similarly, Mahoney (1997) reports on the usefulness of hobbies, reading for pleasure, vacations, attendance of movies or artistic events, physical exercise, peer supervision, recreational games, meditation or prayer, therapeutic massage, keeping a diary, and personal therapy. Walker and Matthews (1997) suggest the need for perspective and stability in life through relaxation, recreation, relationships, and other meaningful activities and challenges beyond work. These findings and recommendations suggest that at a minimum a two-pronged approach needs to be taken to prevent professional burnout and distress: appropriate management of professional life and maintenance of balance in personal life. Effective and important variables in these two categories of prevention behaviors will be addressed in some detail in the following pages and are summarized in Table 9-1.

PROFESSIONAL SELF-CARE

When Coster and Schwebel (1997) queried a group of well-functioning mental health care providers about factors they thought contributed to their lack of impairment symptoms, one answer that arose (among many, of course) was the cost of impairment. Practitioners were deterred from behaviors indicating impairment (i.e., symptoms such as substance use or sexual misconduct) because they feared their consequences (e.g., the loss of licensure). Although this may indeed be an important factor in deterring a clinician from engaging in inappropriate behavior in the moment, it certainly appears that, if this is the only reason care providers refrain from a dangerous action (to self or other), they are already impaired. A number of professional self-care skills are available that serve to keep the counselor or therapist from reaching the point at which the decision about professional behavior is ruled by external sources, such as licensing boards, ethics committees, or credentialing agencies.

Continuing Education

An excellent way to maintain skills at the highest and most professional level is by seeking out opportunities for continuing education (CE). The field of mental health care appears to be ever

TABLE 9-1   Self-Care Behaviors to Maintain Well-Functioning

CATEGORY	RECOMMENDED BEHAVIORS
PROFESSIONAL SELF-CARE SKILLS	
Continuing Education	■ conferences ■ workshops ■ seminars ■ reading books and journals ■ participation in didactic experiences (e.g., classes, lectures) ■ provision of didactic activities
Consultation and Supervision	■ individual supervision ■ group supervision ■ individual consultation ■ group consultation ■ provision of supervision and consultation
Networking	■ association memberships ■ social meetings with colleagues ■ professional meetings with colleagues ■ identification of a mentor ■ service as a mentor
Stress Management Strategies	■ debriefing sessions with colleagues after crises ■ acquisition of time management skills ■ referral ■ professional activity other than client work ■ diversification of clientele ■ scheduling of regular intervals of free time
PERSONAL SELF-CARE SKILLS	
Healthy Personal Habits	■ healthful diet and nutrition ■ healthful physical activity (e.g., exercise, yoga, tai chi) ■ physical self-awareness ■ restful sleep ■ time outdoors, time with nature
Attention to Relationships	■ empathy, tolerance, acceptance, and respectfulness ■ laughter at oneself ■ defenselessness and detachment from one's personal point of view ■ friendships ■ intimate relationships ■ family relationships
Recreational Activities	■ hobbies and interests (e.g., outdoor activities, gardening, painting, writing poetry) ■ travel ■ group memberships (e.g., Sierra Club, religious/spiritual group) ■ volunteer work ■ entertainment (e.g., artistic events, cinema, dinners out)

Continued

TABLE 9-1   Self-Care Behaviors to Maintain Well-Functioning (continued)

CATEGORY	RECOMMENDED BEHAVIORS
PERSONAL SELF-CARE SKILLS (continued)	
Relaxation and Centeredness	■ breathing exercises ■ mindfulness ■ relaxation strategies ■ guided imagery
Self-Exploration and Awareness	■ meditation ■ personal therapy ■ journaling ■ dream work ■ reading ■ self-acceptance

evolving. New research findings are reported every day, and someone somewhere always finds a new and creative solution to old problems. Keeping up-to-date about the most recent and most scientific interventions in the health care field ensures that the client receives optimal care. However, it also keeps the practitioner challenged and enthusiastic by preventing routines and by making sure that clinicians do not get into ruts. New learning obtained through continuing education can be exciting, eye opening, entertaining, and challenging. It is a great way to retain enthusiasm and to discover newness in the same old job. Broadening one's horizons by seeking out new information is extremely satisfying because it dispenses with feelings of stagnation and boredom. The sources of continuing education are almost limitless. The easiest and quickest, but not necessarily most exciting, way to learn is to read professional journals and books. Subscribing to the most relevant journals for one's practice is an excellent means of keeping up-to-date with the field. For example, some of my favorite journals that I read regularly are *Professional Psychology: Research and Practice, Clinical Psychology Review, Journal of Personality Assessment, Psychotherapy*, and the *American Journal of Psychotherapy*. There are many others I read occasionally, but these are my favorites because they are more applied, pragmatic, and realistic than purer research journals (e.g., the highly respected *Journal of Consulting and Clinical Psychology* is not one of my favorites because it is often esoteric and the studies reported tend to have limited external validity for the sake of enhancing their internal validity). Keeping track of new books on the market in one's field of specialization is also an excellent strategy that can be fun as well. I love books and places like the Psychotherapy Book Service or Behavioral Science Book Club are important sources for me. I also regularly visit the exhibits at conventions to look at new books that have been published and do my purchasing then. At some conventions, the book publishers will sell the texts they brought to the convention on the last day at reduced prices so they do not have to ship them back to their publishing house. This is a great opportunity to stock up on terrific books. Another great way to get exposure to new books (and sometimes to get them for free) is to volunteer to review professional books for various journals or newsletters. For example, the American Counseling Association (ACA) regularly reviews books in its newsletter "Counseling Today" and usually the book editor is eager to find professionals to review these books.

Although reading is one of my favorite things to do, I do not always just want to read professional books. Sometimes I prefer to get my continuing education in a more active manner, and that is when I attend conferences and workshops or seminars. There are many ways to find out about CE activities of this sort. The newsletters of most of the large professional associations (e.g., "Counseling Today," "NASW News," "Family Therapy News," "APA Monitor") have monthly listings of CE activities across the world. Most state associations (whether psychological, social work, or counseling) have newsletters that are mailed to members that list all CE activities in the respective state and sometimes across the United States. Online services are also available that offer comprehensive listings of workshops and seminars across the world in the upcoming months. One such source is PsychScapes Worldwide at http://www.mental-health.com. With regard to conventions, all professional associations have at least annual meetings that are designed to share research and practice findings among its members. These can be exciting events that not only deliver CE opportunities but also serve to establish a network of colleagues. Personally, I prefer conventions for networking, and workshops or seminars for CE purposes. This preference is an outgrowth of my research background, which makes me too skeptical about findings disseminated at conventions where scientific review of submissions tends not to be all too rigorous.

One final way of creating CE opportunities is up to the practitioner. The creation of didactic study groups or consultation groups can be great fun (and a means of maintaining professional competence). Attending advanced classes on specific topics at a local university can expand horizons and give exposure to totally new research and practices. Teaching courses and giving lectures is similarly enhancing of one's personal education. No one (I hope) gives a lecture or teaches a class without first running to the library to get completely up-to-date on a topic. Thus, making the commitment to speak about a topic to a group of professionals or students is an excellent way to motivate oneself to get newly immersed in an old topic and to be inspired by what others have to say about the topic. It is my experience (and my opinion) that one never learns anything better than by teaching one's peers. The idea of embarrassing oneself in front of a group of professionals by not being knowledgeable about the topic is greatly motivating for optimal preparation and detail work. Clients will ultimately benefit from the care that has gone into preparing a lecture on a certain clinical topic because the care provider has thoroughly reviewed the literature and certainly is now aware of most anything of relevance to the issue. The fun of learning this way is also not to be underestimated. For me, there is a certain pleasure in collecting, organizing, and expanding a body of literature and then sharing it with others. If that is not quite as pleasurable for the reader, then taking a lecture (rather than giving it) can be almost as much fun and stimulating (provided the speaker is good, of course).

In availing oneself of CE opportunities such as lectures, courses, seminars, and workshops, one needs to be discriminating. Not everything that is out there is of equal quality. I have been to some pretty poor workshops and have chosen to walk out of them! One way to judge the quality of a CE event is to look at who is sponsoring the event. For example, the American Psychological Association (APA) approves CE sponsors and such approval is contingent upon assuring the quality of the event by reviewing presenter credentials and doing outcome evaluation. Although this does not guarantee that a speaker will be inspiring and entertaining, it does guarantee that the person has some basic background knowledge and credentials. Another way to attempt to anticipate the quality of an event is to go to a literature search database and look up the presenter (e.g., Social Work Abstracts, PsycLit). If the person has never published anything, I am skeptical. If presenters have published, I go read some of their work and decide based on that whether the workshop is something I am willing to risk my time on. Word of mouth is helpful too. Asking colleagues about workshops they have attended and enjoyed can be a useful way of making choices among the many options out there. Once good thing about conventions is that one can listen to mini-lectures by many professionals and then can pick the most enjoyable speakers as ones whose workshops or seminars to attend in the future.

Discriminating consumerism is also important in terms of journal and book reading. Some basic knowledge about how to read research reports is very helpful in deciding which strategies have merit and which are based on flawed research. This knowledge is imparted in most mental health related graduate programs. If a research methodology class was not part of a clinician's graduate degree, it may be helpful to seek out such a course to become a better consumer of journal articles in one's field.

Consultation and Supervision

Maintaining competence is a critical aspect of good mental health care practice. One of the best ways to achieve this goal is through ongoing supervision and consultation (Wheeler, 1997). These measures ascertain that the clinician maintains a certain standard of practice and becomes accountable to another professional with regard to the work that is done with the client. Nothing affords clients better protection than the regular seeking of supervision by their provider. Supervision has moved from being a luxury to being a requirement in a profession in which keeping up-to-date with new developments is critical to optimal client care. Supervision and consultation have multiple purposes, of which consumer protection is of course the most important and obvious one. However, supervision also is an excellent gatekeeper and monitor of the entire mental health care profession and creates a feedback loop to individual providers that results in better quality service and enhanced self-development, both personally and professionally (Carroll, 1996). I believe that by far the most helpful form of supervision is delivered in a one-on-one setting. In individual supervision, more than in any other form of consultation, the focus is clearly on the clinician who is in the role

of supervisee, in the same way that therapy is focused on the client. Such an intense focus is critical to professional self-development and maintenance of professional competence.

I believe that supervision needs of clinicians parallel the developmental needs of human beings in life in general. According to self-psychology, all human beings have mirroring needs (which I have also called nurturance and feedback needs), idealization needs (which I have also called guidance and coping needs), and twinship needs (which I have also called companionship and modeling needs [Brems, Baldwin, & Baxter, 1993]). These developmental needs tend to emerge in the order of mirroring needs first, idealization needs second, and twinship needs third. All needs remain active throughout the life span although healthy human beings begin to become able to meet all of these needs independently over the course of their early developmental years. The entry of graduate school or the beginning of a new career in the mental health field is generally very challenging, even for the psychologically healthy individual. As previously indicated, many signs of fragmentation and distress have been identified in psychology trainees and among care providers that are directly linked to these experiences of challenge. Further, when a trainee enters a graduate program or when a new graduate enters the professional workforce and begins to conduct psychotherapy or counseling independently, this is often done before the clinician has had sufficient time and experience to establish a professional self. This can mean that this professional is faced with a challenge similar to that faced by the infant: adjusting to a world that is new and exciting, as well as new and frightening. Supervision can greatly facilitate this process and can prevent impairment and distress.

Feelings of being overwhelmed, inadequate, unable to cope, and alienated from others often ensue as a new professional starts out or as an experienced professional meets excessive challenges. These feelings can be viewed as the fragmentation products of an individual struggling to develop or maintain a professional sense of self in the face of adversity. This process parallels the developmental process all human beings go through when they first develop a sense of self, and the environment must function as the same interpersonal matrix as it does for the infant. Supervision can become the interpersonal matrix that keeps the struggling professional on a healthy trajectory. In other words, just as a clinician is subjected to becoming a selfobject for a client, a supervisor becomes a selfobject for the supervisee. This process in therapy or counseling is always one reflective of transference developed in the attempt to heal or develop a self. The intensity of this relationship is best developed in an individual setting.

The supervisor becomes a selfobject for the supervisee and will need to respond to the trainee from that framework. If the interpersonal process is one primarily of expressed mirroring needs, the supervisor will emphasize mirroring techniques in the session; if it is one of expressed idealization needs, the supervisor will rely on idealization techniques; if it is one of expressed twinship needs, the supervisor will focus on twinship techniques. For all trainees, as for all infants, it appears that the first needs that emerge tend to relate to the establishment of a mirroring pole, second to the establishment of an idealizing pole, and lastly, to the establishment of a twinship function of skills and interests. However, all supervisees will need to be exposed to all types of techniques at some point in the supervision process. Choice of a specific technique therefore needs to be tailored to the expressed need of the supervisee.

Types of transferences and needs expressed by the supervisee toward the supervisor parallel developmental and/or therapeutic processes. The supervisee, in the struggle to develop a healthy professional nuclear self, will express normal developmentally appropriate needs for mirroring, idealization, and twinship. These normal, healthy expressions of needs must be recognized and addressed by the supervisor through acknowledgement, understanding, and occasionally through gratification. They are viewed in the context of

the supervisee's struggle to develop a healthy nuclear professional self; that is, they are viewed as developmentally appropriate and necessary (not always as transferential).

Further, because of possible weaknesses in the supervisee's personal (as opposed to professional) nuclear self, transference needs may arise in the relationship with the supervisor as well. Transferences, emanating from an unhealthy (or less well-developed) aspect of the supervisee's personal self, may occur in the areas of mirroring, idealization, and twinship, depending upon which structure of the personal self is most compromised or unhealthy. These needs, recognized by the supervisor as transferential, are acknowledged, understood, and explained but not gratified. They are discussed in the context of how they enter the therapy or counseling process with the client and how they influence treatment; that is, they are addressed from a countertransference perspective (supervisee toward client). Although the supervisor, in my opinion, never takes on the role of clinician with the supervisee, some exploration of the history of the transference needs may be appropriate, as may be a referral for psychotherapy or counseling for the supervisee. The supervisor will find that transferences in the supervisory relationship may be equally irrational, archaic, and taxing as transferences that develop in a clinician–client relationship. Countertransference reactions (supervisor to supervisee) are to be avoided, but they should be explored and occasionally may require consultation with a colleague.

Depending upon the needs and transferences that emerge in the supervisory relationship, the supervisor has a range of techniques that may be used to deal with the supervisee most effectively. All supervisory strategies have the purpose or goal of helping the supervisee internalize a healthy nuclear professional self. These goals and purposes are summarized in Table 9-2. Important tools that are used in the supervision process to achieve these goals and purposes are the same strategies that are used in psychotherapy or counseling with clients, namely internalization

TABLE 9-2   Purpose and Goals of Supervision Techniques as Adapted to Supervisee Needs

NEED AND STRATEGY TYPE	GOAL OR PURPOSE
Mirroring	■ develop or maintain self-esteem and self respect as a therapist or counselor
	■ gain a realistic self-appraisal of the supervisee's capacity to function as a therapist or counselor
	■ understand and accept shortcomings as a therapist or counselor
	■ develop or maintain realistic and self-assertive ambitions with regard to the supervisee's career in the mental health field
Idealization	■ develop or maintain a set of guidelines and values that provides a structure for the therapy or counseling process
	■ develop or maintain self-confidence and strength about the ability to conduct treatment in a manner that is safe for the client
	■ develop or maintain a sense of calmness and strength that sustains the supervisee during difficult interactions with clients
	■ develop or maintain a strong observing ego that can serve a source of strength and calmness during difficult session
Twinship	■ discover talents and skills that aid in the conduct of treatment
	■ develop or maintain a large set of skills and techniques that are varied and adaptable to clients' needs
	■ develop or maintain a feeling of affinity with other therapists or counselors
	■ develop or maintain a sense of belonging to the chosen professions, e.g., through joining a professional organization

(creating external structure, facilitating transmuting internalization, and managing projective identification), empathy, understanding, and explaining (see Brems, 1999a). Specific supervisory strategies that are chosen depend on the goal that is to be achieved and on the needs of the supervisee. Recommended strategies are summarized according to their function and purpose (i.e., mirroring, idealization, twinship, and general) in Table 9-3.

Although any and all of these supervision experiences can be created and used in group settings, it is less likely that this will happen. Group supervision has its role and purpose but it is different from individual supervision in its intensity and individual focus. Group supervision is preferable for the more experienced clinician who is not facing any great challenges that are creating distress. The truly distressed or impaired provider is better off in individual supervision, where more individualized attention can be paid to very specific needs.

Consultation is another excellent way of maintaining professional competence and is often cheaper than supervision. It is less intense and thus less appropriate in cases of existing distress or impairment. Consultation focuses less on the individual needs of the clinician and more on case management of the clients. Consultation is about the clinician's clients, not about the clinician. Supervision, on the other hand, is primarily about the clinician and only relatedly about the clients. Although a supervisor clearly reviews client cases, his or her focus is on what the cases say about the supervisee. In consultation, the focus is on the client and the case management of each client. The clinician seeking consultation is under less scrutiny and the whole process is less intense. Consultation is an excellent means of keeping focused and oriented in one's clinical work when there is little concern and challenge. Group consultation can add a component of collegiality and joint learning that can be stimulating and exciting. Hiring a leader for a consultation group also

TABLE 9-3   Supervision Strategies by Function or Purpose

MIRRORING TECHNIQUES

1. Teaching jargon to improve the supervisee's self expression as a clinician

2. Providing labels for interventions conducted by the supervisee when the supervisee is not aware of or knowledgeable about these labels

3. Providing assistance to the supervisee in the endeavor to identify a theoretical/conceptual framework for treatment by aiding the supervisee in the exploration, application, and practical understanding of various available theories of psychotherapy or counseling

4. Positive reinforcement of therapeutic skills exhibited by the supervisee

5. Differential reinforcements of existing skills that are incompatible with nontherapeutic habits or interventions expressed or notions held by the supervisee

6. Extinction of nontherapeutic interventions that are fairly minor or inconsequential, and therefore need not be strongly criticized or redirected

7. Regular or routine discussion of and commenting upon the supervisee's positive versus negative skills, habits, and interventions

8. Regular written feedback on a predeveloped form in a manner that can be anticipated by the supervisee

9. Discussion of the supervisee's realistic limitations, and discussion or identification methods for gaining acceptance of these limitations and of working effectively as a clinician despite them and around them

10. Use of videotapes for Self-as-a-Model learning, both in and outside of formal supervision sessions

11. Conduct conjoint problem-solving whenever possible in an attempt to instill confidence in the supervisee about ability to arrive at conceptualizations and solutions independently

12. Offering of understanding for and acceptance of the supervisee's affects and needs arising around the process of becoming a clinician and conducting therapy or counseling, more so, than explaining them

13. Use of active listening to explore the supervisee's perceptions and feelings about her of his ability to conduct treatment, and to help her or him gain acceptance of her or his difficulties

14. Tracking of a supervisee's affects, needs, and behavior to ascertain awareness on the part of the supervisor of fragmentation symptoms on the part of the supervisee

15. Exploration and awareness of mirroring needs in the supervisee, as well as exploration and use of mirroring transferences that develop with the supervisor

16. Discussion of the supervisee's mirroring needs with regard to their impact on treatment

17. Discussion and interpretation of the supervisee's mirroring needs regarding their history and background within ethical confines that avoid dual relationships

18. Recommendation of psychotherapy or counseling for the supervisee if fragmentation is severe and the addressing of mirroring needs would move beyond ethically acceptable limits of the supervisory relationship

IDEALIZING TECHNIQUES

1. Setting rules and boundaries for the supervision process similar to those set for therapy or counseling:
   a. 50-minute hour
   b. no tardiness or unannounced absences
   c. setting of a fee and payment plan (where appropriate)

2. Providing structure for the supervisory process similar to that set for therapy or counseling:
   a. explanation of expectations in the first meeting
   b. definition of supervision and the roles of the supervisor and supervisee
   c. outline of the supervisor's theoretical framework and approach to psychotherapy or counseling
   d. discussion of limits of the supervisory relationship
   e. outline of the required paperwork and its deadlines

Continued

TABLE 9-3 Supervision Strategies by Function or Purpose (continued)

IDEALIZING TECHNIQUES (continued)

3. Use of modeling by the supervisor as an instructional technique

4. Availability of the supervisor for guidance, such as through questions and answers and through the teaching of strategies

5. Providing support to the supervisee with regard to identifying a personal style as exemplified by a theorist or clinician respected or admired by the supervisee

6. Use of videotapes of the therapy or counseling process by either well-known theorists or by the supervisor to facilitate learning through modeling for the supervisee

7. Providing support and guidance for the supervisee when she or he encounters difficulties or voices doubts about certain interventions or the ability to conduct treatment

8. Willingness of the supervisor to serve as a role model as appropriate to the supervisory process and as needed by the supervisee

9. Ascertaining of a focus and direction for supervision, perhaps through setting of clear goals and objectives for the process

10. Encouragement of treatment planning and ascertaining that the supervisee has a sense of direction and focus for the treatment with each client

11. Offering of explanations for, as well as acceptance of, a supervisee's affects and needs arising around the process of becoming a clinician and conducting therapy or counseling, more so than merely understanding them

12. Exploration of the supervisee's need for direction and guidance, as well as encouragement of the development of values and ideals that will help the supervisee guide her or his behaviors and interventions as a clinician

13. Tracking of the supervisee's needs and behaviors to ascertain awareness on the part of the supervisor of fragmentation symptoms on the part of the supervisee

14. Exploration and awareness of idealization needs in the supervisee, as well as exploration and use of idealizing transferences that develop with the supervisor

15. Discussion of the supervisee's idealizing needs with regard to their impact on treatment

16. Discussion and interpretation of the supervisee's idealizing needs regarding their history and background within ethical confines that avoid dual relationship

17. Recommendation of psychotherapy or counseling for the supervisee if fragmentation is severe and the addressing of idealizing needs would move beyond ethically acceptable limits of the supervisory relationship

TWINSHIP TECHNIQUES

1. Use of peer supervision and peer consultation

2. Use of peer observation, either through live observation of sessions conducted by peers or through the joint viewing of videotaped sessions by one of the two peers

3. Staff meetings that require case presentations and viewing of videotaped sessions

4. Use of modeling as a primary instructional strategy to enhance and facilitate skill development and number of therapy or counseling techniques and interventions that the supervisee is familiar with

5. Normalization of difficulties that emerge through discussion of normal responses of novice clinicians

6. Encouragement of peer interaction around treatment issues and personal response themes, perhaps even in the form of a formal peer support group or therapy or counseling group

7. Exploration of the supervisee's need for feeling part of a group of individuals with whom she or he can identify

8. Exploration and awareness of twinship needs in the supervisee, as well as exploration and use of twinship transferences that develop with peers or the supervisor

9. Discussion of the supervisee's twinship needs with regard to their impact on treatment

10. Discussion and interpretation of the supervisee's twinship needs regarding their history and background within ethical confines that avoid dual relationships

11. Movement toward a twinship relationship between supervisor and supervisee as supervision nears its end

TABLE 9-3   Supervision Strategies by Function or Purpose (continued)

GENERAL TECHNIQUES

1. Awareness of transferences established by the supervisee toward the supervisor

2. Awareness of projective identification cycles established between supervisee and supervisor, perhaps as a parallel occurrence of the same phenomenon between the supervisee and her or his client

3. Awareness of countertransference reactions on the part of the supervisor toward the supervisee

4. Awareness of projective counteridentifications on the part of the supervisor that inhibit the supervisory process

5. Thorough familiarity with and knowledge of the supervisee to understand her or his responses to treatment and to be able to empathize

6. Awareness of the issue of dual relationships and avoidance thereof, yet without sacrificing a positive and caring relationship with the supervisee

7. Openness to feedback by the supervisee and willingness to explore the supervisor's own shortcomings

8. Continuing education on the part of the supervisor with regard to the therapy or counseling and the supervisory process, both in terms of keeping up-to-date on research and in terms of theoretical advancements

9. Provision of consistency in all supervisory strategies by adhering to those rules, guidelines, and structures that have been imposed on the supervisee

10. Ethical conduct and uncompromised professionalism in the relationship with the supervisee and the supervisee's clients

---

greatly reduces expense, while still yielding excellent results in terms of new learning and exposure to new material.

One final means of maintaining professional skills and competence is the provision of supervision and consultation. The reasons why the provision of these services can be just as professionally enhancing to the provider as the receiver are the same ones outlined above for giving lectures and workshops. Personal skills have to be honed to consult with others and to serve as the expert. Being a consultant means having expert knowledge in at least one, but often more than one, area. To achieve and maintain such expert status means keeping up with developments in the field, reading and learning constantly. In other words, being a consultant keeps a professional alert and up-to-date.

Networking

Seeking out continuing education activities and supervision or consultation opportunities is an excellent means of meeting yet another professional self-care goal, namely, that of networking with other professionals to reduce isolation and stagnation. Alexander (1997) warns that clinicians need to beware not to be sucked into the mere tediousness of dealing with managed care companies or particularly difficult clients, suggesting that care providers instead need to transcend the impersonal nature of private practice and work toward connectedness with colleagues. Making time for "stimulating and enjoyable encounters with . . . colleagues" is considered essential to maintaining well-functioning (Alexander, 1997, p. 25). One way to do this is through active attempts at networking. This requires that each practitioner take responsibility for reaching out; no one can wait for invitations from others. There are infinite ways of doing this, only limited by each individual clinician's creativity. Networking sessions can be formal or informal, personal or professional, in a work setting or a social setting, individual or group based, and on and on. A few ideas are presented here that are fairly standard ways of networking. These ideas are not particularly creative, though they still can result in enjoyable and entertaining interactions. I do urge each clinician, however, to try to come up with additional networking opportunities that are more creative and stimulating. Sometimes it can be great fun and even professionally enhancing to go on a two-week kayaking trip with colleagues (I have tried this) or to go bowling together (I have not tried that). The possibilities are endless.

Becoming a member of a professional association is an excellent way of introducing oneself into a professional network that is guaranteed to address some of the concerns the individual clinician has felt. Actually attending the meetings of this professional association is even better than just being a passive group member. Too many professionals join associations and pay their dues but never get involved. Having served in numerous capacities for a state psychological association (including as committee member for CE, newsletter editor, president, executive director, etc.), I know that it is much more satisfying and professionally stimulating to be actively involved than to just receive newsletters and visit an occasional convention. I also know that in every professional organization there is always a need for more members to become actively involved. Serving on committees is a great way to network and is usually easily achieved with a simple phone call to the executive office of the association (except for the large national professional associations like APA, NASW, and ACA, where membership on committees is a highly selective process). Aside from being great networking settings, professional associations often provide a wonderful outlet for volunteer and political activities. Social work associations, of course, have a long history of public service agendas and involvement. However, even psychological and counseling associations these days have political and public service agendas that allow interested members to be involved with political leaders in their communities. Thus, they not only gain the collegial relationships that come from being a member but also feel as though they can truly make a difference.

Another great way to network is to participate in social meetings with colleagues. These meetings can come in many varieties. In our state association, we tried once-monthly Friday luncheons that were great fun and often educational because we invited speakers to deliver a brief lecture on a topic of interest. Having visited New Zealand and witnessed the local practice of collegial teatime, Kottler (1997) recommends that we "spend time with friends and colleagues. Have a cup of tea. Reach out to your colleagues for the recognition and support you so richly deserve" (p. 167). Again, only individual creativity limits what these social meetings involve. Professional meetings with colleagues are another avenue of networking that can be stimulating and fun. These overlap greatly with supervision and consultation groups and need no further elaboration.

A final networking strategy involves mentorship: both on the receiving and giving end. Some of the most enjoyable interactions in my career have come from serving as a mentor to young women striving to "make it" in psychology. It is an incredible pleasure to watch a "mentee" succeed and grow. Being a mentee can be a very calming and secure experience because it affords the individual an opportunity to seek counsel in a friendly, nonsupervisory atmosphere. It also provides a target to strive for and a role model to emulate. The identification of a mentor can be a difficult process that requires some initiative and motivation. However, the rewards are great, and reaching out and looking for mentors has been recommended as a prime strategy for preventing isolation and for enriching a counseling or psychotherapy practice and respective skills (Brinson, 1997). Being a mentee is not a passive event; becoming a mentee is an experience that has to be actively pursued and created.

Stress Management Strategies

Developing stress management techniques can easily be combined with quests for networking, CE, and consultation. Many workshops are offered for mental health care providers that address the issue of burnout prevention. These range from personal awareness seminars and workshops to lectures on how to manage a private practice more efficiently. The acquisition of time management skills is a central concern in terms of stress management and is also the topic of many lectures and workshops, especially at conventions and conferences. Books have been written on this topic as well and can greatly facilitate the smooth functioning of a clinician's life. Even online services are available to assist people with stress management (e.g., CATSCo's Help-Stress and Help Think programs at http://www.catsco.com). It is no great difficulty for a mental health care professional to identify sources of information about this topic. However, since many readers may not be inclined to run to the library or a workshop right now or to turn on their computers, I will briefly outline a few helpful strategies that can easily be incorporated into any professional's life at no cost and with minimal effort or time investment.

One of the most important stress management strategies I can think of is debriefing with colleagues after crises. The most likely events to induce professional crises and a clinician's self doubt are crises with clients. The experience of a client's suicide or the threat of homicide by a client can leave a clinician shaken and self-doubting for quite some time. Debriefing with a valued colleague (this is where a mentor comes in handy) can be a lifesaver in these circumstances. Even if a clinician is not well networked in a community, the experience of a major client crisis may become the motivator to seek out a network. If a free debriefing session with a colleague is not available (signaling that the provider is isolated!), then a paid supervision or consulting session is the next best thing. Walking through the actions taken with a challenging client with a colleague can be highly reassuring as well as educational. It is important to have some faith in the person who is sought out for the debriefing. Debriefing is not just about being told that the clinician did everything right; usually everyone makes a mistake somewhere along the road. It is about receiving honest, but sensitive feedback about what was right and not so right. Only then does true learning take place. A clinician does not want to debrief with a colleague who has an axe to grind and

who feels better when pointing out flaws in others. This will only lead to greater self-doubt. A clinician who does not know anyone in the community with whom to debrief may consider calling the local state psychological, social work, or counseling association to ask for a few possible referrals for supervision or consultation. That clinician thereafter also needs to evaluate carefully if all the professional self-care skills described above are in place. If they are not, now is the time to seek them out.

Referral is another good stress management strategy. Not every clinician is equally skilled for all clients and circumstances. Trying to work with clients who prove difficult is bound to create distress and self-doubt. It is best to establish a good referral network and to refer clients with certain characteristics to a provider who thrives with these types of clients. The wonderful reality is that all clinicians are different. The type of client one person cannot work with without despairing is the same client who helps another clinician feel positively challenged and enthusiastic about clinical practice. Getting to know a variety of colleagues through networking facilitates the development of a good referral network, which the clinician can use to send clients on their way when they fit into a pattern that is sure to activate the clinician's countertransference or self-doubt. Admitting to being unable to work with certain clients is no shame; it is a healthy signal of well-functioning and professional and personal self-awareness. Referral is of course also important in the context of not letting caseloads get too high. Especially in private practice, it is tempting to take on every client who calls so as not to miss out on the income. This is not a good reason for client acceptance. Sometimes caseloads are high enough and even if the client appears interesting and within one's expertise it may be important to draw a limit and refer. Each clinician will have to decide how many clients are enough and how many clients are too many. It may be true that everyone will have to exceed their limits at least once or twice before knowing what the right number of clients truly is.

Once that number has been identified, however, it is important to stick with it. One final reason to refer may be incompatible schedules. Rather than accommodating every client's time management needs, sometimes clinicians have to refer a client to another provider if schedules would make the clinician's own life unmanageable. Being contacted by a client who needs to be seen on a Saturday does not mean that the clinician now has to work on Saturday. Referral to a clinician who maintains a practice on weekends may be the best solution, even if it means a loss in income.

Another excellent strategy for stress management to maintain well-functioning is diversification of as many sorts as possible. The two most obvious ways in which a clinician can diversify are to seek professional activities other than client work and to diversify the clientele served. Mental health care providers have many other work options besides direct client contact. They can provide public lectures or specialized seminars and workshops for a variety of groups, ranging from lawyers to physicians to nurses to teachers—the list is endless. They can be adjunct instructors at local universities, colleges, or community colleges; they can give workshops to consumers that are didactic rather than therapeutic. They can serve as consultants, supervisors, and program evaluators. With sufficient research skills, they can be researchers or grant writers who plan and implement new programs. Getting out of the office on a regular basis for some of these other activities can be very freeing and rejuvenating. It is also bound to stimulate the clinician to stretch in new directions and learn new things. The main idea here is not to get locked into a certain self-image of oneself as only one type of care provider who never does anything else. Another means of diversification that is also helpful but not quite as rejuvenating as the previous suggestion is to diversify one's clientele. Burnout appears to be the greatest threat among those practitioners who have specialized in an area to the exclusion of ever seeing any other type of client. If this activity cannot be broken

up by engaging in some other professional activities that get the clinician out of the office, then at a minimum the mental health care provider may want to consider loosening up on the criteria for what types of clients to accept. Only working with sexually abused children day in and day out can be very stressful and emotionally draining. Interspersing that caseload with a few acting-out adolescents or adults in simple existential crises can be rejuvenating and stimulating. Planning how to schedule clients of various types can be another important and helpful point. For example, it is best not to schedule the same types of clients back-to-back, all day long. Interrupting the same routine with an unusual type of client in the middle of the day can recharge and de-stress the clinician.

The final and perhaps most important and potentially easiest strategy for stress management is the scheduling of regular intervals of free time in the daily, weekly, monthly, and yearly schedule. This is no exaggeration; often clinicians will do just fine scheduling some free time in one arena, but never attend to making enough time in another. For example, although they may be faithful about taking a lunch hour each day, they may not have taken a vacation in 10 years. It is important to schedule some free time during work-hours every day. Optimally this would be free time above and beyond the lunch hour. Even half an hour extra each day for a quick walk, some stretching exercises, meditation, or other centering activities can be extremely refreshing and reenergizing. At a minimum, mental health care providers need to be very good about keeping to their 50-mintue hours to have at least 10 minutes between sessions to center and take time for themselves. Anyone who has ever uttered the statements "I'm too busy to go to the bathroom" or "I am too busy to eat" is not scheduling enough free time during the workday. Free time during the work week refers to time to run errands without having to rush, meeting a friend or colleague for an extended lunch, taking a long walk, reading a good (nonprofessional) book, scheduling time for hobbies or interests, or just treating oneself to an unusual activity. Keeping two days per week completely clear of client contact appears to be a minimum of time necessary to reenergize and rejuvenate each week. Being able to work even fewer days in direct client contact would be even better.

Keeping time clear every month means taking time out for a special family event, for a brief extended weekend getaway, for throwing a party or other type of get-together at one's own home, for a meditation retreat, and for similar more time-consuming but rejuvenating activities. Planning these opportunities for well-functioning into the monthly calendar is the only way they will receive the attention they deserve. The yearly schedule refers to planning free time for vacations. It is absolutely critical to take time out away from home or at least away from work. Even a vacation at home can be immensely pleasurable as long as work truly is left completely behind. One colleague of mine once expressed the belief, when I left on yet another backpacking trip, that real researchers do not take vacations. If this is true, I do not desire to be a "real researcher." Life is too short not to schedule free time into my work schedule. Although I firmly believe that work is an important part of life, the emphasis for me is on "part." Work is not all of life and life is not all work. Life is so much more than work; I believe a well-functioning professional schedules time into the work schedule for life to take place every day, week, month, and year.

One common comment that I have encountered in response to this recommendation is that people do not have the time to schedule free time. I heartily disagree. Everyone I have ever known, after some honest deliberation, has come to realize that the free time is possible if some nonessential, time-wasting activities are given up in turn. The worst time-wasting culprits seem to be watching TV and nonpurposeful shopping. Everyone has heard the statistics about how many hours the average person spends in front of the television set or, more recently, on the Internet. My advice is to kill your television set, at least every now and then. If you are very addicted, you may miss it at first, but

after a while you will feel quite liberated. It is much quicker to scan the news in a newspaper or to listen to them on the radio as you commute to and from work or other errands. Beyond that, other forms of entertainment appear much more valuable and lend themselves better to combining them with being sociable or having family time. Shopping, the other culprit, is best done purposefully and with a clear shopping list. Grocery shopping can be accomplished this way with a once-a-week trip to purchase perishables and once-a-month trip to purchase staple goods. All other shopping can be reduced to times when an item is truly needed rather than wanted. Shopping just to be doing something appears wasteful on so many levels. Of course every clinician will have to decide what is the most valuable commodity in his or her own life. However, if stress starts to mount, mental health care providers need to take a look at where they can find extra time to take care of themselves. Making free time is really the bridge between professional and personal self-care. Often clinicians take time away from one of these areas (e.g., the professional) to make time for the other (e.g., the personal) and then feel guilty about their choice. Creating free time by inspecting which activities are nonessential may be a better way to create more time for the essential things in life.

PERSONAL SELF-CARE

Swami Chetanananda expressed the belief that "growing is the most important and essential endeavor that a human being can undertake." This aphorism deserves to be prominently displayed on every clinician's wall. It holds true, not just for clients who are seen in therapy or counseling, but also for the clinician who attempts to help them. Taking care to attend to one's own growth is an important and essential part of being a good counselor or therapist. To facilitate this growth and to prevent common personal problems that are experienced by therapists (e.g., exhaustion, loneliness, illness, substance use [see Mahoney, 1997]), several ideas are presented below that will assist clinicians in incorporating positive self-care skills into their daily lives. These ideas must be rendered individually relevant to each practitioner and are best practiced often and regularly to be maximally useful. Healthy practitioners take time for themselves and attend carefully to personal emotional and physical needs.

Over the years I have developed a set of self-care skills and have attempted to instill a set of values in my students, mentees, and supervisees that appears to have had some success at preventing burnout and distress. I will share these self-care practices below in the hope that some of these values will resonate with other mental health care providers and will prove helpful in their lives. I have found that many articles and books on impairment prevention have stressed the need for a balanced personal life. A simple listing of useful skills generally follows this recommendation with little elaboration. The assumption appears to be that, as mental health care providers, we know how to translate such lists of recommendations into behavior. However, it has been my experience that this is not so. In my years as a teacher and supervisor, I have observed appalling self-care skills among clinicians (including in myself, at times). We espouse healthy diets and exercise with our clients but personally leave home without breakfast every day, hardly ever take time to exercise, and get insufficient sleep. Thus, following are very concrete recommendations about how to take care of one's personal life. This disquisition is, of course, much more personal than any other section of this book because it reflects my personal beliefs and practices about what keeps a person healthy. However, I hope that all readers will find some information that will resonate with them and will be worthy of use. It is my hope that no reader will take offense with any of these recommendations; none *has* to be accepted if it does not appear desirable. I certainly take no offense if people do not incorporate these strategies into their lives. Just because these techniques work for me does not mean they will work for everyone else. I do believe, however, that not incorporating *any* self care skills into one's life will ultimately lead to burnout, distress, and poor client care.

Healthy Personal Habits

There is virtually no limit to the number and variety of personal health care habits a person can develop and engage in. I have chosen the most obvious and essential ones for a brief discussion. Nothing is more personal than personal health care habits. There are no definitive answers or recommendations, and advice that has been published elsewhere is often contradictory and confusing. I have attempted to distill this advice down to the components that tend to be fairly universal to all research findings and suggestions I have reviewed. I will address diet and nutrition, physical activity, rest, physical self-awareness, and awareness of nature.

Nutrition. Nothing stimulates defensiveness more easily than talking to people about their food choices and eating habits. Very few people I have encountered, including clients, supervisees, friends, family members, and others, feel completely comfortable with the choices they have made in this regard. Despite the dangers inherent in raising the topic, I keep talking about diet and nutrition because they are an absolutely essential part of personal well-being and well-functioning. Food choices affect physical and emotional health, a connection not many people seem to make (Null, 1995). There are many types of diets people can choose; a simple differentiation is the omnivorous versus vegetarian diet. Omnivores eat animal (flesh and dairy) and plant foods; vegetarians avoid flesh foods, but not necessarily all animal products. Specifically, the vegetarian lifestyle has large variation within it, including:

- part-time-vegetarianism (people who claim to be vegetarians but consume flesh products on occasion)
- pesco-vegetarianism (vegetarians who eat fish; a contradiction in terms since I have never seen a fish that is not an animal)
- ovo-lacto-vegetarianism (vegetarians who eat eggs and dairy products)
- veganism (vegetarians who shun all animal products; some even reject honey)

- macrobiotics (a special vegetarian diet that is largely grain and vegetable-based with many fermented products, and limited use of spices and herbs) and

- raw food diets (vegetarians who only eat raw foods, including a large variety of nuts, seeds, and sprouts)

Most commonly when people think of a vegetarian diet, they think of the ovo–lacto–vegeterian lifestyle. A vegetarian diet in and of itself is not more or less healthy than a diet that includes meat (i.e., an omnivorous diet). Although research has identified longer life spans and fewer medical problems for certain population groups who live vegetarian lifestyles (e.g., Seventh Day Adventists), these populations have *healthy* vegetarian lifestyles. What makes a vegetarian lifestyle healthy is the conscious choice of wholesome, life-sustaining foods. Junk food vegetarians are no healthier than junk food omnivores. Although I clearly endorse (and follow) a vegetarian diet (of the vegan variety), I also accept omnivorous choices. There are a number of cautions that apply, however. The choice to live a healthy vegetarian lifestyle has many implications, ranging from health concerns to political and social statements. Vegetarianism is a choice that can be made for several reasons:

- health: this type of diet is healthier in terms of reducing the number of medical problems and extending the life span

- spiritual: this type of diet is considerate of animal life and animal well-being

- environmental: this type of diet supports a more sustainable economy that is easier on the planet in terms of pollution and resource use

- financial: this type of diet can be cheaper than a meat-based diet

- global: this type of diet produces more food per acre than a meat-based diet and could ascertain ample food supplies for all the people on earth if practiced universally

An omnivorous diet can also meet these criteria under certain conditions. Specifically, a subsistence lifestyle that used to be practiced by indigenous populations can be respectful of the land, the earth, the animals who are hunted, and the people who practice it. This lifestyle, however, is quickly fading, even in remote areas of the planet. The mainstream American omnivorous diet encourages appalling conditions for animals that are raised for the mere purpose of slaughter. Although I would love to expound on this issue, it is beyond my purpose here. Readers are referred to vegetarian John Robbins's (the could-have-been heir of Baskin-Robbins) *Diet for a New America* (Robbins, 1997), an eye-opening book about the living and dying conditions of animals in the meat and dairy industry.

As alluded to above, however, even a vegetarian diet can be unhealthy and disrespectful. A maximally healthy and respectful vegetarian diet requires that the practitioner of the lifestyle make certain healthy and important choices in addition to the choice to avoid flesh products. Most importantly, the healthiest diet is one that is whole-foods based and organic. Non–whole-foods-based vegetarian diets can be junk food diets that have no healthful impact. Typical American processed foods, such as white flours, salts, and refined sugars and unhealthy fats, such as hydrogenated oils and trans-fatty acids, are the greatest obstacles to health next to toxic and polluted foods, such as nonorganic vegetables grown with unhealthful fertilizers, herbicides, and pesticides and nonorganic, non–free-range meats and dairy products. Although the United States has some restrictions on the use of certain particularly hazardous fertilizers, herbicides, and pesticides, these chemicals are exported to other countries whose foods then in turn are imported again. Thus, even though these chemicals have been banned for use, they are present in the foods we eat if nonorganic choices are made.

There is a political statement in choosing organic foods as well. The use of hazardous chemicals is extremely detrimental and toxic to the health of the farm workers who work with and around them. Further, nonorganic agriculture is large-scale agriculture that is slowly crowding out family-run farms and that tends to be extremely harmful to the environment. Organic farming is sustainable agriculture that does not introduce toxic chemicals into the soil or the groundwater; it protects farm workers and supports family businesses. The problem with nonorganic, non–free-range meat and dairy products stems from two sources in addition to the chemicals introduced into the animal through grazing on nonorganic feed. First, the animals are kept in such unhealthful living conditions that their only means of survival is to be inoculated with large doses of antibiotics. Second, to encourage quick growth (before the animals die from the abhorrent living conditions), they are injected with large doses of hormones. Both of these chemicals, antibiotics and hormones, are then passed along the food chain to the humans who consume the meat or dairy products. Both have been linked to major health problems among humans (e.g., the increased estrogen levels in mainstream American dairy products are thought to contribute to the rising incidence of breast cancer). Early and prolonged use of dairy products has been linked to common health problems among children such as ear infections, hyperactivity, and attention deficits. Finally, the processing of animal products is sufficiently unhygienic and poorly monitored that the introduction of dangerous bacteria can result in life-threatening incidences of food poisoning.

I have summarized the nutritional guidelines and choices that I recommend for use or avoidance in Table 9-4. Although this type of diet may seem to require an inordinate amount of time for food preparation, it really does not require much more energy and time than any other type of diet once a routine has been developed. The transition to this way of eating can be mastered in steps, making the adjustment more acceptable and easier for the body.

In addition to making conscious choices about the foods that are consumed, it is helpful to put some thought into eating habits per se. The average American eats more calories per day than are needed to sustain life. In fact, some researchers are beginning to suggest that calorie restriction (which by others has been reframed as a return to normal calorie levels) is the single most important variable in predicting length of the life span and the only variable that may actually extend the human life span. Eating on the run is not a healthful habit, and it can be extremely relaxing just to eat food slowly and consciously. Making time to sit down for meals, as opposed to eating in the car, in front of the television set, or while talking on the phone or running errands is a centering activity that also facilitates proper digestion. Chewing food well is an important and often-overlooked component of the digestive process, and its absence can account for a variety of health problems. Eating slowly and enjoying the food that is eaten rather than just wolfing it down results in greater relaxation and better health (Millman, 1998). Taking time between bites and swallowing each bite before taking the next one ascertains that food is eaten in proper quantities and can be digested well (Reid, 1994). Making eating an occasion to be relished is a nice way of centering oneself through an already necessary activity.

Physical Activity. Almost as touchy as the topic of food, is the issue of exercise. Everyone knows that being active seems to be related to better health. Nevertheless, only a small proportion of the American public actually follows this advice. The type and level of activity that is optimal for the average person appears less clear-cut than the generic advice to seek out an active lifestyle. Agreement also appears to center around the need for daily exercise. The incorporation of daily exercise is best accomplished by building a routine of exercise into daily life that can be followed even, or especially, on tiresome and stressful days. For physical activity to be healthful, overexercise needs to be avoided as much as underexercise. In other words, it is important to avoid compulsive exercise that stresses the body even further. It is best to choose activities that

TABLE 9-4   Suggested Guidelines for Healthy Nutrition

CATEGORY	SPECIFICS
Healthful Dietary Lifestyle Choices	▪ favor a vegetarian diet ▪ if following an omnivorous diet, favor organic and free-range flesh and dairy products ▪ eat consciously, slowly, and with gratitude ▪ make eating a special occasion that is relaxing and centering ▪ enjoy meal times as a time to center and come together with others ▪ reduce overall caloric intake ▪ eat a variety of foods and rotate foods ▪ follow as many of the suggestions provided below as is reasonable and enjoyable given your preferences ▪ enjoy experimenting with new, wholesome foods (e.g., tempeh, tofu, miso, tahini, dulse, nori, kelp, shiitake)
Foods to Favor	▪ if animal products are used, choose organic and free-range varieties ▪ unprocessed, whole foods (e.g., whole grains; wholesome sweeteners; fresh, not canned or frozen, vegetable and fruits; home-baked breads and pastry products; cold-pressed oils) ▪ a wide variety of organic fruits and vegetables, preferably locally grown and in season ▪ a wide variety of organic grains, rotating for maximizing nutrition and minimizing allergic reaction (e.g., amaranth, quinoa, spelt, kamut, oats, brown rice, wild rice, barley, wheat) ▪ a wide variety of organic legumes, rotating for maximizing nutrition and minimizing allergic reaction (e.g., black beans, garbanzo beans, kidney beans, white beans, adzuki beans, soy beans, pinto beans, lentils, split peas) ▪ daily portions of soy-based foods (e.g., tofu, tempeh, miso, shoyu sauce) ▪ nondairy "milks" (e.g., rice milk, soy milk, almond milk, grain milk) ▪ raw foods (e.g., uncooked vegetables and fruits, sprouts, rolled grains) ▪ small daily amounts of healthful fats (e.g., flax seed oil, nuts, seeds, small quantities of cold-pressed extra virgin olive oil or cold-pressed sesame oil sold in opaque glass containers) ▪ freshly extracted vegetable juices ▪ sea vegetables (e.g., kelp, dulse, nori) ▪ mushrooms (e.g., shiitake, maitake) ▪ pure water in large quantities ▪ herbal teas and green tea
Foods to Avoid	▪ processed foods (e.g., ready-made freezer or microwave meals; foods with white sugars, salts, or flours; foods with hydrogenated fats and oils or trans-fatty acids) ▪ empty foods (e.g., sodas, candy bars, candies, many cereals, coffee) ▪ junk foods (e.g., fast food, fried food, chips) ▪ polluted foods (e.g., produce from countries that do not limit use of certain fertilizers and pesticides; nonorganic, hormonally treated flesh products; nonorganic, hormonally treated dairy products) ▪ toxic foods (e.g., foods containing additives such as food colors, foods containing chemical stabilizers or preservatives) ▪ excessively concentrated foods (e.g., fruit juices, sweetened drinks)

Continued

TABLE 9-4   Suggested Guidelines for Healthy Nutrition (continued)

CATEGORY	SPECIFICS
Helpful Hints	■ practice proper food combining (i.e., keep meals simple, not combining too many different foods at once) ■ eat fruits on an empty stomach and by themselves ■ grind your own whole grain flours to avoid rancidity; if buying whole grain flours, choose a store that grinds them fresh or stores the flours in the refrigerated section or in vacuum-packed bags ■ flake your own grains for cereals to avoid rancidity ■ sauté vegetables in a bit of shoyu, water, or liquid aminos instead of oil ■ eat the bulk of your foods early in the day, not in the evening ■ listen to your body in terms of how best to distribute your food intake over the course of the day (i.e., several small meals versus few big meals) ■ go to your local farmer's market for fresh, organic produce ■ talk to your favorite grocery store manager about stocking organic foods and about special orders of wholesome foods they do not carry regularly ■ bake all of your own pastries, cookies, cakes, and breads to avoid the unhealthful ingredients of packaged products ■ learn to read and always read labels of any pre-packaged foods you buy ■ buy bulk foods whenever possible to save on packaging and cost ■ grow an organic garden and use as many home-grown foods as possible ■ always have a cooked grain and a cooked legume in the fridge ready to combine with fresh vegetables for quick meals ■ make your own convenience or expensive foods such as grain and nut "milks," cereals, pancake mixes, salad dressings, and spice mixes ■ plan consumption of grains and legumes by soaking them during the day and cooking them in the evening for use the next day ■ sprout your own sprouts and grow your own herbs indoors all year long ■ stock a pantry with wholesome foods and do not buy the foods to avoid so as not to be tempted ■ make your teas from bulk herbs and bulk tea leaves; collect your own herbs (e.g., red clover, nettles, dandelion) when in season and dry for year-round use ■ have your water analyzed and install filters as needed in your home
Read and Learn about Nutrition	■ Balch, J. F., & Balch, P. A. (1997). *Prescription for nutritional healing* (2nd edition). Garden City Park, NY: Avery. ■ Crayhon, R. (1994). *Nutrition made simple: A comprehensive guide to the latest findings in optimal nutrition*. New York: M. Evans. ■ Haas, E. M. (1992). *Staying healthy with nutrition*. Berkeley: Celestial Arts. ■ Null, G. (1995). *Nutrition and the mind*. New York: Seven Stories. ■ Pitchford, P. (1993). *Healing with whole foods: Oriental traditions and modern nutrition*. Berkeley: North Atlantic. ■ Pizzorno, J. (1998). *Total wellness*. Rocklin, CA: Prima. ■ Reid, D. (1994). *The complete book of Chinese health and healing*. New York: Barnes and Noble. ■ Robbins, J. (1987). *Diet for a new America*. Walpole, NH: Stillpoint. ■ Roehl, E. (1996). *Whole food facts: The complete reference guide*. Rochester, VT: Healing Arts. ■ Rohe, F. (1983). *The complete book of natural foods*. Boulder: Shambala.

result in enjoyment and that do not hurt physically. A good balance of aerobic exercise and stretching appears to be useful, and the incorporation of the enjoyment of the outdoors can add a relaxing and rejuvenating component. It is helpful to seek variety in terms of exercising too, because this keeps motivation up and repetitive stress injuries down. In designing a physical fitness plan it is best to keep in mind four aspects of physical activity and health: strength (or muscular power); stamina (or aerobic capacity and endurance); suppleness (or flexibility); and sensitivity (or balance, rhythm, and timing). All four components need to be conditioned and attended to. Incorporating exercise that involves strength, stamina, suppleness, and sensitivity into vacations is another excellent way to unwind and keep the body healthy. Hiking, backpacking, kayaking, and similar activities fit the category of exercise while also being extremely pleasurable and relaxing in an outdoor setting that can be healing to mind, spirit, and body.

The three foremost aerobic forms of regular (if not daily) exercise that are relatively safe to engage in and that build strength as well as stamina are walking, swimming, and cross-country skiing. Walking is an excellent exercise that is easy to fit into busy schedules, inexpensive, and possible anywhere and in any weather. Swimming is another form of activity that has a low incidence of injury and can be greatly enjoyable while giving maximum benefits. It is somewhat more limiting than walking in that it requires a special setting and in most parts of the world cannot be engaged in outside throughout the year. The same drawbacks are true for cross-country skiing, another aerobic exercise that is easily learned and that has great physical benefits. Of the three, it is the most expensive form of exercise, requiring an initial investment in gear and probably a few lessons. Beyond the initial investment, however, it is an inexpensive sport, though of course greatly limited by weather. Many other forms of aerobic exercise exist (dancing, tennis, racket ball, running, step classes, and on and on),

and it is important to sample several before settling into a routine. Varying these forms of exercise is also a good idea to keep up motivation and enjoyment. Running, one of the more popular forms of aerobic exercise because of its time efficiency, has many drawbacks, as does high-impact aerobics. It is best to consult with a physician before engaging in any of the potentially more injurious forms of aerobic exercise.

The best supplements to aerobic regimens to build suppleness and sensitivity are stretching exercises, mild workouts with free weights, and systematic exercise routines such as tai chi, chi gong, and yoga. Most of these require some initial learning curve and lessons until a basic routine has been mastered. Once learned, however, they can be carried out anywhere anytime and are extremely flexible. They can easily be engaged in at the office, even for ten minutes between clients. Additionally, exercises such as tai chi, chi gong, and yoga have a strong relaxation or even meditative component and can rest the mind while working out the body. Specific yoga and chi gong routines can be developed with an instructor for specific healing purposes and can be extremely useful for dealing with particular ailments or disease/recurrence prevention.

If a person is currently sedentary, it is important to check with a physician before starting a rigorous exercise program. Once the decision is made to incorporate exercise into daily life, it is a good idea to start exercising slowly, gradually increasing length and intensity. The "main principle in physical activity is gradually and comfortably to be more active" (Bayne, 1997, p. 188). Jumping in full force can lead to injury and quick disillusionment. Thirty minutes of exercise per day most days of the week is a good goal to aim for. Some research has suggested that this span of time does not have to be used in a continuous manner. Instead, exercise can be incorporated into a busy schedule by doing a little bit at a time, several times a day. For example, Thayer, Newman, and McClain (1994) have shown that just 10 minutes of brisk walking had a positive effect for up to two hours in terms of increasing energy and reducing tension! It is important to let the body dictate what is comfortable, rather than to use some external standard of having to work out for a certain length of time or to go for a certain criterion of distance, weight, or similar standards. Some people can easily engage in strenuous aerobic exercise without negative effects, whereas others are better served by less stressful activity. One thing to remember while engaging in all forms of exercise is that very strenuous activity increases the number of free radicals in the body and thus increases the need for proper nutrition, especially the intake of antioxidants. Prolonged strenuous exercise can actually damage the body significantly and can leave the exerciser with a weakened immune system that increases the likelihood of catching colds and other minor ailments. Some sports physicians, for example, indicate that running a marathon can leave the immune system compromised for up to six weeks (K. W. Klingler, personal communication, June 8, 1998).

Rest. Often, when lives get very full with activity and commitments, time is taken away from sleep. This is a bad idea because the cost of reducing sleep can be high. The lack of alertness during the day after a sleep-deprived night alone may undo the time savings of that extra hour of staying awake the day before. Getting plenty of sleep each night is a most important self-care habit. The number of hours required varies greatly from individual to individual. Some people can function well on seven hours; most need eight; a few need as many as nine hours. Everyone has to try to find out personal needs for sleep and must not allow societal rules to influence the choice of how long to sleep. Sleep routines are also best such that a person would awaken in the morning without an alarm clock. This may require going to bed earlier than one used to or starting the day later than most other people. Again, it is most important to listen to the body. Then, the development of routines best follows the dictates of the body rather than some

external criterion of what should or should not be done. Some routines, unfortunately, are dictated. If children have to be at school at a certain time or an employer expects an employee at work at a certain hour, then routines have to be adjusted. However, the number of hours of sleep does not need to be compromised even then. Some people like to incorporate a rest period (i.e., naps) in the middle of the day. Although this practice has never worked for me because it leaves me feeling sluggish and drowsy, it is excellent for some. Experimentation with what works is again the best judge of whether to incorporate this routine into one's life.

Physical Self-Awareness. It is very important to learn to recognize warning signals and pleasure signals that come from within one's body. Bodies often know before the conscious mind does that stress is mounting and changes need to be made. Paying attention to the body and letting it decide when to slow down is an important self-awareness skill and health issue. Bodily symptoms and changes can give an indication when someone may need medical attention, when there is a need to change exercise and nutrition routines, or when the need for relaxation has to be met. Regular physical checkups can be useful even when the body appears to be symptom-free. It is my preference to consult naturopathic physicians for this type of medical care, because they are more attuned to subtle symptoms and bodily changes. Allopathic medicine is still preoccupied with disease and illness as signaled by strong discomfort and overt symptoms. Nonmainstream (and non-Western) medicine instead looks for patterns and changes in functioning that can be early warning signs about disease or less-than-optimal health (e.g., Pizzorno, 1998; Reid, 1994). The definition of health in allopathic medicine is merely the absence of disease; in naturopathic medicine (and in traditional Chinese medicine for that matter), the definition of health is much broader, involving the well-functioning of body, mind, and spirit. Being in touch with one's body is a good idea, and a few critical health monitoring routines can be incorporated by everyone, based on need and risk factors. For example, monthly breast self-exams, daily blood pressure checks, and similar simple health care routines can be life-saving skills for those with risk factors for particular diseases (see Pizzorno [1998] for an excellent book on taking charge of one's own health monitoring).

Taking an interest in one's health care during times of disease or symptoms is another important aspect of physical self-awareness. Blind faith in physicians is often misplaced faith because allopathic medicine often pretends to have the answers to a problem when the suggestion really is at best an educated guess. Being informed about treatment alternatives and not simply accepting every recommendation ever made by a physician is responsible self-care. Medicine has very few definitive answers and many mild physical problems have a wide range of possible solutions. The overuse of certain medications (e.g., antibiotics) has resulted in health care crises of frightening proportions (see Garrett, 1994). The treatment of symptoms rather than causes is another problem of allopathic medicine that can only be overcome by responsible self-care and active involvement in one's personal health care. Taking a pill to alleviate a symptom may be easier in the short term than searching for the cause of the symptom but in the long term tends to create more problems. Symptoms are often signals that the body (or mind) is in distress. Covering the symptom merely prolongs the exposure to the cause of the problem. For example, taking medications for heartburn is a sure way of inviting disaster if the root cause of the problem is not addressed (e.g., a change in diet and eating habits [Pizzorno, 1998]). Even more invasive medical procedures are often confused with being healing when they are really only palliative. Bypass surgery is an important example. The bypass may fix the clogged artery but it will not extend life unless substantial lifestyle changes are made that address the root cause of the disease. Removing the symptoms

(i.e., the arterial clog) does not deliver health. Making responsible choices about risk-benefit ratios of prescribed treatments is another important self-care skill. Chemotherapy for cancer is an example of the cure sometimes being worse than the disease (Moss, 1996). The exploration of treatment alternatives is often left to the patient. This generally means that the patient who is not self-aware will blindly follow the allopathic physician's lead, taking whatever is prescribed without exploring whether preferable alternatives exist (e.g., herbal remedies, nutritional interventions, physical therapies).

The relevance of this issue to therapists and counselors rests in the fact that health care is an active skill and perfect health is only obtained if active involvement is part and parcel of physical self-care. Good physical health, in turn, is an important prerequisite to good mental health. If clinicians cover their own physical symptoms with palliative methods and fail to seek root causes of their ailments, they most likely will model the same passive consumption of physical and mental health care for their clients. This is dangerous practice and goes against the idea of empowering consumers to take an active and deciding role in their own health care (physical and emotional). The lesson is to practice what is preached.

Involvement with Nature. The final personal care habit is somewhat related to relaxation and stress reduction. Being in touch with nature is a centering aspect of life. Too many people in modern life spend hardly any time outdoors. The fact that indoor environments can be perfectly adjusted in terms of warmth, humidity, and similar variables has made it tempting for many to avoid the outdoors when conditions are perceived as less than perfect. This leads to an alienation from the rest of the world and from the very environment that is the natural ecology of our species (Burns, 1998). Humans did not evolve indoors; they are potentially closely tied to nature and have a strong relationship with it. Many people could obtain incredible physical and emotional healing power from nature if only they exposed themselves to it, given that "unthreatening natural environments tend to promote faster more complete recuperation from stress than do urban environments" (Pigram, 1993, p. 402). Health care for centuries was naturally based, not only in the sense of physical health, but also in the sense of mental, emotional, and spiritual well-being. Close communion with nature was perceived as both maintaining health, as well as healing in and of itself. Natural phenomena, such as sacred sites or environments where healing had taken place, interactions with nature such as natural rituals or bathing in natural bodies of water, and natural medicines, such as herbs and foods, were the primary healing and preventative forces that human beings relied on for most of their evolutionary process (Burns, 1998).

Modern life has largely superceded humanity's interaction with raw, natural environments. This is unfortunate given the healing power of natural environments that can reduce stress, enhance positive affect, improve parasympathetic nervous system functioning, and enhance self-esteem (Ulrich, Dimberg, & Driver, 1991). Having the knowledge that being part of nature can be healing in and of itself gives clinicians a powerful tool for self-rejuvenation, relaxation, disease prevention, and healing. Spending some time outdoors every day in as natural an environment as possible has enormous positive consequences for mind, body, and spirit. This time outdoors does not have to be reserved for weekends or vacations, nor should it be guided by weather. The experience of walking in the rain can be immensely pleasurable and healing. The silence of a remote piece of land that allows the person to escape the hectic pace and noisy background of modern society cannot be surpassed in terms of its positive effects by any therapeutic or counseling intervention in the repertoire of today's mental health care providers. Availing oneself of this incredible resource for health and healing means taking care of oneself. Using this resource for clients is, of course, another possibility and one that is beautifully explored in Burns' (1998) book entitled *Nature-Guided Therapy*.

Attention to Relationships

As important as personal health habits, paying attention to relationships is also crucial to personal well-being. Most humans are highly social creatures who feel best if firmly embedded in an interpersonal matrix of meaningful and sustaining relationships (see Kohut, 1984). There are a number of personal traits that can greatly facilitate positive relationships that sustain health as opposed to creating stress. Although it is impossible to outline all of these traits here or to even come close to attempting to teach them, I will simply outline a few suggestions, which when incorporated into personal life tend to be extremely helpful. All of these simple suggestions can greatly enhance relationships in general, as well as closer or intimate relationships. It is beyond the scope of this chapter to help readers fix their relationships with everyone around them. I merely want to draw attention to the fact that relationships play a large role in our lives and can be an important factor in distress and impairment if they become sources of conflict. Paying careful attention and taking responsibility in relationships can reduce stress and create a source of support that is emotionally sustaining at times when work life becomes difficult.

The most important relationships to attend to are friendships and intimate relationships. However, the few guidelines that follow really apply to all relationships, even those with strangers, clients, bosses, and colleagues. To make positive relationships happen, active work has to take place. This work is applied to self and relationships to harmonize interactions with others. Working on the self is the first step in creating positive relationships. Empathy, tolerance, acceptance, and respectfulness are essential ingredients of healthy relationships and largely depend on personal attitudes and self-awareness (which will be dealt with in more detail below). Thus, working on the self to become a better and more tolerant person is a prerequisite to learning to respect others and to allow them to be different from the self. It is important to not place personal rights ahead of those of others; to be considerate of others' needs and respect their right to be themselves; and never to try to form anyone in one's own image. These simple rules ascertain that we respect others and usually that they will respect us in return. If we treat others politely, they will treat us politely. Polite and respectful interactions reduce stress in day-to-day life, making interactions with those in our interpersonal matrix more positive and enjoyable.

Another helpful trait, which tends to keep relationships positive and sustaining, is the ability to laugh at oneself. Those individuals who have learned to laugh at themselves and their own shortcomings, rarely jump to conclusions about others and seldom place blame. They are able to put things in perspective with the recognition that everyone makes mistakes and has flaws that are played out in relationships. The ability to laugh at oneself, however, must not be mistaken as an excuse not to change and grow; being able to laugh at oneself is the beginning. The next step is to look at whether the situation can be avoided in the future through self-change and growth. The ability to laugh at oneself in and of itself, however, often suffices to defuse potentially conflictual situations in relationships since the other people involved will not feel attacked or blamed.

Very similar to the concept of not taking oneself too seriously and being able to accept one's shortcomings is the notion of defenselessness and detachment from one's personal point of view. People who are not defensive and not overly attached to convincing everyone of their personal point of view tend to have more positive and fewer conflictual relationships. It is rare that any issue is so important that a relationship needs to be sacrificed over it. There is usually room for compromise and everyone involved can learn to adjust their point of view or opinion somewhat. Further, the ability not to get defensive when challenged can serve to make interactions with other less conflictual and more collaborative or cooperative. All too often people get locked into perceiving a difference in opinion as a personal attack, followed by the need to defend

themselves and their point of view. Learning to accept differences of opinion as the expression of different, but equally human, ways of looking at the same situation can free us to let go of the need to defend ourselves. This does not mean that we always have to agree with others. It merely means that we recognize that we neither have to get upset nor angry if we perceive a challenge of our actions or beliefs.

One final note is necessary about relationships and their role in burnout and impairment prevention. Most simply put, just simply *having and valuing* relationships is important! Their quality counts, of course. But first of all, we have to develop them. The traits that were discussed earlier are helpful for maintaining relationships, but also for initiating them. Attention to building a close circle of friends is critical, as is the forging of at least one intimate relationship. Attending to family relationships cannot be overemphasized because this is the setting where many of us seek support first and most frequently. Again, it is impossible to deal with issues of developing and maintaining positive intimate and personal relationships in these few pages. It must suffice here to note that relationships are a critical component in clinicians' self-care and must not be ignored. Relationships help balance our lives and often ensure that we keep events around us in the proper perspective, not overreacting to "small stuff." One colleague of mine has a sign up in his office that reads "1/3 1/3 1/3." The first 1/3 refers to relationships, the second to self, and the third to work. This is a wonderful symbol and great model to emulate. Readers who do want to spend more time reading about and working on relationships are referred to one of my other books, *Between Two People: Exercises toward Intimacy* (Johnson, Fortman, & Brems, 1993). There are of course many other resources and books that can easily be found in any popular bookstore.

Recreational Activities

Another highly balancing aspect of life is paying attention to leisure and recreation. Being able to have fun and to enjoy life is often most easily accomplished through recreational activities. The range of activities is literally infinite, and I will merely point out some of the possibilities. The main lesson of this section is to attend to leisure and to learn to view it as important and life sustaining. Ignoring leisure for the sake of making more time for work is not a good idea. Life is too short, and who wants to die thinking, "If only I had tried . . ."? Making time for hobbies and interests is an enjoyable way of adding spice to life and preventing boredom and burnout. It is best to make leisure skills as different from work as possible. There are unlimited possibilities as far as hobbies are concerned and the only limiting factors are motivation, desire, creativity, interest, and physical capabilities. It is never too late to pick up new hobbies or to try to acquire new skills. Hobbies can include outdoor activities (hiking, backpacking, kayaking, skiing, gardening, and so on), introverted activities (painting, writing poetry, composing music, playing music, and so forth), sociable activities (dancing, playing music for others, leading guided hikes, joining a theatre group, and so on), sedate activities (reading, knitting, stamp collecting, etc.), active activities (sports, performing arts, volunteering, etc.), and many more. The most important thing for clinicians appears to be to develop interests and hobbies outside of the mental health care field.

Travel is another way of recreating and spending quality leisure time. Not all vacations have to be lengthy or expensive. Vacations can consist of extended weekend trips and can even be taken at home as long as work is honestly avoided. The definition of vacation is really a broad one that is only limited by personal imagination and preferences. The greatest vacation for some people may be a three-day backpacking trip; for another, it may consist of two weeks in Europe. Making vacations fit personal preferences and finances is the most critical factor, so that they do not end up creating more stress than pleasure. If a vacation is so expensive that it subsequently requires many extra hours of work, it may not be worthwhile. There are also ways to

travel that are more inexpensive, such as using youth hostels instead of hotels, trains instead of planes, and similar time-saving techniques (Luhrs, 1997). Similarly, a vacation that is so full of entertainment and activity that the traveler arrives back home feeling overwhelmed and exhausted does not serve its purpose fully. Striking a balance between novelty and relaxation, excitement and meditation, learning and stress reduction is best considered while planning a trip.

Another aspect of recreation and leisure is group memberships. This is not for everyone, but can be extremely rewarding for more sociable types. Group membership is best chosen away from the mental health profession. In other words, I am not referring to association membership here, which is important to professional self-care. I am referring, instead, to making friends and acquaintances in settings that have nothing to do with one's profession. This type of interaction is stimulating and exciting because it provides exposure to a broad range of healthy human beings who come together for a shared purpose or interest. These purposes or interests can range widely, including group memberships in organizations such as environmental groups (e.g., Sierra Club), religious/spiritual groups, political groups, clubs organized around sports or special interests (e.g., bowling leagues, team sports, choirs), and any other interesting groupings of people who come together on a regular basis for some joint endeavor. Some groups also help people reach beyond themselves as they come together for a greater cause. Such volunteer work can be life enhancing and extremely gratifying.

Finally, also in the category of recreation, there is entertainment. To me this is the leisure category of least importance, though it too should not be entirely neglected. Entertainment can consist of artistic events, cinema, dinners out, and similar activities. It can be an event for one or two, a family, or a whole group of friends. Most importantly, however, entertainment does not refer to watching TV or going shopping. Entertainment for recreation and leisure refers to an activity that is rejuvenating, stimulating, relaxing, comforting, or otherwise self-enhancing, not dulling, passive, or mind numbing. Entertainment can even consist of playing board games with children or party games with a group. In my family of origin, get-togethers often involve games and similar activities that make the meetings more lively, active, and enjoyable.

Relaxation and Centeredness

Another important set of self-care skills is that of relaxation strategies. Of all people, mental health care providers are probably best informed about this particular approach to self-care because it is often incorporated into client care. Application in the personal life of therapists and counselors, on the other hand, may not be as common as would be hoped. The four most common strategies to achieve a state of relaxation are breathing exercises, mindfulness, relaxation strategies proper, and guided imagery. Of these four strategies, most clinicians are well versed in the latter two, but often less familiar with the former two. I will cover breathing and mindfulness in some detail and only provide a quick review of relaxation strategies and guided imagery.

Deep and conscious breathing is a very simple and straightforward means of achieving relaxation. On the other hand, improper breathing is a quick and straightforward way to stress, tension, and uptightness. Many people hold their breath during periods of great stress, and most fail to breathe consciously at any point in their lives. In fact, most of us pay no attention to breathing at all, given that it is an autonomic response we tend to rely on to just happen. However, the truth of the matter is that breathing is very much affected by tension and stress, and these situations cause changes in it, or even cause it to seize momentarily. This sets up a vicious cycle where uptight or improper breathing in turn increases stress in the body even further. Learning to breathe deeply and consciously can be one of the most important components of achieving stress-reduction and mental and physical health. In my

own practice, I have learned to induce a state of deep relaxation merely through the alteration of my breath. No other relaxation strategy works better or faster for me. I use breathing exercises as a prelude to all relaxation strategies, including progressive muscle tension-relaxation exercises, guided imageries, and other visualizations, both in my personal practice, in teaching, and with clients.

To learn healthful deep breathing for relaxation, it is best to watch a baby. Babies automatically breathe in the most efficient and relaxing manner. Their trick is to use their diaphragm as a means of pulling air more fully into the lungs. Most of us breathe using strictly our chest muscles (i.e., the intercostal muscles between the ribs), an inefficient and less relaxing mode of breathing. A person who breathes in this way has no detectable motion of the abdomen during inhalation; merely the chest expands. Diaphragmatic breathing, which is associated with states of relaxation, serenity, and peace, on the other hand, is detected by observing the rise and fall of the abdomen with each inhalation and exhalation, respectively. In diaphragmatic breathing, the diaphragm (i.e., the large muscle between the thoracic and abdominal cavity) flattens with inhalation, moving the abdomen forward and creating space for the lungs to expand with air in the thoracic cavity.

The best way to practice diaphragmatic breathing is to lie flat on one's back with a hand (or both hands) on the stomach. As in most, if not all, breathing exercises, it is best to place the tongue on the alveolar ridge (the soft tissue between the roof of the mouth and the upper front teeth) and to exhale once, forcefully and deeply (as well as noisily) through the mouth. All subsequent breathing is done through the nose exclusively (unless engaging in a special breathing exercise that specifies otherwise). To breathe diaphragmatically, on inhalation, the focus is on flattening the diaphragm against the abdominal cavity, allowing the belly to rise gently. As this movement occurs, the lungs will naturally fill with air. On exhalation, the diaphragm relaxes, the stomach recedes, and the lungs empty. This type of breathing can be engaged in in any rhythm. The slower the breathing the greater the state of relaxation it induces. If one gasps for air between breaths, however, the breathing has slowed too much. It is important to find a comfortable speed and rhythm and then to just observe the breath. Once diaphragmatic breathing has been successfully practiced lying down, it can be used in any body position and can be applied as a quick and easy relaxation strategy anytime, anywhere. It can be used in the middle of a stressful lecture, while sitting with a client, or during a job interview, and no one will even know. Its power of relaxation, however, will take hold immediately, especially after prolonged practice in a relaxed setting.

Because diaphragmatic breathing is strongly associated with a state of relaxation, it is not the preferred way of breathing during physical activity or exertion. While active, breathing is best conducted using the entire breathing apparatus. In this type of breathing, the breath starts with the flattening of the diaphragm and expansion of the abdominal cavity. However, the breath then moves up to the chest with the continued inhalation expanding the lower thoracic cavity and then the upper chest cavity, finally raising the shoulders. Exhalation works in the reverse, with the shoulders dropping first, the lungs contracting, and the diaphragm relaxing to flatten the stomach. This very rejuvenating breath is a good breath to use at the end of a relaxation practice to become alert again. It can also be an excellent means of overcoming fatigue and becoming reenergized anywhere, anytime. A variety of other breathing exercises exist (e.g., Schiffmann, 1996; Weil, 1995), but these two basic forms will get anyone started on the road to relaxation and better vital capacity.

Breathing can easily be combined with the practice of mindfulness, in fact, can be a powerful way to begin to learn being mindful. Mindfulness, in turn, is a helpful precursor for relaxation in general, and relaxation exercises, in particular. Its application is so constant through-

out life that it can be practiced anytime, anywhere, with anyone. It is a mindset, which develops and ultimately becomes second nature, that is very useful in coping with stressors of daily life and in keeping focused on what really matters. Mindfulness is the practice of stillness, centeredness, and full awareness in the present moment. The easiest way to begin to learn mindfulness is to sit still and center attention on breath. Centering attention on breath means observing each inhalation and exhalation and their effects on the body—physical sensations in the nose, the larynx, the lungs, the chest, the abdomen. Full concentration is placed on observing the subtle changes in the body as breath moves in and out. Each small change in the body is noted and attended to with great awareness, but total calm and stillness. Getting this involved with each breath is true mindfulness of the breath, a centering and calming experience that brings along a very peaceful feeling of being in the moment.

Mindful slowing down and centering does not have to involve sitting still and breathing, although this is the easiest way to experience mindfulness and practice it until it takes a hold. Mindfulness can be practiced through any familiar skill or activity by simply placing all attention and concentration on the activity. As in the mindful breathing exercise, every subtle movement and change in the body is noted and appreciated for its complexity. If possible, the activity can be slowed down to appreciate the many subtle muscular movements and incredible coordination more fully. My favorite time to practice mindfulness is when I grind flour. I can fully concentrate on each movement and the smells of fresh grains. It is helpful to the generalization of mindfulness into everyday life to pick a regular activity that is coupled with the practice of mindfulness. It provides a model for the feeling of serenity, centeredness, and peacefulness that is so helpful to relaxation.

The opposite of mindfulness is absent-mindedness and lack of attention. Anyone who has ever left their house and later wondered whether they turned off the stove or locked the door has experienced the opposite of mindfulness. Activities were engaged in without attention to them and hence they were not even consciously registered in awareness. Mindfulness brings a peaceful feeling that clearly focuses the mind on what is important: the present. In mindfulness, attention is concentrated on the moment, and all thoughts of the past and future disappear and along with them the anxiety and stress they may create. Whatever activity one is engaged in at the moment of mindfulness becomes the center and reason of existence and is done for itself, not for an end goal. What becomes important is the process, not the outcome. Thus, when I grind my grains, I try to do it not to make flour, but to grind. If mindfulness is applied to breathing, the focus is to breathe, not to survive or relax. If mindfulness is applied to eating, the purpose is to savor the food, to enjoy every aspect of eating, from chewing to swallowing, not to satisfy hunger or get dinner over with so that something else can be done. Whole books have been written on the topic of mindfulness and it certainly deserves that much attention (Fields, 1985; Hanh, 1975). The few guidelines provided here, however, suffice to initiate the novice into its practice.

Anyone who has achieved relaxed breathing and mindfulness is fully prepared to begin relaxation through specific relaxation strategies or guided imagery. Most health care professionals are well versed in these more concrete strategies, and I will not spend much effort on them here. The following quick review of relaxation strategies is meant mainly to remind clinicians to avail themselves of the very strategies they recommend to clients. No great detail will be provided with regard to how-to because it is assumed that this is a critical skill most mental health care providers already possess but merely need to learn to practice in their own lives.

The idea of using relaxation strategies per se is based on the principle that a body cannot be relaxed and anxious at the same time. Relaxation strategies have been traditionally employed to reduce general anxiety, induce sleep, produce relax-

ation to facilitate coping with a specific anticipated event, or reduce phobic reactions in specific situations. The two primary strategies that have been developed are muscle tension relaxation and pure relaxation. Muscle tension relaxation uses the contrast between tense and relaxed muscles as a means of teaching a person the difference between tension and relaxation for the purpose of inducing relaxation. In muscle tension relaxation, muscle groups are first tensed through specific suggested motions, and then relaxed; the difference between the two states is then noted in a mindful manner. Simple or pure relaxation exercises will focus attention on the same muscle groups but without the initial experience of tension. In other words, the muscle group is focused on with the desire or direction to relax it. In using relaxation, a few cautions apply, both for personal use and use with clients. Most importantly, it is necessary to remember that:

- if pain occurs, the chosen strategy may need to be discontinued (e.g., take a break from tensing a certain muscle group)
- if a floating feeling occurs, this is usually no problem unless the person relaxing is prone to dissociation (which may be a contraindication for use of relaxation)
- a feeling of heaviness is not only normal, but generally desirable unless the person relaxing has a physical problem that is exacerbated by this (e.g., fibromyalgia)
- if the person relaxing sees colors or shapes, this is normal, but can be stopped easily if perceived as unpleasant or disruptive by opening the eyes
- if the person relaxing falls asleep (and this was not the purpose of the relaxation exercise), it is best to awaken them (this is of course difficult if the person has no guide; in such cases it is best to practice relaxation sitting up)

Relaxation is best started with deep breathing, combined with mindfulness of the breath. This slows down, centers, and focuses in the present the person relaxing. It generally works best to work from the periphery of the body to the center. It is best to begin with attention to the feet, calves, thighs, hands, forearms, upper arms, neck, shoulders, scalp, face, and torso, in that order. Relaxation is contraindicated with individuals who are known to have temporal lobe epilepsy (complex partial seizures) and other seizure disorders. It needs to be used cautiously with people who are prone to dissociation and with individuals who suffer from PTSD. Debriefing with the person relaxing (i.e., the client or oneself) after a relaxation exercise is a helpful process that can make subsequent sessions more efficient and useful through appropriate modification or changes. Various resources, including books, tapes, and scripts, are available to persons who would like to incorporate relaxation strategies into their everyday life.

Guided imagery is also known under various other labels, including mental rehearsal, covert modeling, and visualization. It is based on the principles of progressive relaxation and social learning theory and can serve many purposes beyond relaxation. Specifically, in addition to being used to induce relaxation, guided imagery has been used for pain management, skill acquisition and enhancement, self-exploration, and healing. The same cautions that apply to the use of relaxation strategies are also relevant to use of guided imagery. Most importantly, caution needs to be applied when using guided imagery with individuals who have a tendency toward dissociation, excessive anxiety, history of trauma, and certain medical conditions (e.g., seizures, fibromyalgia, arthritis). For relaxation purposes, the procedure of guided imagery usually starts with deep breathing, mindfulness, and a few simple relaxation commands. Visualizations of pleasant scenery are then used to induce a deeper state of relaxation. It is important to remember that imagery is best not limited to the sense of sight, but should also involve the senses of hearing, smelling, touching, and tasting. Unique scenes (including smells, tastes, etc.) are best developed for each individual using the guided imagery be-

cause they work best if highly individualized. However, prepared scripts and tapes are also available, as are numerous books on the subject (e.g., Adair, 1984; Borysenko, 1987; Fogelsanger, 1994).

Self-Exploration and Awareness

"The sense of who one is, and of one's empowering life vision, seems to be at the core of long and creative living" (Jevne & Williams, 1998, p. 5). Emotional stability of the care provider is an important aspect of good therapy as well as of burnout prevention. Therapists are confronted with difficulties by clients on a daily basis; in addition, they have to be able to cope with their own life challenges. Being able to maintain self-esteem, self-efficacy, and a basic sense of personal competence is critical during periods of challenge (Wheeler, 1997). One way to achieve emotional stability is through the development of self-awareness. Self-awareness is of course also critical in the sense that it may prevent or reduce the impact of imposing stereotypes upon clients, forcing them into a provider's value system and worldview. The more self-aware, congruent, and centered counselors are, the less likely it becomes that they will judge their clients and the more likely it becomes that they will be empathically accepting of and positive about them (Johns, 1996). Self-awareness can lead to better professionalism; however, it is also recommended to help therapists "flourish as human beings, who then bring more than the minimum to their therapeutic work" (Johns, 1997, p. 61). Figuring out what really matters in life is an important endeavor that some of us take more seriously than others (Schwartz, 1995). Clinicians can hardly choose to ignore this very important issue if they want to be aware, unbiased, and effective mental health care providers who are neither prone to burnout nor impairment. There are many possible avenues toward self-awareness and personal self-development from which clinicians can choose. I have chosen a few to focus on here, but this list is by no means all inclusive.

One obvious strategy that creates self-awareness is, of course, personal therapy or counseling. One may wonder if this strategy even needs mention in a book for mental health care providers Actually, it appears that it does. Although over 80% of mental health care providers report having been in personal therapy in the past, only 22.2% of doctoral level providers and 38.8% of masters level providers report being currently in therapy. The remainder, a surprisingly large percentage (61.2 79.8%), reject the idea of current personal therapy for several reasons, including concern about the value of personal therapy (15–17%), embarrassment about being a client (13–15%), concern about confidentiality (12 16%), and prior negative experiences with therapy (7–8%) (Mahoney, 1997). Arguments have been made that all therapists in training should be required to be in personal therapy as well. This is a difficult issue that has not been empirically investigated. One can hardly argue, however, that a counselor or therapist will not be more effective and persuasive if personally convinced of the value of personal therapy. To hold such a belief may then translate into practice even without an imposed requirement from a training program or licensing agency. Since readiness for treatment is a critical issue in outcome of therapy, imposed personal therapy may be less effective than personal therapy that is initiated at the discretion of the practitioner. As Johns (1997) points out, therapy's or counseling's "optimal value is likely to emerge at a time of readiness, through reflective self-awareness which identifies a need, triggered by discomfort, uncertainty, the presence or absence of expected or unexpected feelings, unresolved personal issues, challenge of expected beliefs, or crisis, or developmental transitions in both professional and personal maturity" (p. 64). Each clinician will have to decide which road to choose. I am strongly in favor of personal therapy or counseling for mental health care providers and have sought it repeatedly and regularly in my own life. If a clinician makes the choice not to seek therapy or counseling, that person does need to question the level of faith placed in the process and whether any lack of faith

in it is communicated to clients. The potential negative consequences of this are obvious.

Another excellent set of strategies for creating self-awareness is looking at oneself through various self-exploratory means, including journaling, dream work, and reading. Reading is so self-explanatory that I will not write much about it here. There is a virtually endless supply of books and manuals that help people explore their lives, relationships, selves, dynamics, pasts, futures, feelings, thoughts, behaviors, and on and on. In fact, it appears pointless even to try to reference some of those resources as particularly helpful because there are too many to choose from and choices tend to be highly idiosyncratic. Anyone interested in pursuing self-exploration through reading merely needs to walk into a neighborhood bookstore and look around. The options will be endless. The main caution in this regard is to be a careful consumer and not to believe every self-help book that has ever been published. There are some pretty bad books out there. Being a careful and discriminating consumer (as described in the CE section) is essential to the successful use of this medium for self-exploration. Finally, even turning on the computer and getting on the Internet can lead people to interactive reading resources that facilitate self-exploration (e.g., http://www.catsco.com).

Journaling was made popular as a legitimate strategy for self-exploration by Pennebaker (1990). He conducted a series of fascinating experiments in which he asked participants to write about things that were stressful for them to help them get their emotions out in the open. He found that writing in and of itself about stressful life events and the emotions surrounding them was, not only stress-reducing, but helped participants improve coping ability as well as overall health. Based on his research findings, Pennebaker has suggested that writing is most helpful if it deals with both the experience and the emotions and deep feeling involved in the experience. Such self-exploration leads to catharsis and insight, and has numerous extremely positive consequences for the writer. It helps improve mood, increases coping ability, enhances the immune system, and in general appears to promote better physical and mental health. Indeed, journaling appears to be a powerful strategy for clinicians wishing to prevent burnout or impairment. It is also an activity that easily fits into a busy schedule as it can be engaged in anywhere and for brief periods of time. There are virtually no drawbacks to this technique, although some have warned that clinicians need to "guard against becoming morbidly introspective or unduly passive by also emphasizing action" (Bayne, 1997, p. 190). This caution, however, applies to most, if not all, self-exploratory strategies.

A final useful self-exploratory strategy is dream work, which tends to be a familiar strategy to most mental health care providers. The specifics of dream work are more complex than can be dealt with in these pages. However, I am enough of an advocate of dream work to want to alert the reader to its powerful impact by at least mentioning it and encouraging clinicians to engage in it. Exploring one's dreams for achieving insight can be accomplished relatively easily. The best strategy to remember dreams is to direct oneself to remember the dream upon awakening. It is helpful to have a tablet of paper and pen or a tape-recorder by the bedside and to record the dream immediately upon waking from it, even if this is in the middle of the night. Analysis of the dream then occurs the next day. Approaches to analyzing dreams vary widely and it is best to read a few books or take a class to decide which approach is best for the individual. Not all dream analysis approaches rely exclusively on Freud's or Jung's work, although these are certainly the thinkers who have inspired the work on dreams (see Freud, 1900; Jung, 1974). Many larger cities have dream work groups that come together for the sole purpose of working on members' dreams. There are several resources that may prove useful starting points for clinicians interested in using this strategy, including books and tapes that can be found in most popular book stores (e.g., Garfield, 1977; Mahrer, 1989; Taylor, 1983; Ullman & Zimmerman, 1979).

The final, but perhaps most excellent strategy for developing self-awareness is meditation, though this is not the primary purpose of meditation. Meditation is a strategy that could have been listed equally appropriately in the section about relaxation and centeredness or in the section about relationships. It is such a comprehensive and powerful strategy that it touches all aspects of a person's life. The choice to place its discussion in this section was therefore somewhat arbitrary. Not surprisingly, given my difficulty deciding in which section of the chapter to include meditation, I also have some trouble defining it. Meditation is not one thing; it is many things. It has many goals and no goals; it is a difficult process and yet it is easy. Meditation is about finding peace and stillness, about quieting the mind and being in the moment. I am not the only person who has trouble defining meditation or outlining why meditation is done. In LeShan's (1974) book on meditation (perhaps the book most cited as helpful in learning the process), he describes the goal of meditation as "access to more of our human potential or being closer to ourselves and to reality, or to more of our capacity for love and zest and enthusiasm, or our knowledge that we are a part of the universe and can never be alienated or separated from it, or our ability to see and function in reality more effectively" (p. 1). This is the most complicated sentence I have ever read in a book that tries to simplify something, but it is entirely accurate. Meditation returns us to a state of quiet and calm and connects us with our roots in the universe. It is the most calming practice of any skill I have ever engaged in and yet can at times leave the practitioner feeling disturbed or confused at the same time.

Although some teachers say that meditation is best learned with a qualified instructor, basic meditation technique is easily learned. Knowing basic technique does not mean knowing how to meditate and even teachers who have meditated for years accede that it is the practice of meditation that is important, not the goal or the outcome. Practiced over the long term, meditation helps the practitioner develop a calmness and serenity that permeates all of life. It is the ultimate technique for finding defenselessness, relaxation, and peacefulness in life—all states of being that are highly related to well-functioning and preventive of burnout and impairment.

There are many forms of meditation (see LeShan, 1974). Regardless which form one chooses, the basic procedure is to sit in a quiet place, either cross-legged on the floor with a cushion for support or on a straight-back chair with feet on the ground. The sitting posture is chosen because alertness is key to successful meditation, and lying down may lead to drowsiness or sleepiness. Most resources recommend closed eyes; some suggest eyes partially open (as one of my teachers called it: eyes at half-mast). Breathing becomes deeper and deeper as one meditates, and diaphragmatic breathing is most conducive to reaching a deeply relaxed meditative state. From this point, different forms of meditation diverge. The simplest form of meditation merely focuses the mind on the breath. This form of meditation falls in the category of meditations of the outer way because attention is focused on a process or item external to the mind. Another meditation of the outer way is to contemplate an object by looking at it intensely from as many angles as possible during the entire meditation time. Meditations of the inner way focus attention on one's own stream of consciousness without altering it. An example of a meditation of the inner way is the bubble meditation described by LeShan (1974). In this meditation, the person envisions the self sitting at the bottom of a pond, observing bubbles floating upward through the water. As thoughts enter the mind, they are placed in a bubble and are allowed to drift upward through, and ultimately out of, consciousness.

Regardless of which type of meditation is chosen, the universal experience is that it is extremely difficult to keep the mind focused, whether the focus is inward as in the observation of one's own thoughts, or outward as in the observation of one's breath or the contemplation of

an object. The key to success is to continuously bring the mind back to the breath, to the object, or the thought bubbles and, most importantly, to maintain calmness. No one can keep the mind quiet for very long; it is the attempt to do so that brings on the positive effects of meditation. Learning to quiet the mind and coming into the moment is the key to becoming self-aware, connected to a larger cosmos, and defenseless and accepting in relationships. There is literally no better impairment prevention technique than meditation because it is likely to effect all of the topics that were discussed so far almost automatically with regular practice. I strongly encourage the reader to begin to meditate and to seek out a meditation teacher. It will be helpful for the interested reader to read more about meditation by accessing the following resources:

1. Kabat-Zinn, J. (1994). *Wherever you go, there you are*. New York: Hyperion.

2. Kornfield, J. (1993). *A path with heart*. New York: Bantam.

3. Kornfield, J. (1994). *Buddha's little instruction book*. New York: Bantam.

4. LeShan, L. (1974). *How to meditate*. New York: Bantam.

5. Walters, D. (1996). *Meditation for starters*. Nevada City, CA: Crystal Clarity.

In closing the section on self-awareness, I would like to indicate that a desired outcome of self-awareness building is not merely burnout prevention and stress-reduction, but also the development of self-acceptance. Self-acceptance means being comfortable with oneself and not having to pretend to be someone else. Trying to pretend to be someone we are not takes a lot of energy and leads to greater likelihood of burnout and impairment. It also is likely to interfere with genuine acceptance of clients and tends to impair empathy. In fact, self-acceptance is synonymous with empathy for ourselves, and, in turn, "empathy for ourselves is the basis of empathy for others" (Breggin, 1997, p. 159). These are truly the ingredients for success in life and success in career.

SUMMARY AND CONCLUDING THOUGHTS

Self-care among mental health care providers is essential to successful therapeutic work. Losing sight of our personal self, or personhood (Jevne & Williams, 1998), as well as of our humanity is a shortcut to burnout, distress, and impairment. If the therapist suffers, the client will sense it, and the development of the therapeutic alliance will be impeded. In other words, self-care is an essential aspect of being a good counselor or therapist; neglecting the self, means neglecting the client. If it is not a good enough reason for clinicians to take care of themselves for their own sake, then they at least must do so for the sake of their clients. Self-care prevents burnout and impairment, helping clinicians lead long and healthy careers. A balanced life means paying equal attention, at a minimum, to work, play, social relationships, family relationships, personal growth, spiritual well-being, and personal physical and emotional health.

If clinicians do encounter symptoms of burnout, impairment, or distress, they need to be prepared to take steps toward helping themselves and their clients. Resources exist for that purpose, as previously mentioned. The recovery process is not an easy one, and it is clearly beyond the scope of this chapter to address burnout or impairment intervention. Overcoming burnout at a minimum requires the clinician to go through the distinct stages of grieving the loss of a dream and experiencing the anger, depression, and despair that may be involved. It also needs to lead to a point of hope and rebuilding that ushers in renewed enthusiasm for the same career or becomes the foundation of a new life challenge (Jevne & Williams, 1998). Helpful readings, amusing books, and insightful self-disclosures have been shared by a number of practitioners. A small, but helpful, selection of these is listed below.

Blackstone, P. (1991). *Things they never taught me in therapy school: Serious problems, surprising solutions.* Kingston, WA: Port Gamble Press.

Guy, J. D. (1987). *The personal life of the psychotherapist.* New York: Wiley.

Jevne, R. F., & Williams, D. R. (1998). *When dreams don't work: Professional caregivers and burnout.* Amityville, NY: Baywood.

Johns, H. (1996). *Personal development in counsellor training.* London: Cassell.

Kilburg, R. R., Nathan, P. E., Thoreson, R. W. (1986). *Professionals in distress: Issues, syndromes, and solutions in psychology.* Washington, DC: American Psychological Association.

Kottler, J. A. (1995). *Growing a therapist.* San Francisco: Jossey-Bass.

Schwebel, M., Schoener, G., & Skorina, J. K. (1994). *Assisting impaired psychologists* (rev. ed.). Washington, DC: American Psychological Association

APPRECIATION

Thank you for having chosen to go on this journey of exploring the challenges in counseling and psychotherapy with my book. I would love to hear from you with any comments about the book, its content, its effect on you and your career, and any other thoughts you may want to share about the therapeutic process. I hope that the book has contributed even just a small amount of enlightenment and pleasure to your work as a mental health care provider and that you will have or continue to have a happy career. Comments and feedback will best reach me via e-mail at afcb@uaa.alaska.edu or via the address of the publisher of this book as listed in the front matter.

References

Adair, M. (1984). *Working inside out: Tools for change.* Oakland: Wingbow.

Adler, L. E., & Griffith, J. M. (1991). Concurrent medical illness in the schizophrenic patient: Epidemiology, diagnosis, and management. *Schizophrenia, 4,* 91–107.

Ahia, C. E., & Martin, D. (1993). *The danger-to-self-or-others exception to confidentiality (ACA Legal Series, Vol. 8).* Alexandria, VA: American Counseling Association.

Alarcon, R. D., & Foulks, E. F. (1995). Personality disorders and culture: Contemporary clinical views (Part A). *Cultural Diversity and Mental Health, 1,* 3–17.

Alexander, M. B. (1997). Caught in the maze, aghast with this phase. In J. A. Kottler (Ed.), *Finding your way as a counselor* (pp. 25–28). Alexandria, VA: American Counseling Association.

Alexander, M. J., Craig, T. J., MacDonald, J., & Haugland, G. (1994). Dual diagnosis in a state psychiatric facility. *American Journal on Addictions, 3,* 314–324.

Allen, J. P., & Columbus, M. (1995). *Assessing alcohol problems: A guide for clinicians and researchers.* Bethesda, MD: National Institute on Alcohol Abuse and Alcoholism.

American Counseling Association. (1995). *ACA code of ethics and standards of practice.* Alexandria, VA: Author.

American Psychiatric Association (1994). *Diagnostic and statistical manual of mental disorders* (4th ed.). Washington, DC: Author.

American Psychological Association. (1987). Resolutions approved by the National Conference on Graduate Education in Psychology. *American Psychologist, 42,* 1070–1084.

American Psychological Association. (1992). Ethical principles of psychologists and code of conduct. *American Psychologist, 42,* 1597–1611.

American Psychological Association. (1993). Record keeping guidelines. *American Psychologist, 48,* 984–986.

American Society of Addiction Medicine. (1996). *Patient placement criteria for the treatment of substance-related disorders* (2nd ed.). Chevy Chase, MD: Author.

Ammerman, R. T., & Hersen, M. (1992). *Assessment of family violence.* New York: Wiley.

Ammon-Cavanaugh, S. (1995). Depression in the medically ill: Critical issues in diagnostic assessment. *Psychosomatics, 36,* 48–59.

Anderson, B. S. (1996). *The counselor and the law* (4th ed.). Alexandria, VA: American Counseling Association.

Appelbaum, P. S. (1985). Tarasoff and the clinician: Problems in fulfilling the duty to protect. *American Journal of Psychiatry, 142,* 425–429.

Arthur, G. L., & Swanson, C. D. (1993). *Confidentiality and privileged communication*. Alexandria, VA: American Counseling Association.

Auld, F., & Hyman, M. (1991). *Resolution of inner conflict: An introduction to psychoanalytic therapy*. Washington, DC: American Psychological Association.

Azar, S. T. (1992). Legal issues in the assessment of family violence involving children. In R. T. Ammerman & M. Hersen (Eds.). *Assessment of family violence* (pp. 47–70). New York: Wiley.

Baird, B. N. (1999). *The internship, practicum, and field placement handbook: A guide for the helping professions* (2nd ed.). Upper Saddle River, NJ: Prentice-Hall.

Balch, J. F., & Balch, P. A. (1997). *Prescription for nutritional healing* (2nd edition). Garden City Park, NY: Avery.

Barnet, D., Manly, J. T., & Cicchetti, D. (1991). Continuing toward an operational definition of psychological maltreatment. In D. Cicchetti (Ed.), *Development and psychopathology* (pp. 19–29). Cambridge: Cambridge University Press.

Barnett, O. W., Miller-Perrin, C. L., & Perrin, R. D. (1997). *Family violence across the lifespan: An introduction*. Thousand Oaks, CA: Sage.

Barrett-Lennard, G. (1981). The empathy cycle: Refinement of a nuclear concept. *Journal of Counseling Psychology, 28,* 91–100.

Bartwell, A., & Gilbert, C. L. (1993). *Screening for infectious diseases among substance abusers* (Treatment improvement protocol series 6). Rockville, MD: Center for Substance Abuse Treatment.

Basch, M. F. (1980). *Doing psychotherapy*. New York: Basic.

Baumrind, D. (1971). Harmonious parents and their preschool children. *Developmental Psychology, 4,* 99–102.

Baumrind, D. (1973). The development of instrumental competence through socialization. In A. D. Pick (Ed.), *Minnesota Symposia on Child Psychology* (Vol. 7, pp. 3–46). Minneapolis: University of Minnesota Press.

Baumrind, D. (1983). Familial antecedents of social competence in young children. *Psychological Bulletin, 94,* 132–142.

Bavolek, S. J. (1984). *Handbook for the Adult-Adolescent Parenting Inventory (AAPI)*. Eau Claire, WI: Family Development Resources.

Bayne, R. (1997). Survival. In I. Horton & V. Varma (Eds.), *The needs of counsellors and psychotherapists: emotional, social, physical, professional* (pp. 183–198). London: Sage.

Beavers, W. R., Hampson, R. B., & Hulgus, Y. F. (1985). Commentary: The Beavers systems approach to family assessment. *Family Process, 24,* 398–405.

Beck, A. T. (1991). *Beck Scale for Suicide Ideation: Manual*. San Antonio: Psychological Corporation.

Beck, A. T. (1993). *Beck Hopelessness Scale: Manual*. San Antonio: Psychological Corporation.

Beck, A. T., Steer, R. A., & Brown, G. K. (1996). *Beck Depression Inventory—II: Manual*. San Antonio: Psychological Corporation.

Beck, J. C. (1987). The potentially violent patient: Legal duties, clinical practice, and risk management. *Psychiatric Annals, 17,* 695–699.

Beck, J. C. (1988). The therapist's legal duty when the patient may be violent. *Psychiatric Clinics of North America, 11,* 665–679.

Beck, J. S. (1995). *Cognitive therapy: Basics and beyond*. New York: Guilford.

Beck, S. J. (1950a). *Rorschach's test. I. Basic processes*. New York: Grune & Stratton.

Beck, S. J. (1950b). *Rorschach's test. III. Advances in interpretation*. New York: Grune & Stratton.

Beck, S. J. (1957). *Rorschach's test. II. A variety of personality pictures* (rev. ed.). New York: Grune & Stratton.

Bellak, L. (1993). *The T.A.T., C.A.T., and S.A.T. in clinical use* (5th ed.). Boston: Allyn & Bacon.

Bellak, L., & Siegel, H. (1983). *Brief and emergency psychotherapy*. Larchmont, NY: C.P.S.

Bender, L. (1938). *A visual motor Gestalt test and its clinical use*. New York: American Orthopsychiatric Association.

Berlin, F., Malin, M., & Dean, S. (1991). Effects of statutes requiring psychiatrists to report suspected sexual abuse of children. *American Journal of Psychiatry, 148,* 449–455.

Berliner, L., & Elliott, D. M. (1996). Sexual abuse of children. In J. Briere, L. Berliner, J. A. Bulkley, C. Jenny & T. Reid (Eds.), *The APSAC handbook on child maltreatment* (pp. 51–71). Thousand Oaks, CA: Sage.

Berman, P. S. (1997). *Case conceptualization and treatment planning: Exercises for integrating theory and clinical practice*. New York: Sage.

Beutler, L. E., & Berren, M. R. (1995). *Integrative assessment of adult personality*. New York: Guilford.

Blackstone, P. (1991). *Things they never taught me in therapy school: Serious problems, surprising solutions*. Kingston, WA: Port Gamble.

Blaine, J. D., Horton, A. M., & Towle, L. H. (1995). *Diagnosis and severity of drug use and drug dependence.* Rockville, MD: National Institute on Drug Abuse.

Block, J. H. (1965). *The child rearing practices report.* Berkeley: Institute of Human Development, University of California at Berkeley.

Bongar, B., Maris, R. W., Berman, A. L., & Litman, R. E. (1992). Outpatient standards of care and the suicidal patient. *Suicide and Life-Threatening Behavior, 22,* 453–478.

Borum, R., Swartz, M., & Swanson, J. (1996). Assessing and managing violence risk in clinical practice. *Journal of Practical Psychiatry and Behavioral Health, 2,* 205–215.

Borysenko, J. (1987). *Minding the body, mending the mind.* Boston: Addison-Wesley.

Bowden, C. L. (1996). Antimanic agents. In D. F. Klein & L. P. Rowland (Eds.), *Current psychotherapeutic drugs* (pp. 59–67). New York: Brunner/Mazel.

Boylan, J. C., Malley, P. B., & Scott, J. (1995). *Practicum and internship: Textbook for counseling and psychotherapy.* Bristol, PA: Accelerated Development.

Brassard, M. R., Germaine, R., & Hart, S. N. (1987). *The psychological maltreatment of children and youth.* New York: Pergamon.

Breggin, P. R. (1997). *The heart of being helpful: Empathy and the creation of a healing presence.* New York: Springer.

Breier, A. (1993). Paranoid disorder: Clinical features and treatment. In D. L. Dunner (Ed.), *Current psychiatric therapy* (pp. 154–159). Philadelphia: W. B. Saunders.

Brems, C. (1989). Dimensionality of empathy and its correlates. *Journal of Psychology, 123,* 329–337.

Brems, C. (1990). *Manual for a self-psychologically oriented parent education program.* Anchorage, AK: University of Alaska Anchorage.

Brems, C. (1993). *A comprehensive guide to child psychotherapy.* Boston: Allyn & Bacon.

Brems, C. (1994). *The child therapist: Personal traits and markers of effectiveness.* Boston: Allyn & Bacon.

Brems, C. (1996). Substance use, mental health, and health in Alaska: Emphasis on Alaska Native peoples. *Arctic Medical Research, 55,* 135–147.

Brems, C. (1998). Implications of Daniel Stern's model of self development for child psychotherapy. *Journal of Psychological Practice, 3,* 141–159.

Brems, C. (1999a). *Psychotherapy: Processes and techniques.* Boston: Allyn & Bacon.

Brems, C. (1999b). Cultural issues in psychological assessment: Problems and possible solutions. *Journal of Psychological Practice, 4,* 88–117.

Brems, C., Baldwin, M., & Baxter, S. (1993). Empirical evaluation of a self-psychologically oriented parent education program. *Family Relations, 42,* 26–30.

Brems, C., & Johnson, M. E. (1997). Co-occurrence of substance use and other psychiatric disorder: Research and clinical implications. *Professional Psychology: Research and Practice, 28,* 437–447.

Brems, C., & Namyniuk, L. (in press). Comorbidity and related factors among ethnically diverse substance using women. *Journal of Addictions & Offender Counseling.*

Brems, C., Thevinin, D. & Routh, D. (1991). History of clinical psychology. In C. E. Walker (Ed.), *Clinical psychology: Historical and research foundations* (pp. 3–35). New York: Plenum.

Briere, J. (1997). *Trauma Symptom Inventory: Professional manual.* Odessa, FL: Psychological Assessment Resources.

Briere, J., Berliner, L., Bulkley, J. A., Jenny, C., & Reid, T. (1996). *The APSAC handbook on child maltreatment.* Thousand Oaks, CA: Sage.

Brinson, J. (1997). Reach out and touch someone. In J. A. Kottler (Ed.), *Finding your way as a counselor* (pp. 165–168). Alexandria, VA: American Counseling Association.

Bromley, M. A., & Riolo, J. A. (1988). Complying with mandated child protective reporting: A challenge for treatment professionals. *Alcoholism Treatment Quarterly, 5,* 83–96.

Brooks, J. B. (1994). *Parenting in the 90s.* Mountain View, CA: Mayfield.

Brooks, J. B. (1996). *The process of parenting* (4th ed.). Mountain View, CA: Mayfield.

Brown, T. A., & Barlowe, D. H. (1992). Comorbidity among anxiety disorders: Implications for treatment and DSM-IV. *Journal of Consulting and Clinical Psychology, 60,* 835–844.

Brown, T. M., & Stoudemire, A. (1998). *Psychiatric side effects of prescription and over-the-counter medications: Recognition and management.* Washington, DC: American Psychiatric Press.

Bukstein, O. G. (1995). *Adolescent substance abuse: Assessment, prevention, and treatment.* New York: Wiley.

Bulkley, J. A., Feller, J. N., Stern, P., & Roe, R. (1996). Child abuse and neglect laws and legal proceedings. In J. Briere, L. Berliner, J. A. Bulkley, C. Jenny, & T. Reid (Eds.), *The APSAC handbook on child maltreatment* (pp. 271–296). Thousand Oaks, CA: Sage.

Burkhardt, S. A., & Rotatori, A. F. (1995). *Treatment and prevention of childhood sexual abuse: A child-generated model.* Washington, DC: Taylor & Francis.

Burns, G. W. (1998). *Nature-guided therapy: Brief integrative strategies for health and well-being.* Bristol, PA: Brunner/Mazel.

Butcher, J. N., Dahlstrom, W. G., Graham, J. R., Tellegen, A., & Kaemmer, B. (1991). *Minnesota Multiphasic Personality Inventory—2 (MMPI-2): Manual for administration and scoring.* Minneapolis: University of Minnesota Press.

Canter, M. B., Bennett, B. E., Jones, S. E., & Nagy, T. F. (1994). *Ethics for psychologists: A commentary on the APA ethics code.* Washington, DC: American Psychological Association.

Cappuzzi, D., & Nystul, M. S. (1986). The suicidal adolescent. In L. B. Golden & D. Cappuzzi (Eds.), *Helping families help children: Family interventions with school-related problems* (pp. 23–32). Springfield, IL: Charles C. Thomas.

Carey, K. B., & Teitelbaum, L. M. (1996). Goals and methods of alcohol assessment. *Professional Psychology: Research and Practice, 27,* 460–466.

Carroll, M. (1996). *Counselling supervision: theory, skills, and practice.* London: Cassell.

Casey, D. E. (1993). Tardive dyskinesia. In D. L. Dunner (Ed.), *Current psychiatric therapy* (pp. 165–169). Philadelphia: W. B. Saunders.

Cassem, E. H. (1995). Depressive disorders in the medically ill: An overview. *Psychosomatics, 36,* S2–S10.

Castillo, R. J. (1997). *Culture and mental illness: A client-centered approach.* Pacific Grove, CA: Brooks/Cole.

Centers for Disease Control and Prevention (1993, February). *HIV/AIDS surveillance report.* Atlanta: Author.

Chappel, J. N. (1993). Training of residents and medical students in the diagnosis and treatment of dual diagnosis patients. *Journal of Psychoactive Drugs, 25,* 293–300.

Chiauzzi, E. (1994). Brief inpatient treatment of dual diagnosis patients. *New Directions for Mental Health Services, 63,* 47–57.

Choca, J. P. (1988). *Manual for clinical psychology trainees* (2nd ed.). New York: Brunner/Mazel.

Chopra, D. (1994). *The seven spiritual laws of success.* San Rafael, CA: Amber-Allen.

Chrzanowski, G. (1989). The significance of the analyst's individual personality in the therapeutic relationship. *Journal of the American Academy of Psychoanalysis, 17,* 597–608.

Clark, D. C. (1995). Epidemiology, assessment, and management of suicide in depressed patients. In E. E Beckham & W. R. Leber, (Eds.), *Handbook of depression* (pp. 526–538). New York: Guilford.

Clark, R. E. (1994). Family costs associated with severe mental illness and substance use. *Hospital and Community Psychiatry, 45,* 808–813.

Clement, J. A., Williams, E. B., & Waters, C. (1993). The client with substance abuse/mental illness: Mandate for collaboration. *Archives of Psychiatric Nursing, 7,* 189–196.

Connors, G. (1995). Screening for alcohol problems. In J. P. Allen & M. Columbus (Eds.), *Assessing alcohol problems: A guide for clinicians and researchers* (pp. 17–30). Bethesda, MD: National Institute on Alcohol Abuse and Alcoholism.

Cooper-Patrick, C., Crum, R. M., & Ford, D. E. (1994). Identifying suicidal ideation in general medical patients. *Journal of the American Medical Association, 272,* 1757–1762.

Corey, G., Corey, M. S., & Callanan, P. (1988). *Issues and ethics in the helping professions* (3rd ed.). Pacific Grove, CA: Brooks/Cole.

Cormier, L. S., & Hackney, H. (1987). *The professional counselor: A process guide to helping.* Englewood Cliffs, NJ: Prentice Hall.

Cormier, W. H., & Cormier, L. S. (1999). *Interviewing strategies for helpers* (5th ed.). Pacific Grove, CA: Brooks/Cole.

Cornelius, J., Salloum, I., Mezzich, J., Cornelius, M.D., Fabrega, H., Ehler, J. G., Ulrich, R. F., Thase, M. E., & Mann, J. J. (1995). Disproportionate suicidality in patients with comorbid major depression and alcoholism. *American Journal of Psychiatry, 152,* 358–364.

Coster, J. S., & Schwebel, M. (1997). Well-functioning in professional psychologists. *Professional Psychology: Research and Practice, 28,* 5–13.

Cowley, D. S. (1993). Generalized anxiety disorder. In D. L. Dunner (Ed.), *Current psychiatric therapy* (pp. 263–268). Philadelphia: W. B. Saunders.

Crane, L. R. (1991). Epidemiology of infections in intravenous drug abusers. In D. P. Levine & J. D. Dobel (Eds.), *Infections in intravenous drug users.* New York: Oxford University Press.

Crawford, R. L. (1994). *Avoiding counselor malpractice.* Alexandria, VA: American Counseling Association.

Crayhon, R. (1994). *Nutrition made simple: A comprehensive guide to the latest findings in optimal nutrition.* New York: M. Evans.

Crowe, A. H. (1990). *A core curriculum on AIDS and HIV disease: Psychosocial aspects* (module 3). Lexington, KY: University of Kentucky.

Crowe, A. H., & Reeves, R. (1994). *Treatment for alcohol and other drug abuse: Opportunities for coordination*

(Technical assistance publication series 11). Rockville, MD: Center for Substance Abuse Treatment.

Cull, J. G., & Gill, W. S. (1989). *Suicide Probability Scale: Manual.* Los Angeles: Western Psychological Services.

Dagadakis, C. S. (1993). Psychiatric emergencies. In D. L. Dunner (Ed.), *Current psychiatric therapy* (pp. 446–456). Philadelphia: W. B. Saunders.

Dana, R. H. (1993). *Multicultural assessment perspectives for professional psychology.* Boston: Allyn & Bacon.

Darling, N., & Steinberg, L. (1993). Parenting style as context: An integrative model. *Psychological Bulletin, 113,* 487–496.

Daro, D., & Gelles, R. J. (1992). Public attitudes and behaviors with respect to child abuse prevention. *Journal of Interpersonal Violence, 7,* 517–531.

Darou, W. G. (1992). Native Canadians and intelligence testing. *Canadian Journal of Counselling, 26,* 96–99.

Davis v. Lhim, 422, N.W.2d., 688 (Mich. 1988).

Davis, D. (1993). Multiple personality, fugue, and amnesias. In D. L. Dunner (Ed.), *Current psychiatric therapy* (pp. 328–333). Philadelphia: W. B. Saunders.

Davis, J. M., Janicak, P. G., & Khan, A. (1993). Neuroleptic malignant syndrome. In D. L. Dunner (Ed.), *Current psychiatric therapy* (pp. 170–175). Philadelphia: W. B. Saunders.

DeBecker, G. (1997). *The gift of fear: Survival signals that protect us from violence.* Boston: Little, Brown.

Department of Health and Human Services (1993). *Maternal substance use assessment methods reference manual.* Rockville, MD: Author.

Derogatis, L. R. (1975). *Derogatis Sexual Functioning Inventory.* Baltimore: Johns Hopkins University Press.

Derogatis, L. R. (1992). *The SCL-90-R: Administration, scoring, and procedures manual.* Towson, MD: Clinical Psychometric Research.

Diamond, R. J. (1998). *Instant psychopharmacology: A guide for the nonmedical mental health professional.* New York: Norton.

Dietzel, L. A. (1995). *Parenting with respect and peacefulness.* Lancaster, PA: Starburst.

Dinkmeyer, D., & McKay, G. D. (1976). *The parent's handbook.* Circle Pines, MN: American Guidance Service.

Dixon, L, McNary, S., & Lehman, A. (1995). Substance abuse and family relationships of persons with severe mental illness. *American Journal of Psychiatry, 152,* 456–458.

Dodds, J. B. (1985). *A child psychotherapy primer.* New York: Human Sciences Press.

Dolan, Y. M. (1991). *Resolving sexual abuse: Solution-focused therapy and Ericksonian hypnosis for adult survivors.* New York: W. W. Norton.

Dolan, Y. M. (1996). An Ericksonian perspective on the treatment of sexual abuse. In J. K. Zeig (Ed.), *Ericksonian methods: The essence of the story* (pp. 395–414). New York: Brunner/Mazel.

Dorfman, R. A. (1998). *Paradigms of clinical social work* (Vol. 2). New York: Brunner/Mazel.

Dunner, D. L. (Ed.) (1993). *Current psychiatric therapy.* Philadelphia: W. B. Saunders.

Eells, T. D. (1997). *Handbook of psychotherapy case formulation.* New York: Guilford.

Egan, G. (1994). *The skilled helper* (5th ed.). Pacific Grove, CA: Brooks/Cole.

El-Guebaly, N. (1990). Substance abuse and mental disorders: The dual diagnoses concept. *Canadian Journal of Psychiatry, 35,* 261–267.

Elkind, D. (1995). *Ties that stress: The new family imbalance.* Cambridge, MA: Harvard University Press.

Erickson, M. F., & Egeland, B. (1996). Child neglect. In J. Briere, L. Berliner, J. A. Bulkley, C. Jenny, & T. Reid (Eds.), *The APSAC handbook on child maltreatment* (pp. 4–20). Thousand Oaks, CA: Sage.

Ewing, J. A. (1984). Detecting alcoholism: The CAGE questionnaire. *Journal of the American Medical Association, 252,* 1905–1907.

Exner, J. E. (1991). *The Rorschach: A comprehensive system; Volume 2: Interpretation* (2nd ed.). New York: Wiley.

Exner, J. E. (1993). *The Rorschach: A comprehensive system; Volume 1. Basic foundations* (3rd ed.). New York: Wiley.

Fabrega, H. (1992). The role of culture in a theory of psychiatric illness. *Social Science and Medicine, 35,* 91–103.

Faiver, C., Eisengart, S., & Colonna, R. (1995). *The counselor intern's handbook.* Pacific Grove, CA: Brooks/Cole.

Fantuzzo, J. W., & McDermott, P. (1992). Clinical issues in the assessment of family violence involving children. In R. T. Ammerman & M. Hersen (Eds.), *Assessment of family violence* (pp. 11–25). New York: Wiley.

Fast, G. A., & Preskorn, S. H. (1993). Therapeutic drug monitoring. In D. L. Dunner (Ed.), *Current psychiatric therapy* (pp. 529–534). Philadelphia: W. B. Saunders.

Fava, G. A., Morphy, M. A., & Sonino, N. (1994). Affective prodromes of medical illness. *Psychotherapy and Psychosomatics, 62,* 141–145.

Fields, R. (1985). *Chop wood, carry water*. Los Angeles: Jeremy P. Tarcher.

Finkelhor, D. (1986). Sexual abuse: Beyond the family systems approach. *Journal of Psychotherapy and the Family, 2(2)*, 53–65.

Finkelhor, D., & Dziuba-Leatherman, A. (1994). Victimization of children. *American Psychologist, 49*, 173–183.

Firestone, R. W. (1997). *Suicide and the inner voice: Risk assessment, treatment, and case management*. Thousand Oaks, CA: Sage.

First, M. B., & Gladis, M. M. (1993). Diagnosis and differential diagnosis of psychiatric and substance use disorders. In J. Solomon, S. Zimberg, & E. Shollar (Eds.), *Dual diagnosis: Evaluation, treatment, training, and program development* (pp. 23–36). New York: Plenum Medical Books.

Fogelsanger, A. (1994). *See yourself well: Guided visualizations and relaxation techniques*. Brooklyn: Equinox.

Folen, R. A., Kellar, M. A., James, L. C., Porter, R. I., & Peterson, D. R. (1998). Expanding the scope of clinical practice: The physical examination. *Professional Psychology: Research and Practice, 29*, 155–159.

Forer, B. R. A. (1950). A structured sentence completion test. *Journal of Projective Techniques, 14*, 15–30.

Forster, P. (1994). Accurate assessment of short-term suicide risk in a crisis. *Psychiatric Annals, 24*, 571–578.

Foster, P., & Oxman, T. (1994). A descriptive study of adjustment disorder diagnosis in general hospital patients. *Irish Journal of Psychological Medicine, 11*, 153–157.

Freimuth, M. (1996). Combining psychotherapy and psychopharmacology: With or without prescription privileges. *Psychotherapy, 33*, 474–478.

Fremouw, W. J., Perczel, M., & Ellis, T. E. (1990). *Suicide risk: Assessment and response guidelines*. Boston: Allyn & Bacon.

Freud, S. (1900). The interpretation of dreams. In J. Strachey (Trans.), *The standard edition of the complete psychological works of Sigmund Freud*, Vol. 485. New York: Norton.

Fujimura, L. E., Weis, D. M., & Cochran, J. R. (1985). Suicide: Dynamics and implications for counseling. *Journal of Counseling and Development, 64*, 612–615.

Galanter, M., Egelko, S., Edwards, H., & Vergaray, M. (1994). A treatment system for combined psychiatric and addictive illness. *Addiction, 89*, 1227–1235.

Ganz, P. A. (1988). Patient education as a moderator variable of psychological distress. *Journal of Psychosocial Oncology, 6*, 181–197.

Garbarino, J., Guttmann, E., & Seeley, J. A. (1986). *The psychologically battered child*. San Francisco: Jossey-Bass.

Garfield, P. (1977). *Creative dreaming*. New York: Ballantine.

Garner, D. M. (1996). *Eating Disorders Inventory—2: Manual*. Odessa, FL: Psychological Assessment Resources.

Garrett, L. (1994). *The coming plague: Newly emerging diseases in a world out of balance*. New York: Farrar, Straus, & Giroux.

Gerard, A. B. (1994). *Parent-Child Relationship Inventory (PCRI): Manual*. Los Angeles: Western Psychological Services.

Gerner, R. H. (1993). Psychiatric effects of nonpsychiatric medications. In D. L. Dunner (Ed.), *Current psychiatric therapy* (pp. 464–470). Philadelphia: W. B. Saunders.

Ghadirian, A. N., Englesman, F., Leichner, P., & Marshall, M. (1993). Prevalence of psychosomatic and other medical illnesses in anorexic and bulimic patients. *Behavioural Neurology, 6*, 123–127.

Gil, E. (1988). *Outgrowing the pain: A book for and about adults abused as children*. New York: Dell.

Gil, E. (1991). *The healing power of play*. New York: Guilford.

Giles, D. E., & Buysse, D. J. (1993). Parasomniacs. In D. L. Dunner (Ed.), *Current psychiatric therapy* (pp. 360–372). Philadelphia: W. B. Saunders.

Gilliland, B. E., & James, R. K. (1993). *Crisis intervention strategies* (2nd ed.). Pacific Grove, CA: Brooks/Cole.

Gillin, J. S. (1993). Clinical sleep-wake disorders in psychiatric practice: Dyssomniacs. In D. L. Dunner (Ed.), *Current psychiatric therapy* (pp. 373–379). Philadelphia: W. B. Saunders.

Giordano, P. J. (1997). Establishing rapport and developing interview skills. In J. R. Matthews & C. E. Walker (Eds.), *Basic skills and professional issues in clinician psychology* (pp. 59–82). Boston: Allyn & Bacon.

Gitlin, M. J. (1997). *The psychotherapist's guide to psychopharmacology* (2nd ed.). New York: Free Press.

Gleser, G. C., & Ihlevich, D. (1969). An objective instrument to measure defense mechanisms. *Journal of Consulting and Clinical Psychology, 33*, 51–60.

Glickauf-Hughes, C., & Mehlman, E. (1996). Narcissistic issues in therapists: Diagnostic and treatment considerations. *Psychotherapy, 32*, 213–221.

Golden, C. J. (1981). A standardized version of Luria's neuropsychological tests: Quantitative and qualitative approach in neuropsychological evaluation. In F. E. Filskov & T. J. Boll (Eds.), *Handbook of neuropsychology* (pp. 608–642). New York: Wiley.

Golden, C. J., Purish, A. D., & Hammeke, T. A. (1980). *Luria-Nebraska Neuropsychological Battery Manual.* Los Angeles: Western Psychological Services.

Good, B. J. (1993). Culture, diagnosis, and comorbidity. *Culture, Medicine and Psychiatry, 16,* 427–446.

Goodenough, F. (1926). *Measurement of intelligence by drawings.* New York: World Book.

Gordon, T. (1970). *PET: Parent effectiveness training.* New York: Wyden.

Gordon, T., & Sands, J. (1978). *P.E.T. in action.* New York: Bantam.

Gough, H. (1996). *California Psychological Inventory manual* (3rd ed.). Palo Alto, CA: Consulting Psychologists Press.

Gould, S. J. (1981). *The mismeasure of man.* New York: Norton.

Greenberg, S. A., & Shuman, D. W. (1997). Irreconcilable conflict between therapeutic and forensic roles. *Professional Psychology: Research and Practice, 28,* 50–57.

Gregory, R. J. (1996). *Psychological testing: History, principles, and applications* (2nd ed.). Boston: Allyn & Bacon.

Grollman, E. A. (1988). *Suicide: Prevention, intervention, postvention.* Boston: Beacon.

Groth-Marnat, G. (1997). *Handbook of psychological assessment* (3rd ed.). New York: Wiley.

Gutheil, T. G. (1980). Paranoia and progress notes: A guide to forensically informed psychiatric record-keeping. *Hospital and Community Psychiatry, 31,* 479–482.

Guy, J. D. (1987). *The personal life of the psychotherapist.* New York: Wiley.

Haas, E. M. (1992). *Staying healthy with nutrition.* Berkeley: Celestial Arts.

Hammond, K. W., Scurfield, R. M., & Risse, S. C. (1993). Post-traumatic stress disorder. In D. L. Dunner (Ed.), *Current psychiatric therapy* (pp. 288–294). Philadelphia: W. B. Saunders.

Hamner, M. B. (1994). Exacerbation of posttraumatic stress disorder symptoms with medical illness. *General Hospital Psychiatry, 16,* 135–137.

Handler, L. (1996). The clinical use of drawings: Draw-A-Person, House-Tree-Person, and Kinetic Family Drawings. In C. S. Newmark (Ed.), *Major psychological assessment instruments* (2nd ed.; pp. 206–293). Boston: Allyn & Bacon.

Hanh, T. N. (1975). *The miracle of mindfulness.* Boston: Beacon Press.

Hart, S. N., Brassard, M. R., & Karlson, H. C. (1996). Psychological maltreatment. In J. Briere, L. Berliner,

J. A. Bulkley, C. Jenny, & T. Reid (Eds.), *The APSAC handbook on child maltreatment* (pp. 72–89). Thousand Oaks, CA: Sage.

Harter, S. (1986). *Manual for the Adult Self Perception Profile.* Denver: University of Denver.

Haugaard, J. J. (1992). Epidemiology and family violence involving children. In R. T. Ammerman & M. Hersen (Eds.), *Assessment of family violence* (pp. 89–107). New York: Wiley.

Haynes-Seman, C., & Baumgarten, D. (1994). *Children speak for themselves.* New York: Brunner/Mazel.

Heaton, R. K., Chelune, G. D., Talley, J. T., Kay, G. G., & Curtiss, G. (1993). *Wisconsin Card Sorting Test manual-revised and expanded.* Odessa, FL: Psychological Assessment Resources.

Hedlund, J. L., & Vieweg, B. W. (1984). The Michigan Alcoholism Screening Test (MAST): A comprehensive review. *Journal of Operational Psychiatry, 15,* 55–64.

Heilbrun, A. B. (1996). *Criminal dangerousness and the risk of violence.* Lanham, NY: University Press of America.

Heiman, J. R. (1993). Sexual dysfunctions. In D. L. Dunner (Ed.), *Current psychiatric therapy* (pp. 346–352). Philadelphia: W. B. Saunders.

Helfer, M. E., Kempe, R. S., & Klugman, R. D. (Eds.). (1997). *The battered child* (5th ed.). Chicago: University of Illinois Press.

Helmchen, H. (1991). The impact of diagnostic systems on treatment planning. *Integrative Psychiatry, 7,* 16–20.

Helzer, J. E., & Przybeck, T. R. (1988). The co-occurrence of alcoholism with other psychiatric disorders in the general population and its impact on treatment. *Journal of Studies on Alcohol, 49,* 219–224.

Herlihy, B., & Corey, G. (1992). *Dual relationships in counseling.* Alexandria, VA: American Counseling Association.

Herlihy, B., & Corey, G. (1996). *ACA ethical standards casebook* (5th ed.). Alexandria, VA: American Counseling Association.

Herlihy, B., & Corey, G. (1997). *Boundary issues in counseling: Multiple roles and responsibilities.* Alexandria, VA: American Counseling Association.

Hills, H. (1995). Dual diagnosis: Evaluating and treating comorbid disorders. *American Psychological Association Division 50 Newsletter,* 17–19, 25.

Hocking, L. B., & Koenig, H. G. (1995). Anxiety in medically ill older patients: A review and update. *International Journal of Psychiatry in Medicine, 25,* 221–238.

Hoff, L. A. (1989). *People in crisis: Understanding and helping* (3rd ed.). Redwood City, CA: Addison-Wesley.

Hoffman, L., & Halmi, K. A. (1993). Treatment of anorexia nervosa. In D. L. Dunner (Ed.), *Current psychiatric therapy* (pp. 390–396). Philadelphia: W. B. Saunders.

Holsopple, J. Q., & Miale, F. (1954). *Sentence completion.* Springfield, IL: Charles C. Thomas.

Hood, A. B., & Johnson, R. W. (1997). *Assessment in counseling: A guide to the use of psychological assessment procedures.* Alexandria, VA: American Counseling Association.

Hornig-Rohan, M., & Amsterdam, J. D. (1994). Clinical and biological correlates of treatment-resistant depression: An overview. *Psychiatric Annals, 24,* 220–227.

Horton, A. M. (1995, October). *Comorbidity of drug abuse treatment: Where and what.* Paper presented at the 3rd Annual Conference on Psychopathology, Pharmacology, Substance Abuse, and Culture, Los Angeles.

Hudson, W. W. (1992). *The WALMYR Assessment Scales manual.* Tempe, AZ: WALMYR.

Hutchins, D. E., & Vaught, C. C. (1997). *Helping relationships and strategies* (3rd ed.). Pacific Grove, CA: Brooks/Cole.

Iijima Hall, C. C. (1997). Cultural malpractice: The growing obsolescence of psychology with the changing U.S. population. *American Psychologist, 52,* 642–651.

Ivey, A. E. (1994). *Intentional interviewing and counseling.* Pacific Grove, CA: Brooks/Cole.

Ivey, A. E. (1995). Psychotherapy as liberation: Toward specific skills and strategies in multicultural counseling and therapy. In J. G. Ponterotto, J. M. Casas, L. A. Suzuki, & C. M. Alexander (Eds.), (1995). *Handbook of multicultural counseling* (pp. 53–72). Thousand Oaks, CA: Sage.

Iwaniec, D. (1995). *The emotionally abused and neglected child: Identification, assessment, and intervention.* New York: Wiley.

Jarvis, P., & Barth, J. T. (1994). *A guide to interpretation of the Halstead-Reitan Battery for adults.* Odessa, FL: Psychological Assessment Resources.

Jevne, R. F., & Williams, D. R. (1998). *When dreams don't work: Professional caregivers and burnout.* Amityville, NY: Baywood.

Johns, H. (1996). *Personal development in counsellor training.* London: Cassell.

Johns, H. (1997). Self-development: Lifelong learning? In I. Horton & V. Varma (Eds.), *The needs of counsellors and psychotherapists: emotional, social, physical, professional* (pp. 54–67). London: Sage.

Johnson, F. (1993). *Dependency and Japanese socialization.* New York: New York University Press.

Johnson, M. E. (1993). A culturally sensitive approach to child psychotherapy. In C. Brems, *Comprehensive guide to child psychotherapy* (pp. 68–94). Boston: Allyn & Bacon.

Johnson, M. E., Fortman, J., & Brems, C. (1993). *Between two people: Exercises toward intimacy.* Alexandria, VA: American Counseling Association.

Jorge, R., & Robinson, R. G. (1993). Organic mood, anxiety, and anxiety disorders. In D. L. Dunner (Ed.), *Current psychiatric therapy* (pp. 73–79). Philadelphia: W. B. Saunders.

Jung, C. G. (1974). *Dreams.* Princeton: Princeton University Press.

Kabat-Zinn, J. (1994). *Wherever you go, there you are.* New York: Hyperion.

Kalichman, S. C., Craig, M. E., & Follingstad, D. R. (1990). Professionals' adherence to mandatory child abuse reporting laws: Effects of responsibility attribution, confidence ratings and situational factors. *Child Abuse and Neglect, 14,* 69–77.

Kamphaus, R. W., & Frick, P. J. (1996). *Clinical assessment of child and adolescent personality and behavior.* Boston: Allyn & Bacon.

Kane, J. M. (1996). Antipsychotic agents. In D. F. Klein & L. P. Rowland (Eds.), *Current psychotherapeutic drugs* (pp. 113–142). New York: Brunner/Mazel.

Kaplan, R. M., & Saccuzzo, D. P. (1997). *Psychological testing: Principles, application, and issues* (4th ed.). Pacific Grove, CA: Brooks/Cole.

Karoly, P. (1993). Goal systems: An organizing framework for clinician assessment and treatment planning. *Psychological Assessment, 5,* 273–280.

Katon, W. J. (1993). Somatization disorder, hypochondriasis, and conversion disorder. In D. L. Dunner (Ed.), *Current psychiatric therapy* (pp. 314–320). Philadelphia: W. B. Saunders.

Katon, W. J. (1996). The impact of depression on chronic medical illness. *General Hospital Psychiatry, 18,* 215–219.

Katz, I. R. (1993). Delirium. In D. L. Dunner (Ed.), *Current psychiatric therapy* (pp. 65–72). Philadelphia: W. B. Saunders.

Katz, I. R., Streim, J., & Parmelee, P. (1994). Psychiatric-medical comorbidity: Implications for health delivery services and for research in depression. *Biological Psychiatry, 36,* 141–145.

Kaufman, A. S., & Kaufman, N. L. (1993). *Kaufman Adolescent Adult Intelligence Test: Manual.* Odessa, FL: Psychological Assessment Resources.

Keinan, G., Almagor, M., & Ben-Porath, Y. S. (1989). A reevaluation of the relationship between psychotherapeutic orientation and perceived personality characteristics. *Psychotherapy, 26,* 218–226.

Kellerman, H., & Burry, A. (1991). *Handbook of psychodiagnostic testing: An analysis of personality in the psychological testing report* (2nd ed.). Boston: Allyn & Bacon.

Kempe, C. H., Silverman, F. N., Steele, B. F., Droegmueller, W., & Silver, H. K. (1962). The battered child syndrome. *Journal of the American Medical Association, 181,* 17–24.

Kessler, R. C. (1995). The national comorbidity survey: Preliminary results and future directions. *International Journal of Methods in Psychiatric Research, 5,* 139–151.

Kessler, R. C., McGonagle, K. A., Zhao, S., Nelson, C. B., Hughes, M., Eshleman, S., Wittchen, H. U., Kendler, K. S. (1994). Lifetime and 12-month prevalence of the DSM-III-R psychiatric disorders in the United States. *Archives of General Psychiatry, 51,* 8–19.

Kilburg, R. R., Nathan, P. E., & Thoreson, R. W. (1986). *Professionals in distress: Issues, syndromes, and solutions in psychology.* Washington, DC: American Psychological Association.

Kitchens, J. M. (1994). Does this patient have an alcohol problem? *Journal of the American Medical Association, 272,* 1782–1787.

Klein, D. F., & Rowland, L. P. (1996). *Current psychotherapeutic drugs.* New York: Brunner/Mazel.

Kleinke, C. L. (1994). *Common principles of psychotherapy.* Pacific Grove, CA: Brooks/Cole.

Klonoff, E. A., & Landrine, H. (1997). *Preventing misdiagnosis of women: A guide to physical disorders that have psychiatric symptoms.* Thousand Oaks, CA: Sage.

Klopfer, B., Ainsworth, M. D., Klopfer, W. G., & Holt, R. R. (1954). *Developments in the Rorschach technique: Volume 1: Technique and theory.* New York: Harcourt, Brace & World.

Klopfer, B., Ainsworth, M. D., Klopfer, W. G., & Holt, R. R. (1956). *Developments in the Rorschach technique: Volume 2.* New York: New World.

Kloss, C. S. (1996). *Home visiting: Promoting healthy parent and child development.* Baltimore: Paul Brookes.

Knobel, M. (1990). Significance and importance of the psychotherapist's personality and experience. *Psychotherapy and Psychosomatics, 53,* 58–63.

Kohut, H. (1984). *How does analysis cure?* Chicago: International Universities Press.

Kohut, H., & Wolf, E. (1978). Disorders of the self and their treatment. *International Journal of Psychoanalysis, 59,* 413–425.

Kolevzon, M. S., Sowers-Hoag, K., & Hoffman, C. (1989). Selecting a family therapy model: The role of personality attributes in eclectic practice. *Journal of Marital and Family Therapy, 15,* 249–257.

Kolko, D. J. (1996). Child physical abuse. In J. Briere, L. Berliner, J. A. Bulkley, C. Jenny, & T. Reid (Eds.), *The APSAC handbook on child maltreatment* (pp. 21–50). Thousand Oaks, CA: Sage.

Koppitz, E. (1968). *Psychological evaluation of children's human figure drawings.* New York: Grune & Stratton.

Koppitz, E. M. (1975). *The Bender Gestalt test for young children (Vol. 2): Research and application, 1963–1975.* New York: Grune & Stratton.

Kornfield, J. (1993). *A path with heart.* New York: Bantam.

Kornfield, J. (1994). *Buddha's little instruction book.* New York: Bantam.

Kottler, J. A. (1992). *Compassionate therapy: Working with difficult clients.* San Francisco: Jossey-Bass.

Kottler, J. A. (1995). *Growing a therapist.* San Francisco: Jossey-Bass.

Kottler, J. A. (1997). *Finding your way as a counselor.* Alexandria, VA: American Counseling Association.

Kottler, J. A., & Brown, R. W. (1992). *Introduction to therapeutic counseling* (2nd ed.). Pacific Grove, CA: Brooks/Cole.

Krull, K. R. (1997). Psychological testing. In J. R. Matthews & C. E. Walker (Eds.), *Basic skills and professional issues in clinician psychology* (pp. 135–154). Boston: Allyn & Bacon.

Krumboltz, J. D., & Krumboltz, H. B. (1972). *Changing children's behavior.* Englewood Cliffs, NJ: Prentice-Hall.

Kuhn, C., Swartzwelder, S., & Wilson, W. (1998). *Buzzed: The straight facts about the most used and abused drugs from alcohol to ecstasy.* New York: W. W. Norton.

Land, H. (1998). The feminist approach to clinical social work. In R. A. Dorfman, *Paradigms of clinical social work* (Vol. 2; pp. 227–256). New York: Brunner/Mazel.

Lawson, C. (1993). Mother-son sexual abuse: Rare or under-reported? A critique of the research. *Child Abuse & Neglect, 17,* 261–269.

Lemma, A. (1996). *Introduction to psychopathology.* Thousand Oaks, CA: Sage.

Leonard, B. E. (1990). Stress and the immune system: Immunological aspects of depressive illness. *International Review of Psychiatry, 2,* 321–330.

LeShan, L. (1974). *How to meditate.* New York: Bantam.

Lewis, J. A., Dana, R. Q., & Blevins, G. A. (1988). *Substance abuse counseling: An individualized approach.* Pacific Grove, CA: Brooks/Cole.

Lewis, K. N., & Walsh, W. B. (1980). Effects of value-communication style and similarity of values on counselor evaluation. *Journal of Counseling Psychology, 27*, 305–314.

Lipkin, G. B., & Cohen, R. G. (1998). *Effective approaches to patients' behavior: A guide book for health care professionals, patients, and their caregivers* (5th ed). New York: Springer.

Litman, R. E. (1992). Predicting and preventing hospital and clinic suicides. In R. W. Maris, A. L. Berman, J. T. Maltsberger, & R. I. Yufit (Eds.), *Assessment and prediction of suicide* (pp. 362–380). New York: Guilford.

Litten, R. Z., & Allen, J. P. (1992). *Measuring alcohol consumption.* Totowa, NJ: Humana Press.

Loebel, J. P., Dager, S. R., & Kitchell, M. A. (1993). Alzheimer's disease. In D. L. Dunner (Ed.), *Current psychiatric therapy* (pp. 59–64). Philadelphia: W. B. Saunders.

Loeber, R. (1990). Development and risk factors of juvenile antisocial behavior and delinquency. *Clinical Psychology Review, 10,* 1–42.

Lorr, M., & Youniss, R. P. (1985). *Interpersonal Style Inventory: Manual.* Los Angeles: Western Psychological Services.

Ludwig, S., & Kornberg, A. E. (Eds.). (1992). *Child abuse: A medical reference* (2nd ed.). New York: Churchill Livingstone.

Luhrs, J. (1997). *The simple living guide.* New York: Broadway.

Lum, D. (1995). Cultural values and minority people of color. *Journal of Sociology and Social Welfare, 22,* 59–74.

Mahoney, M. J. (1997). Psychotherapists' personal problems and self-care patterns. *Professional Psychology: Research and Practice, 28,* 14–16.

Mahrer, A. R. (1989). *Dreamwork in psychotherapy and self-change.* New York: W. W. Norton.

Maris, R. W. (1992). The relationship of nonfatal suicide attempts to completed suicides. In R. W. Maris, A. L. Berman, J. T. Maltsberger, & R. I. Yufit (Eds.), *Assessment and prediction of suicide* (pp. 362–380). New York: Guilford.

Maris, R. W., Berman, A. L., Maltsberger, J. T., & Yufit, R. I. (Eds.), (1992). *Assessment and prediction of suicide.* New York: Guilford.

Mark, D., & Faude, J. (1997). *Psychotherapy of cocaine addiction: Entering the interpersonal world of the cocaine addict.* Northvale, NJ: Jason Aronson.

Matsumoto, D. (1994). *People: Psychology from a changing perspective.* Pacific Grove, CA: Brooks/Cole.

McGoldrick, M., & Gerson, R. (1985). *Genograms in family assessment.* New York: W. W. Norton.

McKay, M. M. (1994). The link between domestic violence and child abuse: Assessment and treatment considerations. *Child Welfare, 73,* 29–39.

McLellen, A. T., Luborsky, L., Cacciola, J., Griffith, J., Evans, F., Barr., H. L., & O'Brien, C. P. (1985). New data from the Addiction Severity Index: Reliability and validity in three centers. *Journal of Nervous and Mental Disease, 173,* 412–423.

Megargee, E. I. (1995). Assessing and understanding aggressive and violent patients. In J. N. Butcher (Ed.), *Clinical personality assessment: Practical approaches* (pp. 367–379). New York: Oxford.

Meier, S. , & Davis, S. R (1997). *The elements of counseling* (3rd ed.). Pacific Grove, CA: Brooks/Cole.

Messick, S. (1980). Test validity and the ethics of assessment. *American Psychologist, 35,* 1012–1027.

Meyer, R. G., & Deitsch, S. E. (1996). *The clinician's handbook: Integrated diagnostics, assessment, and intervention in adult and adolescent psychopathology.* Boston: Allyn & Bacon.

Meyers, J. E., & Meyers, K. R. (1996). *Rey Complex Figure Test and Recognition Trial: Professional manual.* Odessa, FL: Psychological Assessment Resources.

Miller, G. A. (1985). *The Substance Abuse Subtle Screening Inventory: Manual.* Spencer, IN: Spencer Evening World.

Miller, N. S. (1995). Psychiatric diagnosis in drug and alcohol addiction. *Alcoholism Treatment Quarterly, 12,* 75–92

Miller, N. S., Belkin, B. M., & Gibbons, R. (1994). Clinical diagnosis of substance use disorders in private psychiatric populations. *Journal of Substance Abuse Treatment, 11,* 387–392.

Miller, W. R., & Brown, S. A. (1997). Why psychologists should treat alcohol and drug problems. *American Psychologist, 52,* 1269–1279.

Millman, D. (1998). *Everyday enlightenment.* New York: Warner Books.

Millon, T., & Davis, R. D. (1996). The Millon Clinical Multiaxial Inventory—III (MCMI-III). In C. S. Newmark (Ed.), *Major psychological assessment instruments* (2nd ed.; pp. 108–147). Boston: Allyn & Bacon.

Milner, J. S. (1980). *The Child Abuse Potential Inventory: Manual* (2nd edition). DeKalb, IL: Psytec.

Mirin, S. M, Weiss, R. D., Michael, J., & Griffin, M. L. (1988). Psychopathology in substance abusers: Diagnosis and treatment. *American Journal on Drug and Alcohol Abuse, 14,* 139–157.

Monahan, J. (1993). Limiting therapist exposure to *Tarasoff* liability: Guidelines for risk management. *American Psychologist, 48,* 242–250.

Moos, R. H. (1993). *Coping Responses Inventory—Adult Form: Professional manual.* Odessa, FL: Psychological Assessment Resources.

Morey, L. (1991). *Personality Assessment Inventory: Professional manual.* Odessa, FL: Psychological Assessment Resources.

Morley, J. E., & Krahn, D. D. (1987). Endocrinology for the psychiatrist. In C. B. Nemeroff & P. T. Loosen (Eds.), *Handbook of clinical psychoneuroendocrinology* (pp. 3–37). New York: Guilford.

Morrison, J. (1995). *The first interview: Revised for DSM-IV.* New York: Guilford.

Morrison, J. (1997). *When psychological problems mask medical disorders: A guide for psychotherapists.* New York: Guilford.

Moss, R. W. (1996). *Cancer therapy: The independent consumer's guide to non-toxic treatment and prevention.* Brooklyn: Equinox.

Murray, H. A. (1943). *The Thematic Apperception Test manual.* Cambridge, MA: Harvard University Press.

Namyniuk, L. (1996, November). *Cultural considerations in substance abuse treatment.* Paper presented at the 3rd Biennial Conference of the Alaska Psychological Association, Anchorage, AK.

Namyniuk, L., Brems, C., & Clarson, S. (1997). Dena A Coy: A model program for the treatment of pregnant substance-abusing women. *Journal of Substance Abuse Treatment, 14,* 1–11.

Narrow, W. E., Regier, D. A., Rae, D. S., Manderscheid, R. W., & Locke, B. Z. (1993). Use of services by persons with mental and addictive disorders. *Archives of General Psychiatry, 50,* 95–107.

National Association of Social Workers (1993). *Code of ethics.* Silver Springs, MD: Author.

National Institute on Drug Abuse (1994). *Mental health assessment and diagnosis of substance abuse.* Rockville, MD: US Department of Health and Human Services.

Newman, M. M., & Gold, M. S. (1992). Preliminary findings of patterns of substance abuse in eating disorder patients. *American Journal of Drug and Alcohol Abuse, 18,* 207–211.

Novick, D. M. (1992). The medically ill substance abuser. In J. H. Lowinson, P. Ruiz, R. B. Millman, & J. G. Langrod (Eds.), *Substance abuse: A comprehensive textbook* (2nd ed.). Baltimore: Williams & Wilkins.

Nugent, F. A. (1994). *An introduction to the profession of counseling.* New York: Macmillan.

Null, G. (1995). *Nutrition and the mind.* New York: Seven Stories.

Nystul, M. S. (1993). *The art and science of counseling.* New York: Macmillan.

O'Hagan, K. (1993). *Emotional and psychological abuse of children.* Buckingham: Open University Press.

Oates, R. K. (1996). *The spectrum of child abuse.* New York: Brunner/Mazel

Okun, B. F. (1997). *Effective helping: Interviewing and counseling techniques* (5th ed.). Pacific Grove, CA: Brooks/Cole.

Olson, K. R., Jackson, T. T., & Nelson, J. (1997). Attributional biases in clinical practice. *Journal of Psychological Practice, 3,* 27–33.

Padilla, A. M. (1988). Early psychological assessments of Mexican-American children. *Journal of the History of Behavioral Sciences, 24,* 111–117.

Patterson, L. E., & Welfel, E. R. (1993). *The counseling process.* Pacific Grove, CA: Brooks/Cole.

Patterson, W., Dohn, H., Bird, J., & Patterson, G. (1983). Evaluation of suicidal patients. The SAD PERSONS Scale. *Psychosomatics, 24,* 343–349.

Pearce, S. R. (1996). *Flash of insight: Metaphor and narrative in therapy.* Boston: Allyn & Bacon.

Pennebaker, J. W. (1990). *Opening up: The healing power of expressing emotions.* New York: Guilford.

Perkinson, R. R. (1997). *Chemical dependency counseling: A practical guide.* Thousand Oaks, CA: Sage.

Peterson v. the State, 671, P.2d., 230 (Wash. 1983).

Pies, A. (1995). Differential diagnosis and treatment of steroid-induced affective syndrome. *General Hospital Psychiatry, 17,* 353–361.

Pigott, T. A., Grady, T. A., & Rubenstein, C. S. (1993). Obsessive-compulsive disorder and trichotillomania. In D. L. Dunner (Ed.), *Current psychiatric therapy* (pp. 282–287). Philadelphia: W. B. Saunders.

Pigram, J. J. (1993). Human nature relationships: Leisure environments and natural settings. In T. Garling, & R. G. Gollege (Eds.), *Behavior and environment: Psychological and geographical approaches* (pp. 400–426). Amsterdam: Elsevier.

Pinderhughes, E. B. (1983). Empowerment for our clients and ourselves. *Social Casework, 64,* 331–338.

Pinderhughes, E. B. (1997). Developing diversity competence in child welfare and permanency planning. *Journal of Multicultural Social Work, 5,* 19–38.

Pitchford, P. (1993). *Healing with whole foods: Oriental traditions and modern nutrition.* Berkeley: North Atlantic.

Pizzorno, J. (1998). *Total wellness*. Rocklin, CA: Prima.

Ponterotto, J. G., Casas, J. M., Suzuki, L. A., & Alexander, C. M. (1995). *Handbook of multicultural counseling*. Thousand Oaks, CA: Sage.

Pope, K. S., & Brown, L. S. (1996). *Recovered memories of abuse: Assessment, therapy, forensics*. Washington, DC: American Psychological Association Publications.

Post, R. M. (1993). Mood disorders: Acute mania. In D. L. Dunner (Ed.), *Current psychiatric therapy* (pp. 204–209). Philadelphia: W. B. Saunders.

Potkin, S. G., Albers, L. J., & Richmond, G. (1993). Schizophrenia: An overview of pharmacological treatment. In D. L. Dunner (Ed.), *Current psychiatric therapy* (pp. 142–153). Philadelphia: W. B. Saunders.

Primavera, A., Giberti, L., Scotto, P., & Cocito, L. (1994). Nonconvulsive status epilepticus as a cause of confusion in later life: A report of 5 cases. *Neuropsychobiology, 30*, 148–152.

Primm, B. (1992). Alcohol and other drug abuse: Changing lives through research and treatment. *Journal of Health Care for the Poor and Underserved, 3*, 1–17.

Quinn, J. R. (1993). Evaluation and assessment for transcultural counseling. In J. McFadden (Ed.), *Transcultural counseling: Bilateral and international perspectives* (pp. 287–304). Alexandria, VA: American Counseling Association.

Quitkin, F. M., & Taylor, B. P. (1996). Antidepressants. In D. F. Klein & L. P. Rowland (Eds.), *Current psychotherapeutic drugs* (pp. 23–58). New York: Brunner/Mazel.

Regier, D. A., Farmer, M. E., Rae, D. S., Locke, B. Z., Keith, S. J., Judd, L. L., & Goodwin, F. K. (1990). Comorbidity of mental disorders with alcohol and other drug abuse: Results from the Epidemiological Catchment Area (ECA) study. *Journal of the American Medical Association, 264*, 2511–2518.

Reid, D. (1994). *The complete book of Chinese health and healing*. New York: Barnes & Noble Books.

Reid, W. J. (1998). The paradigm and long-term trends in clinical social work. In R. A. Dorfman, *Paradigms of clinical social work* (Vol. 2; pp. 337–351). New York: Brunner/Mazel.

Reitan, R. M., & Wolfson, D. (1985). *Halstead-Reitan Neuropsychological Test Battery: Theory and clinical interpretation*. Tucson: Neuropsychology Press.

Reynolds, W. M. (1991). *Adult Suicidal Ideation Questionnaire: Professional manual*. Odessa, FL: Psychological Assessment Resources.

Reynolds, W. M., & Mazza, J. J. (1992). *Suicidal Behavior History Form: Clinician's guide*. Odessa, FL: Psychological Assessment Resources.

Richelson, E. (1993). Review of antidepressants in the treatment of mood disorders. In D. L. Dunner (Ed.), *Current psychiatric therapy* (pp. 232–239). Philadelphia: W. B. Saunders.

Ries, R., Mullen, M., & Cox., G. (1994). Symptom severity and utilization of treatment resources among dually diagnosed inpatients. *Hospital and Community Psychiatry, 45*, 562–568.

Riley, J. A. (1994). Dual diagnosis: Comorbid substance abuse or dependency and mental illness. *Nursing Clinics of America, 29*, 29–34.

Rivera-Arzola, M., & Ramos-Grenier, J. (1997). Anger, ataques de nervios, and la mujer puertorriquena: Sociocultural considerations and treatment implications. In J. G. Garcia & M. C. Zea (Eds.), *Psychological interventions and research with Latino populations* (pp. 125–141). Boston: Allyn & Bacon.

Robbins, J. (1987). *Diet for a new America*. Walpole, NH: Stillpoint.

Roberts, T. W. (1994). *A systems perspective of parenting: The individual, the family, and the social network*. Pacific Grove, CA: Brooks/Cole.

Roehl, E. (1996). *Whole food facts: The complete reference guide*. Rochester, VT: Healing Arts Press.

Rogers, C. R. (1961). *On becoming a person*. Boston: Houghton Mifflin.

Rogers, M. P., White, K., Warshaw, M. G., & Yonkers, K. A. (1994). Prevalence of medical illness in patients with anxiety disorders. *International Journal of Psychiatry in Medicine, 24*, 83–96.

Rogers, T. B. (1995). *The psychological testing enterprise*. Pacific Grove, CA: Brooks/Cole.

Rohe, F. (1983). *The complete book of natural foods*. Boulder: Shambala.

Rotter, J. B., & Rafferty, J. E. (1950). *Manual: The Rotter Incomplete Sentences Blank*. New York: Psychological Corporation.

Rowe, C. E., & MacIsaac, D. S. (1986). *Empathic attunement: The technique of psychoanalytic self psychology*. Northvale, NJ: Aronson.

Roy-Byrne, P. P. (1996). Generalized anxiety and mixed anxiety-depression: Association with disability and health care utilization. *Journal of Clinical Psychiatry, 57*, 86–91.

Ryan, L. (1977). *Clinical interpretation of the FIRO-B*. Palo Alto, CA: Consulting Psychologists Press.

Salvia, J., & Ysseldyke, S. (1988). *Assessment in special and remedial education* (4th ed.). Boston: Houghton Mifflin.

Sanderson, C. (1996). *Counselling adult survivors of child sexual abuse* (2nd ed.). London: Jessica Kingsley.

Sansone, R. A., & Shaffer, B. (1997). An introduction to psychotropic medications. In J. R. Matthews & C. E. Walker (Eds.), *Basic skills and professional issues in clinician psychology* (pp. 195–224). Boston: Allyn & Bacon.

Schechter, M., & Roberge, L. (1976). Sexual exploitation. In R. E. Helfer & C. H. Kempe (Eds.), *Child abuse and neglect: the family and the community* (pp. 128–149). Cambridge, MA: Ballinger.

Schiffmann, E. (1996). *Yoga: The spirit and practice of moving into stillness.* New York: Pocket Books.

Schlesinger, E. G., & Devore, W. (1995). Ethnic sensitive social work practice: the state of the art. *Journal of Sociology and Social Welfare, 22,* 29–58.

Schmidt, L. (1992). A profile of problem drinkers in the public mental health system. *Hospital and Community Psychiatry, 43,* 245–250.

Schuckit, M. A (1993). Keeping current with the DSMs and substance use disorders. In D. L. Dunner (Ed.), *Current psychiatric therapy* (pp. 89–91). Philadelphia: W. B. Saunders.

Schuckit, M. A. (1998). *Educating yourself about alcohol and drugs: A people's primer* (rev. ed.). New York: Plenum.

Schuckit, M. A., & Monteiro, M. G. (1988). Alcoholism, anxiety, and depression. *British Journal of Addiction, 83,* 1373–1380.

Schultz, S. C., & Sajatovic, M. (1993). Typical antipsychotic medication: Clinical practice. In D. L. Dunner (Ed.), *Current psychiatric therapy* (pp. 176–182). Philadelphia: W. B. Saunders.

Schutz, W. (1962). *FIRO-Cope.* Los Angeles: Consulting Psychologists Press.

Schwartz, T. (1995). *What really matters: Searching for wisdom in America.* New York: Bantam.

Schwebel, M., Schoener, G., & Skorina, J. K. (1994). *Assisting impaired psychologists* (rev. ed.). Washington, DC: American Psychological Association

Sedlak, A. J. (1990). *Technical amendment to the study findings—National incidence and prevalence of child abuse and neglect: 1988.* Rockville, MD: Westat.

Seligman, L. (1996). *Diagnosis and treatment planning in counseling* (2nd ed.). New York: Plenum Press.

Sellers, E. M. (1996). Antianxiety agents: Benzodiazepine derivatives. In D. F. Klein & L. P. Rowland (Eds.), *Current psychotherapeutic drugs* (pp. 143–160). New York: Brunner/Mazel.

Selwyn, P. A., & Merino, F. (1997). Medical complications and treatment. In J. H. Lowinson, P. Ruiz, R. B. Millman, & J. G. Langrod (Eds.), *Substance abuse: A comprehensive textbook* (2nd ed., pp. 597–628). Baltimore: Williams & Wilkins.

Selzer, M. L. (1971). The Michigan Alcohol Screening Test: The quest for a new diagnostic instrument. *American Journal of Psychiatry, 127,* 1653–1658.

Sheehan, D. V., & Ashok, B. A. (1993). Panic disorder. In D. L. Dunner (Ed.), *Current psychiatric therapy* (pp. 275–281). Philadelphia: W. B. Saunders.

Sherman, M. D. (1996). Distress and professional impairment due to mental health problems among psychotherapists. *Clinical Psychology Review, 16,* 299–315.

Sherman, M. D., & Thelen, M. H. (1998). Distress and professional impairment among psychologists in clinical practice. *Professional Psychology: Research and Practice, 29,* 79–85.

Shneidman, E. (1985). *Definition of suicide.* New York: Wiley.

Shneidman, E. S. (1997). The suicidal mind. In R. W. Maris, M. M. Silverman, & S. S. Canetto (Eds.), *Review of suicidology 1997* (pp. 22–41). New York: Guilford.

Shulman, L. A. (1988). Groupwork practice with hard to reach clients: A modality of choice. *Groupwork, 1,* 5–16.

Sigelman, C. K., & Shaffer, D. R. (1995). *Life-span human development* (2nd ed.). Pacific Grove, CA: Brooks/Cole.

Silverman, M. M. (1997). Introduction: Current controversies in suicidology. In R. W. Maris, M. M. Silverman, & S. S. Canetto (Eds.), *Review of suicidology 1997* (pp. 1–21). New York: Guilford.

Silverman, M. M., Berman, A. L., Bongar, B., Litman, R. E., & Maris, R. W. (1994). Inpatient standards of care and the suicidal patient Part II: An integration with clinical risk management. *Suicide and Life-Threatening Behavior, 24,* 152–169.

Simon, G. E., VonKorff, M., & Barlow, W. (1995). Health care costs of primary care patients with recognized depression. *Archives of General Psychiatry, 52,* 850–856.

Skinner, H. A. (1990). Spectrum of drinkers and intervention opportunities. *Canadian Medical Association Journal, 143,* 1054–1059.

Slaikeu, K. A. (1990). *Crisis intervention: A handbook for practice and research* (2nd ed.). Boston: Allyn & Bacon.

Solomon, A. (1992). Clinical diagnosis among diverse populations: A multicultural perspective. *Families in Society: The Journal of Contemporary Human Services, 73,* 371–377.

Sommers-Flanagan, J., & Sommers-Flanagan, R. (1995). Intake interviewing with suicidal patients: A systematic approach. *Professional Psychology, 26,* 41–46.

Spielberger, C. D. (1995a). *State-Trait Anxiety Inventory (STAI): Professional manual.* Palo Alto, CA: Mind Garden.

Spielberger, C. D. (1995b). *State-Trait Anger Expression Inventory (STAXI): Professional manual.* Palo Alto, CA: Mind Garden.

Stack, S. (1992). Marriage, family, religion, and suicide. In R. W. Maris, A. L. Berman, J. T. Maltsberger, & R. I. Yufit (Eds.), *Assessment and prediction of suicide* (pp. 540–552). New York: Guilford.

Steiner, J. R., & Devore, W. (1983). Increasing descriptive and prescriptive theoretical skills to promote ethnic-sensitive practice. *Journal of Education for Social Work, 19(2),* 63–70

Stelmachers, Z. T. (1995). Assessing suicidal clients. In J. N. Butcher (Ed.), *Clinical personality assessment: Practical approaches* (pp. 367–379). New York: Oxford.

Stern, D. N. (1985). *The interpersonal world of the infant.* New York: Basic Books.

Stern, D. N. (1990). *Diary of a baby: What your child sees, feels, and experiences.* New York: Basic Books.

Stevens, D. E., Merikangas, K. R., & Merikangas, J. R. (1995). Comorbidity of depression and other medical conditions. In E. E. Beckham & W. R. Leber, (Eds.), *Handbook of depression* (pp. 147–199). New York: Guilford.

Strupp, H. H. (1996). Some salient lessons from research and practice. *Psychotherapy, 33,* 135–138.

Suzuki, L. A., Meller, P. J., & Ponterotto, J. G. (1996). *Handbook of multicultural assessment.* New York: Jossey-Bass.

Swenson, L. C. (1997). *Psychology and the law* (2nd ed.). Pacific Grove, CA: Brooks/Cole.

Taylor, J. (1983). *Dream work.* New York: Paulist.

Terre, L., & Ghiselli, W. (1995). Do somatic complaints mask negative affect in youth? *Journal of American College Health, 44,* 91–96.

Teyber, E. (1997). *Interpersonal process in psychotherapy: A relational approach* (3rd ed.). Pacific Grove, CA: Brooks/Cole.

Thayer, R. E., Newman, J. R., & McClain, T. M. (1994). Self-regulation of mood: strategies for changing a bad mood, raising energy, and reducing tension. *Journal of Personality and Social Psychology, 67,* 910–925.

Thorndike, R. L., Hagen, E. D., & Sattler, J. M. (1986a). *Technical manual, Standard-Binet Intelligence Test: Fourth edition.* Chicago: Riverside.

Thorndike, R. L., Hagen, E. D., & Sattler, J. M. (1986b). *Guide to administering and scoring the Standard-Binet Intelligence Test: Fourth edition.* Chicago: Riverside.

Tollefson, G. D. (1993). Major depression. In D. L. Dunner (Ed.), *Current psychiatric therapy* (pp. 196–203). Philadelphia: W. B. Saunders.

Tomb, D. A. (1995). *Psychiatry.* Baltimore: Williams & Wikins.

Tzu, L. (1961). *Tao teh ching* (trans. by John C. H. Wu). New York: Barnes & Noble Books.

Ullman, M., & Zimmerman, N. (1979). *Working with dreams.* Los Angeles: Jeremy P. Tarcher.

Ulrich, R. S., Dimberg, U., & Driver, B. (1991). Psychophysiological indicators. In B. Driver, P. Brown, & G. Peterson (Eds.), *Benefits of leisure* (pp. 73–89). State College, PA: Venture.

U.S. Department of Health and Human Services (1988). *The health consequences of smoking: Nicotine addiction. A report of the Surgeon General.* Rockville, MD: USDHHS Office of Smoking and Health.

Valencia, R. R., & Guadarrama, I. (1996). High-stakes testing and its impact on racial and ethnic minority students. In L. A. Suzuki, P. J. Meller, & J. G. Ponterotto (Eds.), *Handbook of multicultural assessment* (pp. 561–610). San Francisco: Jossey-Bass.

Vandecreek, L., & Knapp, S. (1997). Record keeping. In J. R. Matthews & C. E. Walker (Eds.), *Basic skills and professional issues in clinical psychology* (pp. 155–172). Boston: Allyn & Bacon.

Velting, D. M., & Gould, M. S. (1997). Suicide contagion. In R. W. Maris, M. M. Silverman, & S. S. Canetto (Eds.), *Review of suicidology 1997* (pp. 96–137). New York: Guilford.

Vissing, Y. M., Strauss, M. A., Gelles, R. J., & Harrop, J. W. (1991). Verbal aggression by parents and psychosocial problems of children. *Child Abuse and Neglect, 15,* 223–238.

Volavka, J. (1995). *The neurobiology of violence.* Washington, DC: American Psychiatric Press.

Vollhardt, L. T. (1991). Psychoneuroimmunology: A literature review. *American Journal of Orthopsychiatry, 61,* 35–47.

Walker, C. E., Bonner, B. L., & Kaufman, K. L. (1988). *The physically and sexually abused child: Evaluation and treatment.* New York: Pergamon.

Walker, C. E., & Matthews, J. R. (1997). Introduction: First steps in professional psychology. In J. R. Matthews & C. E. Walker (Eds.), *Basic skills and professional issues in clinical psychology* (pp. 1–12). Boston: Allyn & Bacon.

Walker, E. A., Roy-Byrne, P. P., & Katon, W. J. (1990). Irritable bowel syndrome and psychiatric illness. *American Journal of Psychiatry, 147,* 565–572.

Walters, D. (1996). *Meditation for starters.* Nevada City, CA: Crystal Clarity.

Warner, L. A., Kessler, R. C., Hughes, M., Anthony, J. C., & Nelson, C. B. (1995). Prevalence and correlates of drug use and dependence in the United States. *Archives of General Psychiatry, 52,* 219–229.

Watson, H., & Levine, M. (1989). Psychotherapy and mandated reporting of child abuse. *American Journal of Orthopsychiatry, 59,* 246–256.

Wechsler, D. (1992). *Wechsler Individual Achievement Test Screener: Manual.* San Antonio: Psychological Corporation.

Wechsler, D. (1997a). *Wechsler Adult Intelligence Scale—III: Technical Manual.* San Antonio: Psychological Corporation.

Wechsler, D. (1997b). *Wechsler Memory Scale—III: Technical manual.* San Antonio: Psychological Corporation.

Weil, A. (1995). *Spontaneous healing.* New York: Fawcett Columbine.

Weiss, J. (1993). *How psychotherapy works.* New York: Guilford.

Weiss, M. G. (1994). Parasitic diseases and psychiatric illness. *Canadian Journal of Psychiatry, 39,* 623–628.

Weiss, R. D., Mirin, S. M., & Frances, R. J. (1992). The myth of the typical dual diagnosis patient. *Hospital and Community Psychiatry, 43,* 107–108.

Werth, J. L., & Liddle, B. J. (1994). Psychotherapists' attitudes toward suicide. *Psychotherapy, 31,* 441–448.

Wheeler, S. (1997). Achieving and maintaining competence. In I. Horton & V. Varma (Eds.), *The needs of counsellors and psychotherapists: Emotional, social, physical, professional* (pp. 120–134). London: Sage.

Wickramasekera, I., Davies, T. E., & Davies, S. M. (1996). Applied psychophysiology: A bridge between the biomedical model and the biopsychosocial model in family medicine. *Professional Psychology: Research and Practice, 27,* 221–233.

Wilkinson, G. S. (1993). *Wide Range Achievement Test, Third Edition: Administration manual.* San Antonio: Psychological Corporation.

Wingerson, D., & Roy-Byrne, P. P. (1993). Review of anxiolytic drugs. In D. L. Dunner (Ed.), *Current psychiatric therapy* (pp. 295–302). Philadelphia: W. B. Saunders.

Winnicott, D. W. (1958). *Collected papers.* London: Tavistock.

Wirth, L. (1945). The problem of minority groups. In R. Linton (Ed.), *The science of man in the world crisis* (pp. 347–372). New York: Columbia University Press.

Wise, M. G., & Griffies, W. S. (1995). A combined treatment approach to anxiety in the medically ill. *Journal of Clinical Psychiatry, 56,* 14–19.

Witztum, E., Greenberg, D., & Dasberg, H. (1990). Mental illness and religious change. *British Journal of Medical Psychology, 63,* 33–41.

Wolf, E. S. (1988). *Treating the self: Elements of clinical self psychology.* New York: Guilford.

Wolmark, A., & Sweezy, M. (1998). Kohut's self psychology. In R. A. Dorfman, *Paradigms of clinical social work* (Vol. 2; pp. 45–70). New York: Brunner/Mazel.

Woodcock, R. M., & Mather, N. L. (1989a). WJ-R Tests of Achievement—Standard and Supplemental Batteries: Examiner's manual. In R. W. Woodcock & M. B. Johnson, *Woodcock Johnson Psycho-Educational Battery, Revised.* Allen, TX: DLM Teaching Resources.

Woodcock, R. M., & Mather, N. L. (1989b). WJ-R Tests of Cognitive Ability, Standard and Supplemental Batteries: Examiner's manual. In R. W. Woodcock & M. B. Johnson, *Woodcock Johnson Psycho-Educational Battery, Revised.* Allen, TX: DLM Teaching Resources.

Yufit, R. I., & Bongar, B. (1992). Suicide, stress, and coping with lifecycle events. In R. W. Maris, A. L. Berman, J. T. Maltsberger & R. I. Yufit (Eds.), *Assessment and prediction of suicide* (pp. 553–573). New York: Guilford.

Zaro, J. S., Barach, R., Nadelman, D. J., & Dreiblatt, I. S. (1977). *A guide for beginning psychotherapists.* Cambridge, MA: Cambridge University Press.

Zaubler, T. S., & Katon, W. (1996). Panic disorder and medical comorbidity: A review of the medical and psychiatric literature. *Bulletin of the Menninger Clinic, 60* (2, Suppl. A), A12–A38.

Zellman, G. L., & Faller, K. C. (1996). Reporting of child maltreatment. In J. Briere, L. Berliner, J. A. Bulkley, C. Jenny, & T. Reid (Eds.), *The APSAC handbook on child maltreatment* (pp. 359–381). Thousand Oaks, CA: Sage.

Ziedonis, D. M. (1992). Comorbid psychopathology and cocaine addiction. In T. R. Costen & H. D. Kleber (Eds.), *Clinicians' guide to cocaine addiction: Theory, research, and treatment* (pp. 335–358). New York: Guilford.

Zimberg, S. (1993). Introduction and general concepts of dual diagnosis. In J. Solomon, S. Zimberg, & E. Shollar (Eds.), *Dual diagnosis: Evaluation, treatment, training, and program development* (pp. 4–22). New York: Plenum Medical Book.

Zwaan, M., & Mitchell, J. E. (1993). Bulimia nervosa. In D. L. Dunner (Ed.), *Current psychiatric therapy* (pp. 383–389). Philadelphia: W. B. Saunders.

Zweben, J. E., Smith, D. E., & Stewart, P. (1991). Psychotic conditions and substance use: Prescribing guidelines and other treatment issues. *Journal of Psychoactive Drugs, 23,* 387–395.

Author Index

Subject Index